Critical Roles of Digital Citizenship and Digital Ethics

Jason D. DeHart
University of Tennessee, Knoxville, USA

IGI Global
PUBLISHER of TIMELY KNOWLEDGE

A volume in the Advances in Social Networking and Online Communities (ASNOC) Book Series

Published in the United States of America by
IGI Global
Information Science Reference (an imprint of IGI Global)
701 E. Chocolate Avenue
Hershey PA, USA 17033
Tel: 717-533-8845
Fax: 717-533-8661
E-mail: cust@igi-global.com
Web site: http://www.igi-global.com

Library of Congress Cataloging-in-Publication Data

Names: DeHart, Jason D., 1982- editor.
Title: Critical roles of digital citizenship and digital ethics / Jason DeHart.
Description: Hershey, PA : Information Science Reference, [2024] | Includes bibliographical references and index. | Summary: "This comprehensive research handbook provides a platform for in-depth discussions and empirical investigations into the philosophical foundations and practical implications of digital citizenship. By examining how digital ethics have been and continue to be shaped, the book offers valuable insights into the ways in which educators and researchers can approach these topics in the realm of education."-- Provided by publisher.
Identifiers: LCCN 2023029601 (print) | LCCN 2023029602 (ebook) | ISBN 9781668489345 (h/c) | ISBN 9781668489352 (eISBN)
Subjects: LCSH: Technology--Moral and ethical aspects. | Digital media--Moral and ethical aspects.
Classification: LCC BJ59 .H336 2024 (print) | LCC BJ59 (ebook) | DDC 174/.9302231--dc23/eng/20230928
LC record available at https://lccn.loc.gov/2023029601
LC ebook record available at https://lccn.loc.gov/2023029602

This book is published in the IGI Global book series Advances in Social Networking and Online Communities (ASNOC) (ISSN: 2328-1405; eISSN: 2328-1413)

British Cataloguing in Publication Data
A Cataloguing in Publication record for this book is available from the British Library.

Advances in Social Networking and Online Communities (ASNOC) Book Series

Hakikur Rahman
Ansted University Sustainability Research Institute, Malaysia

ISSN:2328-1405
EISSN:2328-1413

Mission

The advancements of internet technologies and the creation of various social networks provide a new channel of knowledge development processes that's dependent on social networking and online communities. This emerging concept of social innovation is comprised of ideas and strategies designed to improve society.

The **Advances in Social Networking and Online Communities** book series serves as a forum for scholars and practitioners to present comprehensive research on the social, cultural, organizational, and human issues related to the use of virtual communities and social networking. This series will provide an analytical approach to the holistic and newly emerging concepts of online knowledge communities and social networks.

Coverage

- Information Policy Overview
- Communication of Knowledge in Organizations
- Social Models to Design and Support Knowledge Intensive Collaborative Processes
- Advanced Researches in Knowledge Communities
- Strategic Management and Business Process Analysis
- Conceptual Role of ICTs in Knowledge Communication and Management
- Framework of a Pragmatic Conception of Knowledge
- Leveraging Knowledge Communication Networks – Approaches to Interpretations and Interventions
- Performance Evaluation and Benchmarking of Deployed Systems
- Application of Social Network Analysis as a Knowledge Management Tool

IGI Global is currently accepting manuscripts for publication within this series. To submit a proposal for a volume in this series, please contact our Acquisition Editors at Acquisitions@igi-global.com or visit: http://www.igi-global.com/publish/.

The Advances in Social Networking and Online Communities (ASNOC) Book Series (ISSN 2328-1405) is published by IGI Global, 701 E. Chocolate Avenue, Hershey, PA 17033-1240, USA, www.igi-global.com. This series is composed of titles available for purchase individually; each title is edited to be contextually exclusive from any other title within the series. For pricing and ordering information please visit http://www.igi-global.com/book-series/advances-social-networking-online-communities/37168. Postmaster: Send all address changes to above address.

Titles in this Series

For a list of additional titles in this series, please visit: http://www.igi-global.com/book-series/advances-social-networking-online-communities/37168

Social Capital in the Age of Online Networking Genesis, Manifestations, and Imlications
Najmul Hoda (Umm Al-Qura University) and Arshi Naim (King Kalid University, Saudi Arabia)
Information Science Reference • © 2023 • 301pp • H/C (ISBN: 9781668489536) • US $215.00

Global Perspectives on Social Media Usage Within Governments
Chandan Chavadi (Presidency Business School, Presidency College, India) and Dhanabalan Thangam (Presidency Business School, Presidency College, India)
Information Science Reference • © 2023 • 310pp • H/C (ISBN: 9781668474501) • US $215.00

Advanced Applications of NLP and Deep Learning in Social Media Data
Ahmed A. Abd El-Latif (Menoufia University, Egypt & Prince Sultan University, Saudi Arabia) Mudasir Ahmad Wani (Prince Sultan University, Saudi Arabia) and Mohammed A. El-Affendi (Prince Sultan University, Saudi Arabia)
Engineering Science Reference • © 2023 • 303pp • H/C (ISBN: 9781668469095) • US $270.00

Community Engagement in the Online Space
Michelle Dennis (Adler University, USA) and James Halbert (Adler University, USA)
Information Science Reference • © 2023 • 364pp • H/C (ISBN: 9781668451908) • US $215.00

Handbook of Research on Bullying in Media and Beyond
Gülşah Sarı (Aksaray University, Turkey)
Information Science Reference • © 2023 • 603pp • H/C (ISBN: 9781668454268) • US $295.00

Handbook of Research on Technologies and Systems for E-Collaboration During Global Crises
Jingyuan Zhao (University of Toronto, Canada) and V. Vinoth Kumar (Jain University, India)
Information Science Reference • © 2022 • 461pp • H/C (ISBN: 9781799896401) • US $295.00

Information Manipulation and Its Impact Across All Industries
Maryam Ebrahimi (Independent Researcher, Germany)
Information Science Reference • © 2022 • 234pp • H/C (ISBN: 9781799882350) • US $215.00

E-Collaboration Technologies and Strategies for Competitive Advantage Amid Challenging Times
Jingyuan Zhao (University of Toronto, Canada) and Joseph Richards (California State University, Sacramento, USA)
Information Science Reference • © 2021 • 346pp • H/C (ISBN: 9781799877646) • US $215.00

IGI Global
PUBLISHER of TIMELY KNOWLEDGE

701 East Chocolate Avenue, Hershey, PA 17033, USA
Tel: 717-533-8845 x100 • Fax: 717-533-8661
E-Mail: cust@igi-global.com • www.igi-global.com

Table of Contents

Detailed Table of Contents

This chapter reports on a multi-year project that the author conducted using a children's picture book titled The Tantrum that Saved the World. Published in 2017 by World Saving Books, the e-version with its 64 colorfully illustrated pages tells of a little girl who stares down a pending climate crisis by channeling her tantrum power into strategies for saving the world. In the first study, the storybook's words and illustrations were analyzed using a Foucauldian genealogy to show how power circulates and why acquiring only that knowledge stops short of what a critical media literacy (CML) filter might reveal. Two additional studies, one focused on an analysis of critical literacy outside its comfort zone, and the other, a critical post-humanist analysis of digital remixing in a licensed Creative Commons website provided a basis for connecting the multi-year project to the research literature on digital citizenship and ethics.

The digital age has presented both advantages and issues in learning, communication, and overall well-being. Digital technology can provide entertainment, information, and connection but can also increase stress levels, and confusion, and weaken resilience. How can youth build the skills needed to stay focused and avoid digital distractions? What support do educators and parents offer in this endeavor? The chapter will conclude by highlighting the importance of balancing resilience and digital engagement in the digital landscape. While digital distractions can pose challenges to youth's focus and concentration, they can also offer opportunities for learning and creativity. The chapter will argue that youth need to develop a critical and reflective attitude toward digital technology and its role in their lives. The chapter concludes by highlighting the importance of balancing resilience and digital engagement and developing a critical and reflective attitude toward digital technology.

The chapter is a mixed-method study of knowledge, skills, and beliefs of pre-service elementary school teachers on digital citizenship. The quantitative data was gathered through surveying participants. The intervention in the form of a 30-hour module was introduced to participants to identify a change in perceptions. The qualitative data was obtained through the analysis of artefacts in the form of lesson plans. The author grounds the theoretical underpinning into the theory and principles of digital citizenship. The result of the posttest shows a positive change in the knowledge, skills, and beliefs of pre-service elementary school teachers on digital citizenship.

The European Union Digital Education Action Plan addresses the imperative of digital citizenship in the landscape of post-pandemic education. This abstract highlights the plan's focus on equipping learners with the skills to navigate the digital sphere responsibly. Emphasizing digital literacy, online safety, and ethical behavior also the plan aims to foster critical engagement with safer technology by promoting inclusivity and equitable access, it seeks to empower individuals to thrive as responsible digital citizens in an interconnected world.

Digital citizenship is a term that has been used globally. However, many schools have also traditionally operated on a perceived need to behaviorally control children's activity. These control mechanisms have been presented in communities using terms such as citizenship–where the children and community are led to believe that strict obedience is for the common good. In this chapter, the authors share the experiences of Green, a child identified with multiple disabilities. The authors address laws in the United States and inclusive ethics and demonstrate the potential harm that is done when digital citizenship is defined around a narrow set of compliance behaviors that are too rigid for many students without support. The authors propose an alternative framing of digital citizenship that accounts for inclusive ethics and that encircles adults and institutions in collateral obligations regarding digital technologies' ethical use. The authors end with suggestions for classroom practices that support these definitions and offer an alternative ending for Green.

The digital transforms and influences the social practices, the mediations, and the quality of the information in the new digital sphere of modern visual cultures. Digital information is as permanent as it is transitory, real and ephemeral, current and obsolete. Therefore, ethical concerns must be able to keep up with the rapid development of the digital. This chapter is based on exploratory research on the digital image used daily in mass media and social media. The relevance of this anchoring in the digital image is due to the increasingly visual information world, raising fundamental ethical implications. There are ethical criteria (e.g., truth, facts, authenticity) that guide good information practices, whether in the mass media or in social media. They are central standards for responsible citizenship. Following a theoretical-conceptual strategy, the purpose is to argue the ethical use of digital images in daily social practices and technological mediations.

The teaching of writing is an ethical endeavor on many fronts. This chapter focuses on aspects of teaching ethical practices that have evolved over the last three decades as teachers of writing have adapted instruction to account for new tools that impact how a writer composes and the nature of the final product created. Many of these aspects are again at a turning point with the explosion in AI technologies in recent years, and it is essential that writing teachers see this moment with "historical consciousness" (Bruce, 2003, p. 13) and help writers to understand the ethical implications of being a writer in the age of AI. Teachers must recognize the history of writing development, view writing as a form of inquiry, and understand the complexities of authorship and attribution from a sociocultural and digital perspective so that they can help young writers develop ethical stances in this new era of writing.

As the field of intelligent libraries continues to evolve, Metaverse technology presents a promising opportunity for driving innovation and improving library services. This chapter delves into the potential of Metaverse to transform libraries into fully-intelligent spaces. The authors will explore the underlying technologies and theoretical foundations of Metaverse, and examine how it can be applied to enhance library collection management, reading spaces, cultural promotion, reading experiences, and user services. By harnessing the power of Metaverse, intelligent libraries have the opportunity to create immersive experiences and achieve significant progress in their digital transformations.

Participatory budgeting is a recent mechanism that is related to open innovation and crowdsourcing in public entities and allows the participation of society in governmental interests. This chapter aims to analyse the use of open innovation and crowdsourcing tools in the public sector in Portugal, in the form of participatory budgets, and also to understand the characteristics of the participatory budget processes used, as well as the motivations for their implementation and the difficulties faced. For these purposes, the investigation uses a quantitative methodology, by means of a questionnaire sent to Portuguese municipalities. The research carried out reveals influences between the size of municipalities and the use of participatory budget and its characteristics. The investigation also highlights the main motivations for its implementation, the greatest difficulties experienced, and the perceived barriers to non-implementation.

This chapter aimed to investigate the predictive role of new media literacy levels and critical thinking dispositions of university students on their digital citizenship levels. The study was carried out with 124 university students studying in different departments of a state university in Türkiye, and the data were gathered using New Media Literacy Scale, Sosu Critical Thinking Dispositions Scale, and Digital Citizenship Scale. It was concluded that university students had high new media literacy levels and critical thinking dispositions while they presented moderate digital citizenship levels. Also, university students' new media literacy levels and critical thinking dispositions had a moderate and positive relationship with their digital citizenship levels. It was also concluded that university students' new media literacy levels and critical thinking dispositions significantly predicted their digital citizenship levels, and they together explained 17% of the total variance on their digital citizenship levels.

In today's digital era, it is important to examine the ways researching in digital spaces differs from the general practice of conducting qualitative research. For this reason, this chapter will consider the work done by researchers to include the ethical considerations, affordances, and constraints of conducting qualitative research in digital spaces. Then, with these points in mind, it will include a specific discussion on the research literature concerning how literacy researchers have studied adolescents' digital work both in and outside of educational spaces. Finally, implications will be shared on the future of research in digital spaces, such as with AI, and what is said and unsaid within the research literature about conducting qualitative research in digital spaces.

It is important to be a role model and guide for the child in the process of acquiring the rules and behaviors that should be followed in order to protect from the risks during the use of technology and not to create risks for others. In this context, as a result of the research conducted with preschool teachers who are the guides and role models of children, it was seen that their understanding of digital ethics was based on protection from the harms of technology, correct use, and the application of moral and ethical rules to the digital environment. It was stated that precautions such as restricting posts about children and controlling the channels they share were taken. They stated that they wanted to receive training on digital ethics to improve themselves, guide and protect children, and that they obtained information various resources. They stated that they shared with children, families, and colleagues on issues such as safe internet and digital citizenship and emphasized that teachers and parents should also be trained for digital ethics education to be given to children.

Digital citizenship has been defined in a multitude of ways in the literature. The definitions promote a range of teaching practices spanning from protectionism to social activism. This chapter examines what the literature and research say about digital citizenship including how it is defined, the theoretical foundations at work in those definitions, and whose responsibility it is for teaching about and supporting digital citizenship. The National Educational Technology Plan (NETP) developed by the U.S. Department of Educational Technology and the International Society for Technology Education (ISTE) has created guidelines for digital citizenship education that incorporate a critical component. This chapter considers the political implications of those definitions and guidelines. Research is presented that examines teacher candidate preparation to provide insight about whether teacher candidates feel prepared to support the digital citizenship goals proposed by both the NETP and the ISTE.

Digital citizenship and digital ethics are closely related concepts regarding how to behave responsibly and ethically in the era of artificial intelligence (AI) in the digital world. The author shared his perspectives on digital citizenship and digital ethics based on his experiences working with PGDs, Master, and MPhil students at Kathmandu University School of Education, Nepal. The author attempted to reflect on how PGDs, Master, and MPhil students understood and acted on these concepts—digital citizenship and digital ethics based on the respecting, educating, and protecting framework of digital citizenship and digital ethics. Further, the author argued that digital citizenship is the responsible and respectful use of technology to engage in online activities: look for trustworthy sources and protect and promote human rights. Similarly, digital ethics are rules and moral principles governing individuals' interpersonal behavior. Overall, the author concluded that digital citizenship and ethics are related concepts that deal with digital technology's moral and social aspects.

Preface

DIGITAL WORLDS AND DIGITAL RULES

This book has its genesis in the theoretical work I have done since 2015, but also in the age in which I was born and grew as a learner, including my experiences as a researcher and classroom teacher. As I consider the words collected in this book, I return to the foundations of thought that have informed my work.

Ontologically, Lankshear and Knobel (2008) suggested that our world is in a state of hybridized flux, with new ways of meaning-making and new tools/materials appearing on a frequent basis, providing a reconceptualization of what we mean by connection and communication. These authors later on wrote of the ways in which new practices and new ways of being would be made possible through digital and remixed environments and compositional methods (Lankshear & Knobel, 2011). I also note the work of Kist (2000) in envisioning classrooms that would incorporate multimodal meaning-making practices, as well as commentary from the New London Group (1996) related to the ways in which informational superhighways would shape learning and lived reality.

Given the age of these citations, at this point, I note the speculative nature of many of these authors; that is, scholars have envisioned and continue to envision possibilities for educational practices, often through the innovations of youth, in juxtaposition to the descriptions of the ways educational institutions approach and embrace or resist social and cultural transformations. The innovation that is often glimpsed in what youth explore is hardly work of the past, and readers can see more about the work that is being documented in educational settings through my edited collection, *Innovations in Digital Instruction Through Virtual Environments*. While the purpose of that project is to capture and distill examples of digital, virtual, and multimodal work and classroom encounter, this book takes a step into the overarching theories and larger conversations about the implications of the digital for conceptualizations of both citizenship and ethics. I have previously written about this topic (Buchholz et al., 2020), and I note the work of Luke (2018) in positioning digital citizenship and digital ethics about more than technical abilities, and more about participation and ideological engagements, including a sense of social justice.

Simply put, the guiding question at the heart of digital citizenship and digital ethics is: What are the guiding principles? I play with this notion at the outset of this introduction with the world "rules." Black et al. (2022) have pointed out the deficit thinking that so often guides approaches to digital citizenship, and I liken this fearful vantage to the ways that society takes up questions of Artificial Intelligence (AI), and its manifestations in ready-made questions to the well-worn types of questions that are often posed in educational settings. Developing a set of rules or guidelines is especially ephemeral work as online environments are ever-changing (Bilki et al., 2022; Forzani & Leu, 2017). Moreover, the interlinking nature of digital environments is considerable, including questions of digital identity (Masiero & Bailur, 2021), concepts of audience (Wargo, 2015), and possibilities for activism (Kaun & Uldam, 2018). The privileged practices of schooling spaces are sometimes renegotiated or conceptualized through the home-based practices of literacies that are part of daily life for youth (Buchholz & Pyles, 2018), including larger questions of how practices and interactions change across spaces (Pyles et al., 2019).

In response to these complicated questions, contributors in this book speak to the roles of educators, the ways composing can be explored, the ways that digital citizenship and digital ethics have been defined and explored in research literature, and the perceptions and ideas that exist in situ as educators and students grapple with ways of being and knowing in virtual and digital spaces and through the tools afforded by these spaces. It is this balance of affordance and restriction, of possibility and danger, that is of central inquiry when approaching the digital – there are boundaries to describe, surely, but also ways in which students can explore and expand both their knowledge seeking and knowledge sharing identities in digital realms.

With shifts in audience, changed in the tools and methods available for communication and networking, possibilities for access to information, and more avenues of exploration, there is an optimism to the shape of digital citizenship. Buckingham (2020) pushes back on binary views of digital spaces as risk versus reward or promise versus peril, and notes a need for continued criticality and exploration of digital spaces, with attention to extending instruction to account for digital and social change. From this more complex and critical viewpoint, there are certainly responsibilities to consider and, as in many cases with questions of social and ethical imperative, it is arguably not the enterprise of the individual teacher or even a single educational institution to create and enforce a set of rules or guidelines. Instead, and ideally, citizens are the primary motion of citizenship and people are the driving force of a thoughtful politic.

There is no simple and pared-down solution or ten commandments of digital interaction. To that problematized end, I note the contributions of the authors in this volume for speaking to research and practice from a global context, and for considering a multitude of questions and sociopolitical realities as they have crafted chapters. From the ways in which students with disabilities have been left out of the conversation to the ways that children's literature can align with concepts of digital citizenship and ethics, the scholars who have agreed to be part of this project share widely and richly.

I now offer an overview of the contents of this book, with appreciation to these authors and the peer reviewers who have helped shaped this work.

A Map of the Book

Invited contributor Donna Alvermann leads the collection with a chapter titled "Critical Media Literacy and Relationships of Power: Implications for Digital Citizenship and Digital Ethics." Throughout this collection, a variety of authors share about the perspectives of educators and students. Providing a unique approach, Alvermann explores the notion of authors' responsibility in building positive citizenship for young audiences, one of the four cornerstones which form the basis of digital ethics. The data in this chapter stems from a multi-year study of a children's text that focuses on the climate crisis, analyzed through a Foucauldian lens. This chapter author speaks to relationships of power, and the vital need for knowledge building when developing critical media literacy (CML). This work speaks aptly to notions of criticality, which I view as an essential element of citizenship and ethics in all environments, including the digital and stemming from the textual.

In Chapter 2, invited contributors Ian O'Byrne and Spencer Ward explore "Balancing Resilience and Digital Distractions: Youth Staying Focused in the Digital Landscape" exploring the positives and drawbacks relative to the large amounts of content that are available through digital media. O'Byrne and Ward comment on the ways in which young people can navigate these diversions, as well as locate information, and the authors further comment on the ways in which educators and families can offer their expertise as guides in the process. Criticality is a central point that the authors circle back to, as well as reflection. The authors ultimately focus on the notion of resilience, in spite of any hurdles that youth face, and speak to the rich affordances that digital spaces offer. This position offers much to consider in terms thinking about how digital citizenship and digital ethics are taken up in educational settings.

Tetiana Holovatenko presents a mixed-methods study with pre-service elementary school teachers in Chapter 3, "Elementary School Teacher Training for Digital Citizenship in English Language Classrooms." Holovatenko's study focuses on the beliefs relative to digital citizenship, and includes findings based on surveys, as well as a 30-hour module intervention process. Qualitative data were generated through lesson plan artifact analysis. The author reports positive results, and grounds the findings in theory relative to digital citizenship.

In Chapter 4, "Empowering Digital Citizenship Through Distance Education: A Technology-Driven Education Action Plan," authors Seda Adıgüzel and Selami Eryılmaz comment on the global shift toward digitalization and the ways in which these changes affect daily interactions. The authors locate their discussion in light of the COVID-19 pandemic, and note the ways in which educational institutions have made use of digital tools and approaches in this time. Framing their work, Adıgüzel and Eryılmaz share about educator perceptions of distance/online education in light of the Council of Europe Digital Education Action Plan 2021-2027. By exploring this movement and context, the authors speak to trends and issues that continue to emerge and shape practices, research, and policies.

Chapter 5 features the work of Mary Rice and Shernette Dunn, and is titled "Challenging Digital Citizenship Discourse Through the Lens of (Dis)ability." Rice and Dunn comment on the global applications of digital citizenship, and juxtapose this global notion with the assumption that is conveyed in school practices and policies which seek to control behaviors and activities among young people. The authors discuss the idea of obedience in light of concepts of citizenship, and share a case study of a child with identified multiple disabilities (Green). I deeply appreciate how the authors explore questions of digital citizenship through a critical lens, and do so in a way that centers stories of youth who might be otherwise marginalized in the conversation, and in the policies and practices that stem from this conversation.

From this consideration, Rice and Dunn comment on the nature of laws in the United States that relate to the ethics of inclusivity, and speak to the need to broaden conceptualizations of what digital citizenship means beyond obedience and compliance. The authors share a revised framework for digital citizenship, composed in light of the ways in which students can be harmed by traditional, limited approaches, and in light of ethical considerations. By exploring digital citizenship in this way, the authors contribute a vision of what is possible in this field, and how educational institutions might provide more developed support for students in their learning and growth.

In Chapter 6, Paulo Barroso considers "Digital Image Ethics: Social Practices and Technological Mediations" with an eye to the ways in which digital work transforms social realities. Barroso writes about the idea of visual cultures in our current context, and speaks to the work of exploring ethics in light of both the fixed and ethereal nature of digital information. This chapter includes discussion of citizenship and responsibility, as well as what it means to be ethical, in a visual world that composed, in part, of social media and other media-based ways of generating and sharing information. Overall, Barroso's project is focused on suggesting ways that ethics can be applied to the digital in terms of images.

Chapter 7 is an invited chapter from Kristen Hawley-Turner and Gary Pankiewicz, and is titled "Developing Writers in the New Digital Age: Ethical Stances for Writers and Teachers." Hawley-Turner and Pankiewicz discuss the ethical implications involved in teaching writing, and notes this changing work in the context of changing tools and practices. The authors locate this discussion in the current context of AI technologies, and return to these ethics questions. They speak well to the linking together of ethical foundations, digital possibilities, and composition work with students and calls on teachers to expand notions of authorship, consider the social and digital implications of technologies, and position practices with a view to the history of writing development. This is a timely chapter that serves as a reminder of why writing is an important aspect of this larger conversation.

In Chapter 8, "Metaverse Libraries: Leveraging Technology for Empowering Digital Citizenship," contributor Mohammad Daradkeh explores the notion of an intelligent library, and comments on the field of library science as an evolutionary one. The author considers Metaverse technology as an innovation with potentially positive implications, including the possibility of further shaping library environments. The notion of the Metaverse is considered in terms of theoretical foundations, technological affordances, and implications for library-based practices, including questions of managing resources, space, sharing about cultural and social knowledge, and more. I appreciate Daradkeh's knowledge in this chapter as it extends my understanding as a literacy scholar and represents the author's expertise relative to reading spaces that I love and support.

Following this consideration of library spaces, Chapter 9 in authored by Sjusan Bernardino and Vera Rocha, and expands on social and cultural considerations of economic realities related to educational implications. These authors comment on "Open Innovation and Crowdsourcing in the Public Sector in Portugal: The Participatory Budget," a topic which speaks to the economic trends, social innovations, and intersections with government that undergird economics education. Given the linking of social and political aspects of meaning-making to the implications of literacy practices, this work holds a relevant thread as a means of thinking through foundational knowledge for the ways digital technologies shape the larger, global world of communication.

Chapter 10 is authored by Ali Orhan, and is titled "The Role of New Media Literacy and Critical Thinking on Predicting Digital Citizenship: A Cross-Sectional Study With Turkish University Students." This chapter once more speaks to the global perspective; in this case, Orhan examines the connections between literacy levels with new media and critical thinking dispositions. In all, 124 university students from a variety of departments participated in this study, which drew on the New Media Literacy Scale, Sosu Critical Thinking Dispositions Scale, and Digital Citizenship Scale. Orhan noted that university students demonstrated considerable levels of both new media literacy, as well strong critical thinking dispositions, with moderate measures related to digital citizenship. Orhan also noted a predictive relationship between literacy levels in new media, critical thinking, and digital citizenship, accounting for seventeen percent of the variance related to digital citizenship. Orhan's contribution suggests possibilities for thinking about the interrelations of the concepts, including the importance of digital literacy in relation to digital citizenship, and the role of critical thinking in approaching these topics.

From a quantitative focus, the next chapter explores qualitative inquiry. In Chapter 11, Crystal L. Beach explores the notion that "The Future Is Now: Considerations of Conducting Qualitative Research in Digital Spaces." Beach shares insights into the ways that research inquiry is shaped and reconsidered when taking place in digital spaces, including the ethics, benefits, and drawbacks of this kind of work, with particular attention to qualitative paradigms. The author discusses studies that have been conducted in digital spaces with adolescents in the literacy field as a means of locating this theoretical discussion in practical applications and concrete examples, and speaks to the ways that further implications may take shape in light of AI and other trends and shifts. Beach links both the theoretical and practical with a view to qualitative work.

The authors of Chapter 12 pose the question: "Do We Support Ethical Behavior in Digital Tool Use in Early Childhood?" This chapter is the result of collaboration between Feyza Aydin Bölükbaş, Kübra Engin, Emine Bozkurt Polat, Kadriey Selin Budak, and İlkay Ulutaş, and focuses on the critical nature of modeling and guiding for young children as they negotiate technology. The authors conducted a study with preschool teachers and noted concern from these educators in relationship to the roles of protection from potential harm, appropriate applications of ethical concepts in digital contexts, and boundaries for restrictions and rule-based approaches. Educators expressed a desire for further professional development, and spoke to the need for working with families in approaching digital contexts. These authors capture the thinking and approach of a group of educators, and speak to the ways in which digital environments are approached through a framing of guidance.

In Chapter 13, Vicki Hosek comments on "The Politics of Digital Citizenship Education," first noting the many ways in which the concept of digital citizenship has been approached through research, and considers questions of how this concept is located in theoretical foundations and how digital citizenship is approached in The National Educational Technology Plan (NETP) and by the International Society for Technology Education (ISTE). Hosek further explores questions of roles and responsibilities for sharing about this construct with students, noting the political ramifications of particular approaches. This chapter considers the political implications of those definitions and guidelines. From this review of extant research and policy, the author then shares information from teacher candidates regarding their comfort with the goals that are shared by the NETP and ISTE. Hosek's work links together the theoretical dimensions of digital citizenship, as well as practical and political implications, and I appreciate the way this author has then linked these ideas to pre-service teacher voice.

Finally, Chapter 14, titled "Digital Citizenship and Digital Ethics: An Educator's Perspectives," presents thoughts from Niroj Dahal. Digital citizenship and digital ethics are closely related concepts regarding how to behave responsibly and ethically in the era of artificial intelligence (AI) in the digital world. The author shared his perspectives on digital citizenship and digital ethics based on his experiences working with PGDs, Master, and MPhil students at Kathmandu University School of Education, Nepal. The author attempted to reflect on how PGDs, Master, and MPhil students understood and acted on these concepts—digital citizenship and digital ethics based on the respecting, educating, and protecting framework of digital citizenship and digital ethics. Further, the author argued that digital citizenship is the responsible and respectful use of technology to engage in online activities, look for trustworthy sources, and protect and promote human rights. Similarly, digital ethics are rules and moral principles governing individuals' interpersonal behavior. Overall, the author concluded that digital citizenship and ethics are related concepts that deal with digital technology's moral and social aspects.

What Do We Craft in the Digital?

As I leave the reader to explore the chapters in this volume, I center my final words in this introduction around the notion of crafting. As a literacy scholar, I am struck by the possibilities for composition and access to meaning-making practices in digital spaces; yet, I have more in mind when considering crafting, and revisit a notion of encounter. I have been drawn to explorations of what adolescents and adults can do in terms of creating with digital tools as a key aspect of my research trajectory, including the visual and comics-based work with which creators and authors engage. In what ways, beyond this visual and material composing, are ideas framed, ideologies reified or challenged, identities affirmed, and institutions critiqued? In what ways is the world shaped and reshaped through movements in the digital?

Centering the human is of utmost importance, and this work holds continued implications for concepts of democratic practice and citizenship. As a researcher and teacher, I emphasize the need to continue envisioning and shaping a more inclusive and thoughtful world through critical practices, both within digital encounter and in the wider social realm.

In what ways do the critical and ethical overlap? In what ways, if any, can these concepts be practiced with more efficacy and with insight to their powers and affordances in digital spaces?

Jason D. DeHart
University of Tennessee, Knoxville, USA

REFERENCES

Bilki, Z., Satar, M., & Sak, M. (2023). Critical digital literacy in virtual exchange for ELT teacher education: An interpretivist methodology. *ReCALL*, *35*(1), 58–73. doi:10.1017/S095834402200009X

Black, R., Walsh, L., Waite, C., Collin, P., Third, A., & Idriss, S. (2022). In their own words: 41 stories of young people's digital citizenship. *Learning, Media and Technology*, *47*(4), 524–536. doi:10.1080/17439884.2022.2044848

Buchholz, B. A., DeHart, J. D., & Moorman, G. (2020). Digital citizenship during a global pandemic: Moving beyond digital literacy. *Journal of Adolescent & Adult Literacy*, *64*(1), 11–17. doi:10.1002/jaal.1076 PMID:32834710

Buchholz, B. A., & Pyles, D. G. (2018). Scientific literacy in the wild: Using multimodal texts in and out of school. *The Reading Teacher*, *72*(1), 61–70. doi:10.1002/trtr.1678

Buckingham, D. (2020). Epilogue: Rethinking digital literacy: Media education in the age of digital capitalism. *Digital Education Review*, (37), 230–239. doi:10.1344/der.2020.37.230-239

DeHart, J. D. (Ed.). (2023). *Innovations in digital instruction through virtual environments*. IGI Global. doi:10.4018/978-1-6684-7015-2

Forzani, E., & Leu, D. J. (2017). Multiple perspectives on literacy as it continuously changes: Reflections on opportunities and challenges when literacy is deictic. *Journal of Education*, *197*(2), 19–24. doi:10.1177/002205741719700203

Kaun, A., & Uldam, J. (2018). Digital activism: After the hype. *New Media & Society*, *20*(6), 2099–2106. doi:10.1177/1461444817731924

Luke, A. (2018). Digital ethics now. *Language and Literature*, *20*(3), 185–198.

Masiero, S., & Bailur, S. (2021). Digital identity for development: The quest for justice and a research agenda. *Information Technology for Development*, *27*(1), 1–12. doi:10.1080/02681102.2021.1859669

Pyles, D. G., Rish, R. M., & Warner, J. (Eds.). (2019). *Negotiating place and space through digital literacies: Research and practice*. IAP.

Wargo, J. M. (2015). Spatial stories with nomadic narrators: Affect, Snapchat, and feeling embodiment in youth mobile composing. *Journal of Language & Literacy Education*, *11*(1), 47–64.

ADDITIONAL READING

Kist, W. (2000). Beginning to create the new literacy classroom: What does the new literacy look like? *Journal of Adolescent & Adult Literacy*, *43*(8), 710–718.

Lankshear, C., & Knobel, M. (2006). *New literacies: Everyday practices and classroom learning*. Open University Press.

Lankshear, C., & Knobel, M. (Eds.). (2008). *Digital literacies: Concepts, policies and practices*. Peter Lang.

The New London Group. (1996). A pedagogy of multiliteracies: Designing social futures. *Harvard Educational Review*, *66*(1), 60–92. doi:10.17763/haer.66.1.17370n67v22j160u

Acknowledgment

Strong academic and thoughtful work can only be accomplished with the help of willing reviewers. In addition to the contributors in this volume who participated in peer review, I gratefully acknowledge the following scholars for their help as reviewers:

Tonya Baker, *National Writing Project, USA*

Christian Goering, *University of Arkansas, USA*

Adam Jordan, *College of Charleston, USA*

Beth Lloyd, *College of Charleston, USA*

Alex Panos, *University of South Florida, USA*

Chapter 1
Critical Media Literacy and Relations of Power:
Connecting to Digital Citizenship and Ethics

Donna E. Alvermann
https://orcid.org/0000-0001-6881-0657
University of Georgia, USA

ABSTRACT

This chapter reports on a multi-year project that the author conducted using a children's picture book titled The Tantrum that Saved the World. Published in 2017 by World Saving Books, the e-version with its 64 colorfully illustrated pages tells of a little girl who stares down a pending climate crisis by channeling her tantrum power into strategies for saving the world. In the first study, the storybook's words and illustrations were analyzed using a Foucauldian genealogy to show how power circulates and why acquiring only that knowledge stops short of what a critical media literacy (CML) filter might reveal. Two additional studies, one focused on an analysis of critical literacy outside its comfort zone, and the other, a critical post-humanist analysis of digital remixing in a licensed Creative Commons website provided a basis for connecting the multi-year project to the research literature on digital citizenship and ethics.

"I live in the actual world, I participate, I act—but one eye gazes toward a world that could be but is not yet. I am armed with an imagination, a hope, a passion and willingness to choose. Look for me in the whirlwind." --Maxine Greene (1988)

DOI: 10.4018/978-1-6684-8934-5.ch001

INTRODUCTION

Releasing one's social imagination through aesthetic experiences in education was American philosopher Maxine Greene's way of engaging participants in her seminars on citizenship education at Teachers College, Columbia. A teacher and social activist until her death in 2017, Greene's words still ring relevant in a chapter focused on digital citizenship and ethics. Her conceptual framing of social imagination also speaks to authors of children's picture books, especially in terms of how they position their young audiences in relation to older youths and adults. It is this scenario that raises ethical concerns among professionals who work with young children.

Given that one could easily envision a young child, while engaging with an e-version of *The Tantrum that Saved the World* (Hiebert & Mann, 2017), taking to heart Greta Thunberg's oft-quoted statement that no one is too small to make a difference. After all, the young protagonist Sophia in *The Tantum* did just that when she stared down a pending climate crisis by channeling her tantrum power into strategies for saving the world. According to Megan Otto (2021), a content specialist and marketing writer for the Literary Hub (lithub.com), picture book authors frequently create story characters who oppose the very rules that adults have previously set (https://lithub.com/what-is-the-point-of-childrens-books-about-the-climate-crisis/). Questioning the ethics of this situation and its unrealistic expectations, Otto shared her preference for books that position young children in ways that enable them to recognize they can influence adults around them by engaging as collaborators rather than as adversarial competitors.

AIM OF THIS CHAPTER

The purpose of this chapter, which integrates findings from three different but related projects that Alvermann was involved in between 2017 and 2022, is focused on seeking connections between her findings and those of researchers currently working within a paradigm that addresses digital citizenship and ethics. Research framed within this new paradigm is intended to address issues of civic importance that accelerated during COVID-19, were fast forwarded by advances in instructional technologies, and now appear in the form of generative artificial technologies (e.g., Chat GPT) that are raising questions among teachers, teacher educators, and students about how reading, writing instruction, and even self-sponsored composition may be impacted (Robinson, 2022).

A Brief Flashback

As early as 2020, Buchholz, DeHart, and Moorman had published a commentary in the International Literacy Association's *Journal of Adolescent & Adult Literacy* in which they argued for moving beyond digital literacy to a more inclusive term—digital citizenship—to account for the inequities and numerous other issues that surfaced as a direct consequence of COVID-19. Earlier publications by Mirra (2018) on educating for critical civic empathy and Hawley Turner's (2019) edited book on the ethics of digital literacy had also signaled a new movement in the field of literacy education. However, it took Buchholz et al.'s focus on broadening the knowledge base for digital literacy that caught the present author's attention. Clearly, changes were afoot.

For example, in a review of Hawley Turner's (2019) edited book, *The Ethics of Digital Literacy: Developing Knowledge and Skills Across Grade Levels*, DeHart (2021) traced the need to address certain ethical concerns associated with digital citizenship. Specifically, his review called attention to Hawley Turner's engagement with the International Society for Technology in Education framework for digital citizenship prior to expanding on those standards. Of upmost importance, he emphasized the necessity for developing "an awareness of the ethical questions facing citizens online that are difficult to encompass as a set of skills or competencies" (n. p.). DeHart also pointed to the chapter authors' careful "consideration of a set of steps for teachers to nurture participatory and social justice-oriented digital citizenship as part of the curriculum" (n. p.) By noting that the authors had acknowledged COVID-19's deepening impact on the existing digital divide and social inequities, DeHart left no doubt but what Hawley Turner's book would be well received in the field of education. And, indeed it has been, judging by the many positive reviews of the book, including a recommendation from *Choice Reviews*.

The time it took for the present author to acknowledge important changes and new research directions in the field of literacy education may have been due in part to her reluctance to let go of digital literacy as the umbrella concept for all things that matter in her professional life as a researcher. Certainly as a teacher, a mentor, and an active service provider at the local, national, and international levels, she senses the value of stepping outside herself and learning alongside others more steeped in the research literature on digital citizenship and ethics. It is toward the goal of learning how that literature connects to, pushes against, and/or offers insights into tried and 'true' findings that excites and beckons onward.

STUDY #1: THE ORIGINAL

The impetus for a multi-year project stemmed from a teacher-researcher project Alvermann had initiated during her development of courses for an online Master's degree program that involved preservice and practicing teachers critiquing texts they read aloud to their elementary students or that older students read for themselves. Follow-up, small-group discussions provided spaces in which to share their critiques. One such discussion of *The Tantrum that Saved the World* prompted what evolved into a multi-year project that first focused on critical media literacy and a picture book's underlying relations of power (Alvermann, 2019).

The Tantrum That Saved the World

This 2017 children's picture book grew from a Kickstarter project, one of several popular crowd-source funding portals on the internet. Herbert described the process in a three-minute video that included the authors' rationale and credentials. The book's protagonist (Sophia) is portrayed as a little girl (possibly a nine- or ten-year-old) who initially views the unexpected arrival of climate refugees on her front porch as an intrusion and an unwelcomed disturbance in her small, peaceful part of the world.

Sophia was minding her business one day,
When quite without warning, a bear came to stay.
The ice that he lived on had ceased to exist.
He hoped that Sophia would kindly assist.
Startled and flustered, Sophia said, "**No.**"
But the bear came right in. He had nowhere to go.

Soon other refugees passed through Sophia's front door: a family from Kiribati, a low-lying island nation in the equatorial Pacific Ocean; a New England fisherman; an Andean flamingo; a Bengal tiger; a young Syrian farm family; a sea turtle, and bees. Each had a story to tell about the increasingly warm temperatures that were affecting them. Each also *'expected Sophia to halt their demise.'* Claiming she is just a kid, Sophia convinces the refugees that they must join forces and go with her to City Hall *'where the powers that be...would sort this mess once and for all.'* But the adults at City Hall proved to be of no help and told the motley crew to *'take a number and wait, if you please.'* Hours passed and no one paid any heed to Sophia's explanation that even if she and the folks at City Hall were not feeling the effects of climate change in their little part of the world, *'everyone loses when one part's neglected.'*

It's at this point that Sophia takes matters into her own hands. A colorful illustration shows her turning red in the face and throwing a tantrum as a result of sensing how it is her future, too, that the adults in City Hall are selling out. As the climax unfolds, Sophia is the source who urges cooperative action. She proposes that teamwork and sheer strength in numbers of people doing practical things (individually and daily) can buy time for Earth's living inhabitants to adapt to climate change. But a happy storybook ending was not the authors' sole intent. Indeed Herbert and Mann envisioned their book being part fiction, part science, and part action plan, consisting of Sophia's six Rs: Refuse (things you don't need); Repair (mend rather than throw away); Rot (compost food waste); Reuse (repurpose and share); Recycle (plastic packaging); Reduce (your impact on Earth)

The relations of power revealed in the so-called turnabout of this picture book on how Earth's inhabitants treat their world is worthy of analysis if for no other reason than it demonstrates how power circulates. According to Foucault 1990/1976), "Power is not something that is acquired, seized, or shared, something one holds on to or allows to slip away." (p. 94). In other words, one cannot 'empower' someone else. To even attempt to do so, Foucault argued, would simply imply that the one holding power initially decides when, where, and how to dispense it. Expanding on Foucault's concept of power relations, Lilja and Vinthagen (2018) introduced the notion of 'dispersed resistance', which they claim can show itself in unremarkable everyday subtleties (e.g., Sophia's six Rs) or more demonstrable and loud outbursts (her tantrum at City Hall).

An Analysis of Underlying Power Relations

For purposes of analyzing *The Tantrum's* content (illustrations and words), the author chose to apply Foucault's (1977) approach to genealogy. In this instance, the term 'genealogy' bears no resemblance to the more common method of researching one's ancestors. Nor to history's chronological listing of events in time-order fashion. Instead, Foucault's genealogy begins with a question posed in the present. For example, How was it possible for Herbert and Mann (2017) to have co-authored an award-winning picture book that ostensibly earns Otto's (2021) general critique of authors who create young story characters (e.g., Sophia) as children who are left to deal with a crisis for which adults were responsible?

In answering that question, the author of this chapter used the conceptual tools of genealogy such as descent, emergence, and deconstruction (cf. Hoffman & Alvermann, 2020) to discover 'accidents' that involved inherently complex networks of power relations (including domination, submission, and resistance). The use of these tools uncovered forces at work that would otherwise have remained hidden or badly flawed. Piecing together such evidence made it possible to answer the study's research question. Basically, relations of power that emerged along parallel pathways were interrupted accidentally at various points along the way. Or as Foucault would explain the process, "People know what they do, they frequently know why they do what they do, but what they don't know is what *what* they do does" (Dreyfus & Rabinow, 1983, p. 187, emphasis added). This somewhat quirky five-word phrase ('what *what* they do does) is akin to thinking about normalizing unstable systems, which Dreyfus and Rabinow are quick to explain in this way:

There is no inherent logic of stability. Rather, at the level of practices there is a directionality produced from petty calculations, clashes of wills, meshing of minor interests. These are shaped and given a direction by the [politics of power]. This directionality has nothing inherent about it and hence cannot be deduced. It is not a suitable object for a theory. It can, however, be analyzed, and this was Foucault's project. (Dreyfus & Rabinow, 1983, p. 188)

STUDY #2: CRITICAL LITERACY OUT OF ITS COMFORT ZONE

In this section, *The Tantrum that Saved the World* appeared in an invited chapter co-authored by George Lovell Boggs, Nerida Spina, Donna Alvermann, and Barbara Comber (2021). Its purpose was to challenge and help move beyond an old cliché of critical literacy scholarship that weakens—'we go along to get along.' The intended audience included educators and researchers who embrace critical awareness as a key part of their responsibility for (and in consequence of) research-based knowledge production.

From the start of their journey to bring critical literacy out of its comfort zone, the four teacher educators/literacy researchers (Boggs and Alvermann from the United States and Nerida Spina and Barbara Comber from Australia) had encountered varying levels of difficulty in obtaining a print copy of *The Tantrum*. For example, Barbara discovered she would need to order a copy and wait for its arrival from the United States, or go against her usual preference and order it as a digital text. When Barbara investigated further, she discovered that one of the book's authors was Megan Herbert, an Australian writer and illustrator who teamed up with Michael Mann, an award-winning American climate scientist, to produce the picture book.

Initially, Barbara's interest was piqued by the fact the picture book was not readily available in Australia. Wondering if it was because she had tried to buy it from an independent bookseller, Barbara checked with a larger Australian book dealer. However, a search by an employee of that establishment proved futile, though other children's books with similar titles surfaced, including *The Worms that Saved the World* (Doyle & Deeley, 2017) and *The Fart that Saved the World* (Gill, 2017). In the end, after she had decided to purchase the online ('carbon-free') version of *The Tantrum that Saved the World*, payment to either Amazon or Pay Pal required Euros rather than Australian dollars.

In reflecting on the access issues accompanying Barbara's search, her co-researchers had to wonder how anyone without her financial and technological resources, mobility, social networks, and linguistic competencies would have navigated a situation that complex. The situation certainly drew attention to how interlocking certain elements of privilege can be—that is, how inextricably tied they are to forces of domination, inequity, and control. The irony of their situation was not lost on the four researchers. Nor did it halt their intention to push critical literacy out of its comfort zone. With this as background, it is no surprise to learn that their next Zoom discussion focused on identifying suitable options for conceptually framing their attempts to alter how critical literacy is currently perceived.

The two options that seemed of greatest fit were Smith's (2005) approach, as an institutional ethnographer, to studying how texts of varying kinds effectively organize and regulate people's work, and

Foucault's (1997) interest in ethics, subjectivity, and truth. Ultimately, the team agreed to frame its write-up (Boggs, Spina, Alvermann, & Comber, 2021) mostly around Smith's approach as a critical ethnographer. However, findings from the earlier Foucauldian analysis of Sophia's tantrum (study #1) were engrained in Boggs' consciousness, largely because of the latter's appreciation of parables and their usefulness to him as a songwriter. Due to the present chapter's focus on digital citizenship and ethics, Boggs' interpretation of the Foucauldian analysis seems most relevant here.

The basic parameters of [Foucault's] critical philosophy offer to challenge the justice of social justice agendas that so readily cling to critical literacy research. If *The Tantrum* were read as a parable of critical literacy research, the story might be darkly humorous. For instance, advanced industrialized states whose civilizations prospered to the extent they were able to efficiently extract and use natural resources of all kinds, including humans, may indeed welcome the world's climate refugees and they may indeed produce the best-known and loudest voices calling for change. But it won't be because of the orderliness of a cosmic narrative of righting wrongs, or because human lives or the lives of polar bears are sacred. It will be because globalization, including its academic apparatus, has conditioned individuals themselves to accept this evolution of power relations. A particularly pointy barb in the imagined parable is that Sophia is only aware of her subjectivity in the most naive and unwieldy way. She senses disturbances in *her* environment, but sees *others* as the problem, yet still sees *herself* alone as the source of remedy (Routledge e-version of Boggs, et al., 2021, p. 6).

While this parable of critical literacy research will likely have its critics, it also falls well within Greene's call for social imaginaries in "…a world that could be but is not yet." A world in which ideologically coded texts are exposed for the role they play in maintaining hidden power relations. Whether or not future critical educators and researchers will be joined by others occupying influential positions in sectors of government, industry, commerce, and the media to stop the ravages of climate change and migration is a question worthy of consideration by passionate, digital citizens equipped with appropriate knowledge and ethical "know-how" to act responsibly and effectively. The time has come (and perhaps passed) when well-meaning authors of children's picture books are made scapegoats. Or when children of Sophia's age are tasked with saving the world they've inherited.

STUDY # 3: DIGITAL REMIXING IN A POSTHUMAN WORLD

In this section, *The Tantrum*'s theme broadens to include technology and social media, in particular *Tik Tok* and *Becoming3lectric*. This time, a new textual analysis of the picture book's protagonist, Sophia, focuses on exposing a text's ideological codes. For that focus to be useful, it is necessary that readers have access to the study's Creative Commons site: http://www.becoming3lectric.com.

The Tantrum's Further Involvement

But first a brief description of how Sophia figured into the third and final analysis of *The Tantrum that Saved the World* is in order. If one were to read that picture book in ways that the institutional ethnographer Dorothy E. Smith fosters (see Stanley, 2018), the first step would be to define the *text*'s ideological code, not Sophia's. So the question becomes one slightly akin to the oft-quoted phrase "mirrors, windows, and sliding glass doors" (Bishop, 1990, p. ix-xi). Who is Sophia? A picture book character who sees herself as part of the world (mirror), through other peoples' eyes (window), or through an experience that connects her to the world (sliding door)? Arguably, she is all three at different points in the story's unfolding, just as Hutcherson Price proved to be in the digital remix study (Beach, Alvermann, Loomis, Wright, & Hutcherson Price, 2023). For more on the Sophia/Hutcherson Price analogy, see s later section in this chapter titled "Exposing Ideological Texts."

For now it is sufficient to focus on how Smith (2005) might have connected Sophia to Hutcherson Price and hence the digital remix project (study #3). Smith would have asked what work is *The Tantrum* text being asked to do? Is it to function as a token of multiculturalism, a climate-change anxiety relaxer, or a meme announcing a serious media campaign to save the world? By asking such questions of the text, it is possible to expose a text's ideological code and its potential impact on a story character such as Sophia. Practice in reading such codes is precisely what Smith would envision as an appropriate take-up of her life's work. Practice level questions also loosely connected *The Tantrum* and Sophia to the questions on entanglement posed by Beach et al. in their study of digital remixing in online classes.

Entanglements

In the term *entanglements* is a clue that points to research designs commonly associated with critical posthuman literacy, a concept that Leander and Burriss (2020) claim is a bridge between critical literacy studies and posthumanist theory. They describe it as a juncture in time when advances in technology, algorithmic computational agents, and artificial intelligence make it possible for readers and composers to 'become' with machines. Noting that critical literacy educators and researchers are trailing professionals in other closely related fields, Leander and Burriss attribute this lag to the prevalence of mediational and representational perspectives held by a majority of scholars in critical literacy studies. Arguably, it is the combination of mediational and representational bias that is proving inadequate for researching within a critical posthuman literacy framework. To be clear, Leander and Burris are careful not to dismiss earlier work of scholars such as Smith (2005) on the grounds that a person's ideological preferences and professional identities are choices one makes for reasons that align with their own beliefs and goals in life.

Entanglements of a different kind that cut across studies involving critical media literacy and posthumanist theories include one on Becoming3lectric.com (Alvermann, Wright, & Wynne, in press) and one on Tik Tok (Wynne, Wright, and Alvermann, 2021). Because the first is an extension of the digital remix project and still in press), the second one on Tik Tok is discussed here. Briefly, Judith Butler's (1990) theory on gender performativity (1990) and queer theory (Butler, 1993) provided a hybridized framework that supported inquiry into how three Gen-Z Tik Tokers perceived themselves as being performed *by* gender rather than as subjects doing the performing. This finding is a twist of sorts in terms of how gender performative theory has typically been studied in the past. What may be if greater interest to readers of this chapter on connecting to digital citizenship and ethics is how Wynne's involvement in the Tik Tok study led to her unexpected interest in designing a social media channel inside Tik Tok that hosted her Master's thesis project—one similar to the digital remixing online study but of shorter duration.

On Being a Digital Citizen Experiencing Digital Ethics

On a practical level the digital remix project (study #3) experienced some difficulties from the start that were findings of a sort Alvermann had not foreseen when choosing to do a long-term study of digital remixing that involved heavy oversight by her university's human subjects office. Findings (boundaries 1- 3 below) were presented as part of an alternative format session at the annual meeting of the Literacy

Research Association (Beach, Alvermann, Loomis, Wright, & Hutcherson Price, 2022). The audience could not have access to Becoming3lectric (the Creative Commons website) because the study was still undergoing blind review. All the audience could know was the purpose of the study—to explore ways in which the cultural practice of digital remix might mediate human participants' and nonhuman participants' (the remixes) means of connecting on a public website using Vagle's (2012) post-intentional approach to phenomenology as the analytic of choice.

First Boundary

The researcher's proposal to conduct a study completely online (including the recruitment of participants, data collection, and analysis using Vagle's 2012 post-intentional approach to phenomenology) under a licensed Creative Commons website for an indeterminate period of time took an unusually long time to get approved, largely because it was the first of its kind at her university in 2014. The IRB staff supported the researcher in meetings that involved outside consulting on legal matters.

Second Boundary

The IRB's decision to withhold the right of potential contributors to upload their remixes to the Creative Commons website resulted in a setback, design-wise, for the study. While we had originally written the proposal with inclusive intentions in mind, we were required by a panel of IRB reviewers to have a member of the research team approve or disapprove each contributor's offering. Undoubtedly, this move discouraged some potential participants from volunteering to join the study. As an aside, it is worth mentioning that never once did any team member have to censor a contribution intended for http://www.becoming3lectric.com.

Third Boundary

From the start of the journal submission process, difficulties arose due to constraints associated with potential reviewer bias and blind review. By definition, remixes are endless hybridizations of other people's remixes; thus, any charge of plagiarism should not apply. A related issue was the fact that as researchers, the team could not include any examples of participants' remixes in manuscripts submitted for review. Not surprisingly, reviewers were reluctant to spend time evaluating a manuscript that had no visible findings (artifacts). This information is shared for the purpose of potentially gaining peer support in adjusting the blind review process to meet the realities of conducting research fully and openly online.

FURTHER CONNECTIONS TO DIGITAL CITIZENSHIP AND ETHICS

As noted earlier in the chapter's purpose statement, the author integrated findings from three different but related analyses of a popular children's picture book on climate change in an attempt to make connections between her findings and those of researchers currently working within a paradigm that addresses digital citizenship and ethics.

Sophia's Centeredness as a Digital Citizen

Verbal and visual characteristics that portrayed Sophia, the protagonist, in an e-version of *The Tantrum that Saved the World* were first analyzed using a Foucauldian genealogy to show how power circulates and why acquiring only that kind of information stops short of what a critical media literacy (CML) filter could reveal. In a subsequent invited chapter for a research handbook on critical literacies (Pandya, Mora, Alford, Golden, & De Roock, 2021), the ethics of Sophia's actions were analyzed by moving critical literacy outside its comfort zone to lay bare some assumptions about its credibility as an analytic frame. A third and final analysis of Sophia's role in *The Tantrum* used the text's ideological codings to pave the way for her eventual appearance in the connections section. Here, Sophia might be described as an apt muse for Maxine Greene and present-day followers of her social imagination project. Or at the very least a hovering, personified influence in the author's naming of three areas of connection between Sophia and the researchers cited in the emergent literature on digital citizenship and ethics.

Educating for Empathy

In Mirra's (2018) earlier contributions to the digital citizenship and ethics realm through her book on civic engagement, there was a broadening of the field that served to alert teachers, teacher educators, and administrators in some instances that conditions were not going to change at the secondary and tertiary levels without a concerted effort on everyone's part to imagine futures where students were included as stakeholders in their own education. That she never promised instant turnarounds, nor situated her work in subfields of education such as content area reading for subject matter teachers, or disciplinary literacy for those who knew their own specialty areas (e.g., how they were structured) sufficiently well to mentor students in a new process. Instead, Mirra picked up on the potential appeal of civics at a time when most educators were doing well to stay connected to their students, complete overloads of paper work, and still have a life. The Pandemic years of course would introduce new challenges, many of which depended on students and teachers alike to develop fairly thick skins while not losing empathy for those who are faring less well.

That both Mirra and Garcia would find common ground in participating in the Digital Democratic Dialogue (3D) (2022) online social network of teachers and students from six demographically distinct communities across the US was a milestone in the digital citizenship and ethics movement. Taking seriously what secondary students accurately defined as the ills of this country and then insisting on action projects to move toward social imaginaries, which perhaps are still a far reach (Greene, 2020), is nothing short of genius (Muhammad, 2020). With "Other suns: Designing for racial equity through speculative education" (Garcia & Mirra, 2020) making the rounds in social media, the blogging community, and required reading lists of academics who find the writing style invigorating, it is almost a guarantee that Mirra's idea of educating for empathy will be around for a long time. After all, who can argue with notions of humaneness, critical empathy, and speculative education (Mirra & Garcia, 2020; Mirra & Garcia, 2022; Thomas & Corbitt, 2019).

In a professional space where media literacy educators, such as De Abreu (2023), rub elbows with media magnates, journalists, and climate scientists effortlessly on a weekly basis is to know that educating for empathy doesn't require a classroom full of young people who are action oriented in a civic-minded way. Indeed, De Abreu's project is shaping dialogue amid broken conversations in the world of media. To do that successfully is to recognize that "we are in a world of cultural climate change" (p. 1) where people talk past each other all too often, or worse, suddenly feel at odds with society. De Abreu offers a way out of hopelessness by practicing what she calls "nuanced speaking" (p. 2), a practice that involves talking in shades of gray—a point Toni Morison (1993) endorsed three decades ago.

Exposing Ideologically Coded Texts

Digital ethics, which are part and parcel of being a digital citizen today, tend to dominate research projects that focus on exposing ideologically coded texts. For instance, Brownell's (2021)work with elementary-age students demonstrates that building on the understanding that no texts are neutral leads naturally into a closer look at the politics of text. Brownell observed that even young children can agentively participate in activities that require applying critical literacies and appropriate rhetoric when they are supported and have opportunities to act on real world political issues. The real world of Hutcherson Price includes adhering to strict time scheduling in a daily work week. As a participant in the online remixing study (Beach et al., 2023), Hutcherson Price's remix "Working 8 – 5" was inspired by Maya Angelou's poem, "I Know Why the Caged Bird Sings." It was the most popular remix on the Creative Commons website. In Hutcherson Price's response to a critic's remix, she wrote with indignation and anguish:

Stung. Slapped. Misunderstood. Chastised. Gaslit. "That's how I felt after reading the critique of my heart-pouring cry for intellectual and creative freedom at work." The intensity of Hutcherson Price's hurt feelings was balanced by another remix that provoked a flood of happy emotions. In that remix, a simple word with flowers around it, made her immediately cry when she saw it. "I experienced comradeship and felt relieved and safe."

Later, while reflecting on the incident, the research team of which Hutcherson Price was a member wondered to what degree author-ownership of a remix is retainable, especially when an ideologically coded text is involved.

Although no answer was forthcoming in Hutcherson Price's group, it is interesting to note that a significant number of other researchers who hail from disciplines outside education are attempting to deal with similar issues. For instance, Wright, Sandlin, and Burdick (2023) concluded from their study of fake news and a number of other kinds of texts in a post-truth era that responsibility falls on the digital citizen to sort out potential answers. In their words, "it is the radical transmission of dissonant affects, inevitably looming traumas, and a world characterized by unending, carnal violence that shape the political views and identities of viewers and learners.

Engaging the Challenges of AI and Algorithmic Literacies

A major concern among researchers working in the new paradigm of digital citizens and ethics is the fact that few digital citizens are engaged with platform research. This is evident from the author's multi-year project as well as from the emergent research literature on digital citizens and ethics. It seems probable that part of the difficulty lies in finding an identity with which those researchers already invested in AI and Algorithmic Literacies can agree on (Garcia & Nichols, 2020; Hendrix-Soto & Nash, 2023; Yildiz, 2023). Platform designs that were popular in the past have typically tended to ignore criticality as a feature of any great virtue. However, with digital literacy scholars now on the scene, Nichols and LeBlanc (2021) are toying with the idea of dropping the word "literacy" and going with "civic media ecology" as a substitute. Their "ecological" stance, according to Nichols and LeBlanc, avoids pointing to previously held attachments to local, representational, and humanist perspectives. All things considered, such a move would seem to invite a sameness that suffocates.

In general, the advent of new literacy practices have historically led to ideological struggles. For instance, a decade ago societal influences outside the classroom led to introducing digital writing instruction to children of elementary age in Australia. This event, which subsequently sparked controversy over user-generated and often co-constructed web content, posed significant challenges to the expert-dominated field of teaching and teacher education (Mills & Exley, 2014). A similar stir is occurring now with the acceleration of AI via natural language generation platforms (e.g., Chat GPT), which could potentially change how we perceive truth, creativity, plagiarism, and any number of other concepts associated with teaching and interpreting written texts (Nash, Hicks, Garcia, Fassbender, Alvermann, Boutelier…2023; Robinson, 2022). For the most part, English educators are no longer fearful of being replaced by AI platforms and devices that come and go in rapid succession. Nor are they prone to dismiss the corporate influence of big tech companies and algorithms on what is available for reading and writing. However, despite an ever growing awareness of the degree to which they are capable of maintaining a professional presence in the lives of children and young people, English educators and critical literacy theorists have yet to take into sufficient account how AI is changing what it means to be critically literate in a digital age (Leander & Burriss, 2020). Until those missing pieces of pedagogical knowledge and know-how are in place, it is doubtful that ELA teachers and teacher educators will be successful in ensuring that equity and quality education are viewed and enacted as basic human rights (Belshaw, O'Byrne, & Salmon, 2023).

OVERVIEW

In the introduction to this chapter, Alvermann noted: It is toward the goal of learning how research on digital citizenship and ethics connects to, pushes against, and/or offers insights into her own work in a digital literacies framework that excites her.

Months later after several self-initiated revisions of this chapter, she can say with certainty that literacy scholars who are committed to the new paradigm are far more action oriented than she is. This is evident not only in their discursive claims but also in the kinds of topics they are researching and where they are choosing to inquire—mostly in diverse community settings outside established institutions of teaching and learning. That's the pushback, and she is inclined to take it seriously—to spend less time 'thinking' in order to have more time for 'doing' action-oriented research.

Where connections between digital citizenry and its associated ethics seem to overlap most clearly with Alvermann's agenda as a digital literacies researcher is in the area of multimodal media (broadly defined) that exert little or no effort to expose relations of power responsible for systematically maintaining the status quo (Rowsell, Morrell, & Alvermann, 2017). This situation may not be hopeless but it definitely aggravates, especially when issues of justice are at the core. Taking two steps backward for every one step forward is not in her definition of progress.

Finally, an important insight gained from writing this chapter lies in Alvermann's recognition that continuing her scholarly fascination with all things posthuman is about to be interrupted. As much as she enjoys engaging with Braidotti's (2018) thinking along the lines of a theoretical framework for critical posthumanities, or 'becoming with machines' (Leander & Burriss, 2020), it is Maxine Greene's (1988) invitation to 'look for [her] in the whirlwind' that beckons.

REFERENCES

Aguliera, E., & de Roock, R. S. (2022). Datafication, educational platforms, and proceduralized ideologies. In L. Pangrazio & J. Sefton-Green (Eds.), *Learning to live with datafication: Educational case studies and initiatives from across the world* (pp. 100–118). Routledge. doi:10.4324/9781003136842-6

Alvermann, D. E. (2019). The power of discourse: CML and "The Tantrum that Saved the World.". *International Journal of Critical Media Literacy*, *1*(1), 128–136. doi:10.1163/25900110-00101011

Alvermann. D. E. (2022a). *Cultural relevancy matters: Intersection of methods and epistemology*. Virtual presentation for Doctoral Summer School at UCLouvain-la-Nueva, Belgium.

Alvermann, D. E. (2022b). *Three Boundaries from Start to Finish.* Annual meeting of the Literacy Research Association, Phoenix, AZ.

Alvermann, D. E., & Hagood, M. C. (2000). Critical media literacy: Research, theory, and practice in "New Times.". *The Journal of Educational Research*, *93*(3), 193–205. doi:10.1080/00220670009598707

Alvermann, D. E., Wright, W. T., & Wynne, E. (in press). Researching media literacy practices using both critical and posthuman inquiry. In P. Fastrez & N. Landry (Eds.), *Media literacy and media education research methods*. [Part of the Routledge Research in Media Literacy and Education Series directed by Pete Bennett and Julian McDougall.]

Beach, C. L., Alvermann, D. E., Loomis, S., Wright, W., & Hutcherson Price, L. (2023). Digital remixing online: Entangled feelings. [Open Access]. *Digital Culture & Education*, *14*(4), 92–108. https://www.digitalcultureandeducation.com/volume-14-4

Belshaw, D., O'Byrne, I., & Salmon, T. (2023). Algorithm and data literacy. Response to UNESCO's call for contribution. *We Are Open.* https://blog.weareopen.coop/ai-literacy-a-cross-sectional-slice-of-new-literacies-1906c8e44abe

Bishop, R. S. (1990). Mirrors, windows, and sliding glass doors. *Perspectives - Gerontological Nursing Association*, *6*, ix–xi.

Boggs, G. L., Spina, N., Alvermann, D. E., & Comber, B. (2021). Critical literacy out of its comfort zone—Productive textual tantrums. In J. Z. Pandya, R. A. Mora, J. H. Alford, N. A. Golden, & R. S. De Roock (Eds.), *The handbook of critical literacies* (pp. 327–334). Routledge. doi:10.4324/9781003023425-38

Bolter, J. D. (2016). Posthumanism. *The International Encyclopedia of Communication Theory and Philosophy.* https://onlinelibrary.wiley.com/doi/10.1002/9781118766804.wbiect220

Braidotti, R. (2018). A theoretical framework for the critical posthumanities. *Theory, Culture & Society*, *36*(6), 31–61. doi:10.1177/0263276418771486

Brownell, C. (2021). Children's rhetoric in an era of (im)migration: Examining critical literacies using a cultural rhetorics orientation in the elementary classroom. *Research in the Teaching of English*, *55*(3), 265–288. doi:10.58680/rte202131186

Buchholz, B. A., DeHart, J., & Moorman, G. (2020). Digital citizenship during a global pandemic: Moving beyond digital literacy. *Journal of Adolescent & Adult Literacy, 64*(1), 11–17. doi:10.1002/jaal.1076 PMID:32834710

Butler, J. (1993a). *Bodies that matter: On the discursive limits of "sex"*. Routledge.

Butler, J. (1993b). Critically Queer. *GLQ, 1*(1), 17–32. doi:10.1215/10642684-1-1-17

Carroll, W. K. (2010). 'You are here': Interview with Dorothy E. Smith. *Socialist Studies/Etudes Socialistes, 6*(2), 9–37.

De Abreu, B. S. (2023). Shaping dialogue amid broken conversations. In B. S. DeAbreu (Ed.), *Media literacy, equity, and justice* (pp. 1–7). Routledge.

DeHart, J. D. (2021, June 23). Review of *The ethics of digital literacy: Developing knowledge and skills across grade levels* by K. Hawley-Turner. *Education Review, 28*. doi:10.14507/er.v28.3111

Doyle, K., & Deeley, S. (2017). *The worms that saved the world*. Kevin Doyle Books.

Dreyfus, H. L., & Rabinow, P. (1983). *Michel Foucault: Beyond structuralism and hermeneutics* (2nd ed.). University of Chicago Press. doi:10.7208/chicago/9780226154534.001.0001

Foucault, M. (1977). Nietzsche, genealogy, history. In D. Bouchard (Ed.), *Language, counter-memory, practice: Selected essays and interviews* (pp. 139–164). Cornell University Press.

Foucault, M. (1990). The history of sexuality: An introduction (Vol. 1; R. Hurley, Trans.). Vintage Books.

Foucault, M. (1997). The hermeneutic of the subject. In P. Rabinow (Ed.), *Michel Foucault: Ethics, subjectivity and truth* (HurleyR., Trans.). The New Press.

Garcia, A., & Mirra, N. (2020). Other suns: Designing for racial equity through speculative education. [published online]. *Journal of the Learning Sciences, 32*(1), 1–20. doi:10.1080/10508406.2023.2166764

Garcia, A., & Nichols, T. P. (2021). Digital platforms aren't mere tools—They're complex environments. *Phi Delta Kappan, 102*(6), 14–19. doi:10.1177/0031721721998148

Garland, D. (2014). What is a "history of the present"? On Foucault's genealogies and their critical preconditions. *Punishment & Society, 16*(4), 365–384. doi:10.1177/1462474514541711

Gill, M. C. (2017). *The fart that saved the world*. M. C. Gill.

Greene, M. (1995). Art and imagination: Reclaiming the sense of possibility. *Phi Delta Kappan, 76*(5), 378–382.

Greene, M. (1998). Introduction. In W. Ayers, J. Hunt, & T. Quinn (Eds.), *Teaching for social justice* (pp. xxvii–xivi). The New Press.

Hawley Turner, K. (2019). *The ethics of digital literacy: Developing knowledge and skills across grade levels*. Rowman & Littlefield.

Hendrix-Soto, A., & Nash, B. (2023). Critical approaches to media literacy in teacher education: Accounting for the challenges of post-truth politics. [JoLLE]. *Journal of Language & Literacy Education, 19*(1), 1–16.

Herbert, M., & Mann, M. E. (2017). *The tantrum that saved the world*. World Saving Books.

Hoffman, J. V., & Alvermann, D. E. (2020). What a genealogical analysis of Nila Banton Smith's *American Reading Instruction* reveals about the present through the past. *Reading Research Quarterly, 55*(2), 251–269. doi:10.1002/rrq.269

Leander, K. M., & Burriss, S. K. (2020). Critical literacy for a posthuman world: When people read, and become, with machines. *British Journal of Educational Technology, 51*(4), 1262–1276. doi:10.1111/bjet.12924

Lilja, M., & Vinthagen, S. (2018). Dispersed resistance; Unpacking the spectrum and properties of glaring and everyday resistance. *Journal of Political Power, 11*(2), 211–229. doi:10.1080/2158379X.2018.1478642

Luke, A. (2012). Critical literacy: Foundational notes. *Theory into Practice, 51*(1), 4–11. doi:10.1080/00405841.2012.636324

Mills, K. A., & Exley, B. (2014). Time, space, and text in the elementary school digital writing classroom. *Written Communication, 31*(4), 434–469. doi:10.1177/0741088314542757

Mirra, N. (2018). *Educating for empathy: Literacy learning and civic engagement*. Teachers College Press.

Mirra, N., & Garcia, A. (2020). I hesitate but I do have hope: Youth speculative civic literacies for troubled times. *Harvard Educational Review, 90*(2), 295–321. doi:10.17763/1943-5045-90.2.295

Mirra, N., & Garcia, A. (2022). Guns, schools, and democracy: Adolescents imagining social futures through speculative civic literacies. *American Educational Research journal, 59*(2).

Morrison, T. (1993) *Toni Morrison: The Nobel Prize in Literature 1993*.

Muhammad, G. E. (2020). *Cultivating genius: An equity framework for culturally and historically responsive literacy*. Scholastic.

Nash, B., Hicks, T., Garcia, M., Fassbender, W., Alvermann, D., Boutelier, S., McBride, C., McGrail, E., Moran, C., O'Byrne, I., Piotrowski, A., Rice, M., & Young, C. (2023). Artificial intelligence in English education: Challenges and opportunities for teachers and teacher educators. *English Education, 55*(3), 201–206.

Nichols, T. P., & LeBlanc, R. J. (2021). Media education and the limits of "literacy": Ecological orientations to performative platforms. *Curriculum Inquiry, 51*(4), 389–412. doi:10.1080/03626784.2020.1865104

Otto, M. (2021). What is the point of children's books about the climate crisis? *Literary Hub*. https://lithub.com/author/meganotto/

Pandya, J. Z., Mora, R. A., Alford, J. H., Golden, N. A., & De Roock, R. S. (Eds.). (2021). *The handbook of critical literacies*. Routledge. doi:10.4324/9781003023425

Robinson, B. (2022). Speculative propositions for digital writing under the new autonomous model of literacy. *Postdigital Science and Education*, *5*(1), 117–135. doi:10.100742438-022-00358-5

Rowsell, J., Morrell, E., & Alvermann, D. E. (2017). Confronting the digital divide: Debunking Brave New World discourses. *The Reading Teacher*, *71*(2), 157–165. doi:10.1002/trtr.1603

Smith, D. E. (2005). *Institutional ethnography: A sociology for people*. Rowman Altamira.

Stanley, L. (2018). *Dorothy E. Smith feminist sociology & institutional ethnography*. X Press.

Thomas, P., & Corbitt, A. (2019). Moving beyond troll rhetoric and facilitating productive online discourse in K. Hawley Turner (Ed.). The ethics of digital literacy: Developing knowledge and skills across grade levels (pp. 81-87). Rowman & Littlefield.

Vagle, M. D. (2014). *Crafting phenomenological research*. Routledge.

Wright, R. R., Sandlin, J. A., & Burdick. (2023). *What is critical media literacy in an age of disinformation?* Open Access Wiley Online Library. doi:10.1002/ace.20485

Wynne, E., Wright, W. T., & Alvermann, D. E. (2021). Creating gaps in understanding: How Genz Z disrupts gender norms on Tik Tok. *The International Journal of Critical Media Literacy*, *3*(1), 1–23. doi:10.1163/25900110-03030001

Yildiz, M. N. (2023). Algorithmic social justice through participatory action research. In B. S. De Abreu (Ed.), *Media literacy, equity, and justice: Media binds or blinds?* (pp. 243–256). Routledge.

KEY TERMS AND DEFINITIONS

Climate Change: Attributes the warming of planet Earth to human activities in the so-called Anthropocene Era that continue to contribute to changes in wind and air currents, rainfall patterns, droughts, storms, melting ice, and the resultant rise in sea level (Herbert & Mann, 2017).

Critical Literacy: Refers to the use of technologies of print and nonprint media as forms of human communication that need analyzing for purposes of critiquing and transforming normative structures, ruling systems, and everyday practices that govern how people live and interpret their lives and the lives of others (Luke, 2012).

Critical Media Literacy (CML): Seeks to understand how the meanings that audiences make of various media texts (both print and nonprint) are negotiated in relation to one's different situations and positionings (e.g., adult, child, teenager, sexual orientation, race, ethnicity, socioeconomic class, and cultural contexts). In this way, CML pushes back on assumptions about consumerism such as who or what controls how humans act (Alvermann & Hagood, 2000).

Critical Posthumanism: Argues that new fields of transdisciplinary knowledge, sometimes referred to as critical posthumanities, are emerging from convergences along a spectrum that includes cognitive capitalism, critical post-humanities, and post-anthropocentrism. Analytically, this spectrum is thought to range from a critique of the humanist 'ideal' to critiques of hierarchies in various species (e.g., human exceptionalism) (Braidotti, 2018).

Foucauldian Genealogy: Invokes French philosopher Michel Foucault's approach to analyzing historical materials for purposes of writing critical history in ways aimed at bringing about a revaluing of values in the present day (Garland, 2014).

Institutional Ethnography: A method of inquiry developed by Marxist, feminist scholar Dorothy E. Smith to initially study the everyday social networking relations among women (but later extended to people in general). Smith fostered inquiry that begins from the standpoint of people, rather than from objectified, discursively constructed positions (Carroll, 2010).

Posthuman World: Designates a society where arguments in favor of the materiality of digital devices has become more convincing with the rise of social media, such that embodied digital practices of their users extend and maintain social relationships—rather than simply displace them (Bolter, 2016).

Relations of Power: A concept French philosopher Michel Foucault views as not something that is acquired, seized, shared—something that one holds on to or allows to slip away. Instead, power circulates and is constantly at play both in non-egalitarian and mobile relations (Foucault, 1990/1976).

Textual Analysis: A method for analyzing texts (both print and non-print media) in institutionally-located contexts that come with implied instructions for reading and establishing authorization. When texts self-constitute as factual, then differences (e.g., contrary viewpoints) are said to add up to a micro politics (Stanley, 2018).

Chapter 2
Balancing Resilience and Digital Distractions:
Youth Staying Focused in the Digital Landscape

W. Ian O'Byrne
College of Charleston, USA

Spencer Ward
Walden University, USA

ABSTRACT

The digital age has presented both advantages and issues in learning, communication, and overall well-being. Digital technology can provide entertainment, information, and connection but can also increase stress levels, and confusion, and weaken resilience. How can youth build the skills needed to stay focused and avoid digital distractions? What support do educators and parents offer in this endeavor? The chapter will conclude by highlighting the importance of balancing resilience and digital engagement in the digital landscape. While digital distractions can pose challenges to youth's focus and concentration, they can also offer opportunities for learning and creativity. The chapter will argue that youth need to develop a critical and reflective attitude toward digital technology and its role in their lives. The chapter concludes by highlighting the importance of balancing resilience and digital engagement and developing a critical and reflective attitude toward digital technology.

INTRODUCTION

The digital age has presented both advantages and issues in learning, communication, and overall well-being. Digital technology can provide entertainment, information, and connection but can also increase stress levels, confusion, and weaken resilience. While digital tools and spaces offer many benefits, such as increased access to information and opportunities for social connection, they can also be intrusive and negatively impact personal relationships. Additionally, the shift to online learning during the COVID-19

DOI: 10.4018/978-1-6684-8934-5.ch002

pandemic has highlighted the digital divide and the challenges faced by low-income students and those with special needs. It is important to examine how digital technologies are playing a role in our lives, particularly in the lives of children and adolescents, and to consider their impact on well-being and academic progress. How can youth build the skills needed to stay focused and avoid digital distractions? What support do educators and parents offer in this endeavor?

Over the last two decades, the Internet has wielded a profound and transformative influence on various facets of human existence, with perhaps no domain more profoundly affected than learning and social behaviors (Garcia et al., 2021; UNESCO, 2022). Research on the impact of Internet-enabled devices in schools remains ongoing (Arnott et al., 2018). However, various educational practices and approaches such as web-based inquiry, multimodal creation, and online communication of ideas have been shown to foster digital literacy (Spires et al., 2019. Over the past two decades, research has consistently highlighted the complexity of reading and writing in digital environments. However, discussions surrounding screen time often overlook the crucial aspects of cognitive and social development. Instead, many articles tend to emphasize concerns about potential harm to developing brains caused by screen usage. For example, in a *Psychology Today* article, Dunckley (2014) opens with the claim, "Addiction aside, a much broader concern that begs awareness is the risk that screen time is creating subtle damage even in children with 'regular' exposure, considering that the average child clocks in more than seven hours a day" (Rideout et al., 2010, para. 2). Although all the data she presents are from studies of youth who are Internet and video-game addicted, she encourages parents to "arm yourself with the truth about the potential damage screentime is capable of imparting—particularly in a young, still-developing brain" (para. 3).

While we acknowledge the medical relevance and validity of the available data, we do not perceive Internet addiction as an inevitable outcome. Furthermore, we believe that the current discourse surrounding this issue fails to provide a comprehensive understanding of the intricate interactions between developing brains and digital devices. Some research suggests that "screentime" is problematic, perhaps even addicting, and arguments that children's use of screens should be limited abound. (Street et al., 2021) There are legitimate concerns about children's physical, intellectual, and emotional well-being; yet conversations about *screentime* focus predominantly on the time spent on devices, often overlooking fundamentally important questions about what youth are learning by using technology. We also believe that media mentorship plays a crucial role in empowering youth to be active, critical, and responsible participants in today's media-rich society (RobbGrieco & Hobbs, 2013). It can equip them with the necessary skills and knowledge to navigate the media landscape effectively, make informed choices, and become creators of media rather than passive consumers (Passey et al., 2018; Cortesi et al., 2020).

This chapter aims to explore these questions by examining the concept of digital resilience to digital distractions, which can be understood as the ability to avoid, cope with, or absorb disturbances caused by digital technology. Digital distractions are intrusions and hinderances that are brought on by the availability of technological sources and content (Wu & Xie, 2018). We posit that through rich, dynamic, engaging, complex, creative, culturally, and socially relevant online learning experiences, youth can thrive as fully participatory digital-age citizens (Jenkins et al., 2015; Buchholz et al., 2020). In short, we seek to position youth as agentive, proactive, and spirited creators, not just consumers. We seek opportunities for them to build identity, and agency, and seek critical consciousness in an age of screentime. We will make the case that the Internet is the dominant text of a globally connected world, and as such, youth need to build the skills, habits, and practices necessary to survive and thrive in current and future digital contexts. The chapter will review the current research on these topics and discuss the implications for youth development and education.

LITERATURE REVIEW

Youth are growing up in a connected world where the web affords unprecedented learning opportunities, has made information plentiful, and put experts, figuratively speaking, at our fingertips. Ubiquitous mobile technologies have made the promise of improved access a realistic, achievable goal. This globally networked landscape has prompted a new vision for education, demonstrating the value of learning anywhere, anytime, and with equal access as a fundamental human right (Collins & Halverson, 2018). Never in the history of civilization have we seen such a potentially transformative literacy technology adopted by so many in so many different places in such a short period of time (Leu et al., 2018). The challenge is that even with these incredible opportunities, serious questions remain about how individuals can effectively negotiate the firehose of information available online as they contend with digital distractions while maintaining personal agency, self-efficacy, and resilience.

This leaves the current generation of educators, researchers, and policymakers with the critical responsibility to help facilitate a cultural transition during this period of profound change, ensuring a successful and positive move from print to digital resources and the hybrid spaces in between (Livingstone & Blum-Ross, 2020). Additionally, it forces us to address print-to-digital shifts in a manner that guarantees equity for all learners across the globe. Simply put, how can parents, guardians, and educators help young people become digitally literate while also being mindful of their well-being and personal development?

Towards Digital Critical Consciousness

As the Internet has become a major part of our lives, digital and social spaces have become the public square where people may interact and communicate. Students today are born into a digital world but still have to learn about digital literacy (Livingstone & Sefton-Green, 2016). To prepare students for the constant change of what it means to be literate (O'Byrne, 2018), we must expand our definition of "text" to include multimodal formats (Serafini, 2012). Text is not only ambiguous (Belshaw, 2012) but also deictic as future needs will be met by newer technologies and discourses (Forzani & Leu, 2017).

Literacy and skills are related but distinct concepts. Literacy focuses on knowledge, skills, and dispositions, while skills refer to work habits, dispositions, and strategies essential for digital literacy. Examples of digital skills include copy and pasting and critical thinking, while digital literacy encompasses the ability to read, write, and participate in online spaces. Digital skills involve the ability to efficiently encode and decode digital information with speed and fluency, while digital literacy relates to the awareness, attitude, and ability to use digital texts and tools (Iordache et al., 2017). It is important that youth build the knowledge, practices, skills, and dispositions necessary to survive and prosper in current and future contexts (Serdyukov, 2017).

Focusing on literacy practices and skills is just one part of preparing students for the future; there must also be opportunities to foster critical consciousness (Maker Castro et al., 2022). Research findings indicate that students have the capacity to develop self-regulatory skills while engaging in online research within an educational setting (Putman et al., 2015). Increasingly, we are seeing young people facing systemic discrimination have increased difficulty in achieving well-being (Kutsyuruba et al., 2015). In this context, wellbeing includes physical, mental, and socioemotional health, and research is still striving to understand the function that digital spaces, particularly in a post-pandemic society, have (Masi et al., 2021). Critical consciousness is a combination of reflection, motivation, and action against oppression (Seider et al., 2020).

Critical consciousness focuses on understanding the world and exposing social and political contradictions and includes taking action against oppressive elements in one's life (Jemal, 2017). Critical consciousness is an understanding of structural inequalities and oppressions in society and encourages questioning of dominant narratives and power dynamics to bring about social justice and equity (Watts & Hipolito-Delgado, 2015). Critical consciousness contributes to youth development by fostering awareness, empowerment, critical thinking, civic engagement, and personal growth and equips individuals with the tools and mindset necessary to navigate and challenge the complexities of the world they live in (Cadenas et al., 2020)

Through the development of digital literacies and interactions in a networked society, youth can gain insight into the sources of social issues (e.g., racism, sexism, poverty, discrimination) critical consciousness gives them insight into the sources of these problems and how power dynamics maintain inequality (Hobbs, 2021). Moving far beyond simple digital literacy practices and skills, youth can achieve a sense of power and confidence by developing critical consciousness, enabling them to challenge oppressive systems and become involved in political and social issues (Kunnath & Jackson, 2019). This in turn gives them the agency to advocate for change and make a meaningful difference in their lives, and the lives of others.

Developing critical consciousness in the use of digital technologies also provides opportunities to build and enhance critical thinking capabilities (Stubbs & Hullinger, 2022). Youth learn to question current ideals, evaluate data and media with a critical eye, and gain a more nuanced understanding of intricate social topics (Mirra et al., 2015). This gives them a larger outlook and enables them to make well-informed choices and drive personal growth and social transformation (Valtonen et al., 2019). These intersections between critical consciousness and digital social contexts encourage youth to cultivate empathy, compassion, and a sense of social responsibility (Barrera et al., 2017).

This involvement encompasses various civic activities, such as voting, community organizing, and advocacy, aimed at addressing injustices and effecting positive change. A systematic review examining the role of digital skills in the lives of young people (aged 12-17) revealed that possessing information, navigation, and processing skills is associated with several favorable outcomes. These include increased civic participation, enhanced online opportunities, improved academic grades, and a greater propensity for seeking information related to school assignments (Livingstone et al., 2021). It is hoped that individuals are then provided opportunities to transform and transcend conceptions of self which lead to more equitable and inclusive relationships, ultimately helping to generate a more just and inclusive society (Khalifa et al., 2016).

Towards Digital Resilience

As indicated earlier, we are presented with one of the paradoxes of digital literacy. Digitally networked spaces allow educational models that are more sustainable and accessible and allow learners to remain engaged, but they also serve as a source of distraction. Digital tools and resources provide access to a wealth of information and educational content from around the world. Learners can explore diverse perspectives, access up-to-date resources, and engage in real-time discussions with peers and experts across geographical boundaries. This enhanced accessibility promotes inclusivity and allows learners to access educational materials and opportunities that may otherwise be unavailable to them. We also need to acknowledge that digital spaces also pose challenges in terms of potential distractions as the abundance of entertaining content, social media platforms, and messaging apps can divert learners' attention from their educational activities.

Digital distractions are avoidance mechanisms derived from technology (Wu & Xie, 2018). In technology-enhanced educational environments, digital distractions are present within instructional and personal technology (Liao & Wu, 2022). Digital distractions can range from games, videos, or other technological activities that serve to avoid the main task (Chen et al., 2020). Students often switch tabs to avoid an assignment with games or videos (Bergdahl et al., 2020; McDaniel, 2020). With unlimited digital distractions, it is challenging for youth to focus on an educational task within a technology-enhanced learning environment (Kearney & Maakrun, 2020). Digital distractions, including notifications, messages, and content, can contribute to the fragmentation of attention and information overload, making it hard to focus on sustained, critical thinking (Gui et al., 2017).

Digital Distractions interfere with youth's education. Bergdahl et al. (2020) found a significant correlation between lower performance grades and in-class multitasking behavior. Students have been shown to use multitasking behaviors as a purposeful method to distract themselves and combat boredom or disinterest (Unsworth & McMillan, 2017). The accessibility of digital distractions poses a significant challenge, particularly when students are encouraged to utilize technology to enhance their learning experiences (Cheong et al., 2016). The allure of these distractions often leads students to seek them out to evade learning, consequently exerting a negative impact on their academic performance. Engaging excessively in digital diversions, such as the overuse of social media, gaming, or entertainment, can have adverse effects on critical thinking abilities (Al-Samarraie et al., 2021). Excessive preoccupation with digital activities can hinder individuals from dedicating sufficient time and focus to engage in critical thinking, reflection, and meaningful involvement in social matters (Cladis, 2020).

Not only are digital distractions negatively influencing academic achievement, but the amount of time youth spends on technology outside of school may also hurt their education. Current studies found a negative relationship between time spent on technology and academic achievement (Anthony et al., 2021; Rozgonjuk et al., 2021; Türel & Dokumaci, 2022). Moreover, Anthony et al. (2021) concluded that more than one hour per weekday and four hours per weekend day over a year could significantly influence academic achievement test scores, attitudes toward education, and concentration. Bergdahl et al. (2020) found significant correlations between time spent on social media and streaming services and low academic grades. Digital distractions may be more significant for students of average and low achievement (Pérez-Juárez et al., 2023). Current research suggests that high-achieving students overcome digital distractions with self-regulated strategies while low-achieving students often use digital multitasking to cope with boredom during lessons (Bergdahl et al., 2020).

Digital distractions have the potential to influence all students' successes; however, with acknowledgment and proper training, students and instructors can be equipped to combat digital distractions better (Haleem et al., 2022). The goal in this should be to build resilience in digital spaces. Resilience to digital distractions is the ability to avoid, cope with, or absorb disturbances caused by digital technology (Bergdahl et al., 2020). The effect digital distractions have on critical thinking can differ from person to person and in different circumstances. Some can utilize digital distractions without hindering their ability to think critically or reflect. Additionally, technology can be used to encourage critical engagement and learning when utilized consciously and appropriately. A constant influx of new information can prevent meaningful contemplation and evaluation as youth need to build the skills needed to stay focused and avoid digital distractions. These skills can be taught and practiced in and out of schools to provide youth with strategies to combat digital distractions (Turner et al., 2020). Digital resilience helps build the critical mindset necessary for comprehending and challenging societal norms, power structures, and inequities (Harris et al., 2020).

Towards Media Mentorship

Media mentorship refers to a relationship in which a more experienced individual guides and supports a younger person in developing their media literacy skills and navigating the media landscape (Campbell & Haines, 2016). It involves providing guidance, sharing knowledge, and fostering critical thinking abilities related to media consumption, creation, and interpretation (Jocius et al., 2021). Media mentorship can include helping youth acquire technical abilities, such as photography, video editing, podcasting, and graphic design, allowing them to express themselves creatively through media (Hobbs et al., 2013). Mentors can empower youth to understand their online presence, privacy, and appropriate behavior through digital citizenship (McGillivray et al., 2016). They teach ethical practices, such as respecting copyright and proper attribution in order to promote respectful online communication as well as guidance in avoiding online risks (Davis et al., 2010).

Media mentors guide others to critically assess media messages, helping them recognize biases, evaluate sources, and identify manipulation and individuals develop a foundation for informed media consumption and decision-making (Höttecke & Allchin, 2020). Media mentorship develops a safe space for youth to discover passions, express opinions, and narrate experiences using different media outlets (Pinkard et al., 2017). This guidance, dialogue, and apprenticeship have the potential to build confidence, strengthen self-worth, and motivate young people to be heard (Resnick, 2017). As individuals proceed through higher education and careers, mentorship can include advice on media-related careers, industry information, and education paths to help them make informed decisions and take advantage of media-related job opportunities (Collins & Halverson, 2018). Overall, media mentorship plays a crucial role in empowering youth to be active, critical, and responsible participants in today's media-rich society. It equips them with the necessary skills and knowledge to navigate the media landscape effectively, make informed choices, and become creators of media rather than passive consumers.

For successful media mentorship in an educational setting, instructors must teach students how to use technology effectively and continuously monitor their technology use to lessen students' multitasking behavior (McDaniel, 2020). Flanigan and Babchuk (2022) concluded that instructors primarily use prevention strategies to cope with students' digital distractions instead of reactive strategies in fear of damaging their teacher-student relationship. In technology-enhanced learning environments, instructors must account for digital distractions and educate students on the role digital distractions have on their academic success (Aaron & Lipton, 2018). Media mentorship can be integrated into the classroom to support students in developing media literacy, technical skills, critical thinking abilities, and a responsible digital presence (Redmond, 2015). Educators can facilitate discussions and activities that promote critical thinking about media messages allowing learners to analyze advertisements, news articles, or social media posts, examining biases, stereotypes, and persuasive techniques employed by media creators (Reeves & Crowther, 2019).

Parents and guardians can also engage in media mentorship with youth to support their development in digital spaces. Parents and guardians can encourage critical thinking and media literacy in their youth by discussing the media they consume, such as movies, TV shows, news articles, and social media posts (De Abreu, 2010). Parents and guardians can work with youth to create media content, such as family vlogs, photo albums, or podcasts. This not only allows the youth to express themselves creatively but also helps them to acquire technical skills while strengthening personal bonds. This can include helping youth explore hobbies such as photography, video editing, coding, and writing. This help may take the form of providing resources and connecting them with educational programs or mentorship opportunities. Most importantly, parents and guardians can set an example for youth by modeling healthy media habits (Schwartz et al., 2014). These include balanced media use, responsible online behavior, and critically engaging with media.

STAYING FOCUSED IN THE DIGITAL LANDSCAPE

Adolescents' media use is complex. The discussion provides youth, parents, caregivers, and educators with ideas to create a digital balance to positively influence all stakeholders' relationships with technology and each other. Furthermore, our goal is to create a digital balance to harness the positive attributes of technology while combating the negative.

Communicate with Youth to Understand and Explain the Effects of Technology

One of the first steps in working with youth to examine the impact of technology and developing digital resilience is to engage in open and informed discussions with youth about screentime and technology (Lee & Hancock, 2023). In an era where digital devices have become an integral part of everyday life, it is crucial to recognize the potential implications and effects they have on the physical, mental, and socioemotional well-being of young individuals. By initiating meaningful conversations, parents, educators, and caregivers can foster a greater understanding of the benefits, risks, and responsible use of technology, thereby promoting healthy digital habits and mitigating potential negative consequences (Chassiakos & Stager, 2020). Effective communication strategies and ongoing dialogue are necessary to guide and empower youth as they navigate the digital landscape successfully.

To prepare for these conversations, adults should conduct a self-reflection to evaluate their relationship with technology (Turkle, 2016). Your views and actions must align with your technological expectations. Authenticity is essential while generating a workable plan to create a digital balance with your youth. There is also a need to situate youth as co-learners, and perhaps as experts in these discussions as you strive for a critical balance as we use digital technologies. When engaging in conversations with youth about screentime and technology, it is vital for adults to approach the discussions without positioning themselves as infallible experts (Turner et al., 2020; O'Byrne et al., 2021).

Adopting a humble and open-minded attitude fosters a collaborative and inclusive environment that encourages the free exchange of ideas (Lee, 2018). Recognizing that technology evolves rapidly, and that youth often possess a deep understanding of its nuances, adults should embrace a mutual learning approach. This allows for the exploration of diverse perspectives, encourages critical thinking, and empowers youth to actively participate in shaping their own digital experiences (D'Amico et al., 2016). By acknowledging that both parties have unique insights to offer, adults can create a more meaningful dialogue that respects the expertise and autonomy of young individuals while still providing guidance and support when needed (Jennings et al., 2006).

Recognizing the Effect of Digital Distractions on Productivity and Mental Health

Digital distraction is a growing issue in our tech-saturated world, with smartphones, social media, and online activities constantly vying for our attention. This can lead to disrupted workflow, reduced productivity, stress, anxiety, and poor overall wellness (Farivar et al., 2022). To combat the negative effects of digital distractions, it's essential to recognize them and devise strategies to minimize their impact (Skulmowski & Xu, 2021).

As shown above, studies have shown that excessive digital distractions negatively impact productivity as e-mail, social media, and instant messaging can interrupt focus and detract from essential tasks (Addas & Pinsonneault, 2015). This lessens both the speed and quality of output, as well as inhibiting deep, creative thinking (May & Elder, 2018). To combat this, individuals can set boundaries, manage notifications, and create dedicated periods of no digital distractions, leading to a more productive work environment. Individuals can also identify opportunities to consider the work that they find valuable and whether productivity and instant gratification are the end goal (Davenport & Ronanki, 2018). Or, instead, do we seek, and perhaps create space for youth to seek deep, insightful, creative solutions in their work?

Digital distractions not only decrease productivity but also impact mental health (Chen et al., 2020). Excessive time on social media and other digital platforms can lead to loneliness, depression, and lowered self-esteem (Hunt et al., 2018) as comparisons with 'highlight reels' of others' lives can be damaging. The addictive nature of online activities such as gaming or binge-watching can disturb sleep (Flayelle et al., 2017). To achieve good mental health and well-being, it's important to prioritize self-care, set digital boundaries, and use technology mindfully and intentionally (Orlowski et al., 2015). This provides a balance between the digital world and personal well-being.

Setting Learning and Personal Goals and Priorities

Prioritizing mental health and well-being is essential for setting learning and personal goals and balancing mental health, productivity, critical consciousness, and digital distractions. Talk with youth about respecting their emotional needs, taking breaks, and seeking support when needed (Souers & Hall, 2016). When setting goals, adults and youth should consider their own well-being, self-care, and work-life balance. Maintaining a balance between productivity and self-care is essential. Setting goals is great for productivity, but honest discussions should focus on the need to sometimes give yourself a break. Taking care of yourself (i.e., exercising, meditating, or spending quality time with family and friends) helps reduce stress and keeps you in a positive frame of mind. Talk with youth about how taking care of yourself is just as important as reaching your goals.

As you discuss goals with youth, help them understand the social, cultural, and political contexts that affect them (Jenkins, 2016). Consider how your objectives will impact yourself, others, and society, and ensure they promote justice and ethical behavior (Giangrande et al., 2019). Technology can spread false information, create echo chambers, and fuel online harassment, which entrenches existing biases (Kozyreva et al., 2020). Additionally, access to technology can be unequal, limiting critical engagement for some. More research is needed to evaluate technology's effect on critical thinking, particularly the perception, knowledge, and involvement of youth in social issues, both positively and negatively (Heberle et al., 2020). Acknowledgment of privilege and biases and striving to pursue goals have the potential to bring about positive change (Lawless & Chen, 2021).

To successfully reach learning and personal goals, create a balanced, mindful approach that emphasizes mental health and well-being (Lewallen et al., 2015). Self-care, including reflection about the broader impact of personal and professional goals, can be used to maintain a healthy work-life balance. Regular reflection with youth provides opportunities to stay focused and reach goals, and while striving to efficiently manage digital distractions.

Controlling Media Use by Limitation or Scheduling

Limiting or scheduling screen time is essential for achieving good mental health, well-being, productivity, critical consciousness, and staying clear of digital distractions (Weinstein & James, 2022). Too much time spent on screens can cause mental health issues, sleep disturbances, and reduce productivity. Establishing rules on media usage helps individuals regain control of their schedule and promote overall well-being. For some, this includes establishing times of uninterrupted work or study and turning off unnecessary notifications. Apps or browser extensions can be used to block or limit access to distracting websites and apps. By controlling digital distractions, adults and youth can be productive and stay focused on their goals.

Schedule specific time slots for digital media such as social media, entertainment, or news. An example of this is identifying activities that are for *screens on*, *screens off*, or *screens on the side*. "Screens on" time refers to time spent with technological devices, or in front of screens. The key factor in this relationship is the interaction with a display device, and a significant amount of attention in the interaction is paid to the content on the screen. This may include passive consumption or active creation of content. This means that it might include watching television or playing video games. It may include surfing the Internet or coding. "Screens Off" time refers to time spent away from technological devices, or not in front of screens. The key factor in this relationship is the lack of a display device in the environment. Most of the attention in these interactions is paid to the self or other individuals in the room. This may involve music playing in the room or listening to an audio podcast. But it should not include video or interactions with a screen. "Screens on the side" refers to times spent working and using devices on the side for assistance, or a resource. The key factor in this relationship is the attention and focus on the self or other individuals in the room. Focus is also usually paid to an activity or project. As an example, this may include using a tablet to display directions, or a video as families cook together. This is identified by targeted, active consumption of content. Passive consumption of content is not usually included as it detracts from the project or task at hand. The use of screens on the side is important as it aids in the completion of the task. This can help maintain a healthy balance by preventing media from consuming the entire day and distracting from important tasks or self-care activities. Setting boundaries around screen time allows for a more intentional engagement while still giving you control.

Limiting or scheduling media consumption is an effective way to maintain positive mental health, achieve goals, nurture critical thinking, and limit distractions. Setting boundaries ensures time is spent on activities that promote well-being. Controlling screen time can lead to more intentional engagement with media and a more informed outlook. Striking a balance between online activities and other aspects of life helps to enhance mental health, productivity, and satisfaction.

Getting Feedback and Help from Teachers, Peers, And Family

To achieve mental health, well-being, productivity, and critical consciousness while managing digital distractions, it is essential to get feedback and support from teachers, peers, and family members. Such individuals can offer valuable advice, aid, and accountability to successfully cope with life's complexities. Teachers are essential for guiding and supporting students in their educational journey. They can provide feedback on academic progress, recommend study strategies, and supply resources to manage digital distractions. Open communication with teachers allows for an understanding of expectations and potential adaptations to meet personal needs. Seeking their feedback can also develop a critical mindset and facilitate conversations on diverse views and social matters.

Peers can offer valuable support and aid in collaboration. Through the discussion of shared goals and difficulties, peers can offer encouragement, guidance on how to resist digital distractions and new ideas on various topics. Collaborative learning environments foster critical thinking as it encourages reflection on different perspectives and promotes empathy. Asking for help and advice from peers benefits learning and also creates a sense of community and mutual development. Families offer invaluable support for maintaining mental health and balance in life. Engaging in honest conversations enables understanding of individual goals and establishes a supportive environment. Families can provide a safe space to discuss challenges, offer helpful advice and set healthy boundaries for digital distractions. They can also prioritize self-care and help individuals manage their work-life balance.

Seeking guidance, support, and perspectives from teachers, peers, and family members is essential in attaining a healthy balance in mental health, productivity, and critical consciousness while managing digital distractions. Through engaging actively with these sources, challenges can be addressed more efficiently, leading to significant personal growth, sharpened critical thinking, and a sense of community. In short, their input is invaluable in successfully integrating these essential aspects of life.

CONCLUSION

Technology offers youth an invaluable platform for critical thinking and awareness of social issues. It provides easy access to information, diverse perspectives, and a medium for expression and activism. The internet, social media, and digital tools can unite young people with like-minded individuals and communities. Technology has the potential to spread knowledge, give a voice to those who have been ignored, and unite people to make positive changes. It can help young people confront existing power structures, fight for social justice and take action to create lasting progress. Building digital resilience and fostering critical consciousness provides the skills necessary to effectively navigate network communities.

With this great power, also comes great responsibility. Youth need opportunities to balance resilience and digital distractions in current and future situations. While digital distractions can pose challenges to youth's focus and concentration, they can also offer opportunities for learning and creativity. Choosing wisely between media and tools is key to maintaining mental health, productivity, well-being, and critical consciousness. Different tasks require different approaches, and mindful decisions can have a big impact on our well-being and productivity. To become more aware, we must intentionally select media and tools that support a variety of views, recognize inclusivity, and challenge our thinking. This entails engaging with reputable news sources, listening to thought-provoking podcasts, or reading relevant blogs, or deliberately looking for content that forces us to reflect on our beliefs and opinions. Doing this helps us contribute to critical discussions and makes a more informed and sympathetic global community.

REFERENCES

Aaron, L. S., & Lipton, T. (2018). Digital distraction: Shedding light on the 21st-century college classroom. *Journal of Educational Technology Systems*, *46*(3), 363–378. doi:10.1177/0047239517736876

Addas, S., & Pinsonneault, A. (2015). The many faces of information technology interruptions: A taxonomy and preliminary investigation of their performance effects. *Information Systems Journal*, *25*(3), 231–273. doi:10.1111/isj.12064

Al-Samarraie, H., Bello, K. A., Alzahrani, A. I., Smith, A. P., & Emele, C. (2021). Young users' social media addiction: Causes, consequences and preventions. *Information Technology & People*, *35*(7), 2314–2343. doi:10.1108/ITP-11-2020-0753

Anthony, W. L., Zhu, Y., & Nower, L. (2021). The relationship of interactive technology use for entertainment and school performance and engagement: Evidence from a longitudinal study in a nationally representative sample of middle school students in China. *Computers in Human Behavior*, *122*. Advance online publication. doi:10.1016/j.chb.2021.106846

Arnott, L., Palaiologou, I., & Gray, C. (2018). Digital devices, internet-enabled toys and digital games: The changing nature of young children's learning ecologies, experiences and pedagogies. *British Journal of Educational Technology*, *49*(5), 803–806. doi:10.1111/bjet.12676

Barrera, D., Willner, L. N., & Kukahiko, K. I. (2017). Assessing the development of an emerging critical consciousness through service learning. *Journal of Critical Thought and Praxis*, *6*(3).

Bergdahl, N., Nouri, J., Fors, U., & Knutsson, O. (2020). Engagement, disengagement and performance when learning with technologies in upper secondary school. *Computers & Education*, *149*. Advance online publication. doi:10.1016/j.compedu.2019.103783

Buchholz, B. A., DeHart, J., & Moorman, G. (2020). Digital citizenship during a global pandemic: Moving beyond digital literacy. *Journal of Adolescent & Adult Literacy*, *64*(1), 11–17. doi:10.1002/jaal.1076

Cadenas, G. A., Cantú, E. A., Spence, T., & Ruth, A. (2020). Integrating critical consciousness and technology in entrepreneurship career development with diverse community college students. *Journal of Career Development*, *47*(2), 162–176. doi:10.1177/0894845318793968

Campbell, C., & Haines, C. (2016). *Becoming a media mentor: A guide for working with children and families*. American Library Association., doi:10.1177/0894845318793968

Chassiakos, Y. L. R., & Stager, M. (2020). Current trends in digital media: How and why teens use technology. In *Technology and adolescent health* (pp. 25–56). Academic Press., doi:10.1016/B978-0-12-817319-0.00002-5

Chen, L., Nath, R., & Tang, Z. (2020). Understanding the determinants of digital distraction: An automatic thinking behavior perspective. *Computers in Human Behavior, 104*, 106195. doi:10.1016/j.chb.2019.106195

Cheong, P. H., Shuter, R., & Suwinyattichaiporn, T. (2016). Managing student digital distractions and hyperconnectivity: Communication strategies and challenges for professorial authority. *Communication Education, 65*(3), 272–289. doi:10.1080/03634523.2016.1159317

Cladis, A. E. (2020). A shifting paradigm: An evaluation of the pervasive effects of digital technologies on language expression, creativity, critical thinking, political discourse, and interactive processes of human communications. *E-Learning and Digital Media, 17*(5), 341–364. doi:10.1177/2042753017752583

Collins, A., & Halverson, R. (2018). *Rethinking education in the age of technology: The digital revolution and schooling in America.* Teachers College Press.

Cortesi, S., Hasse, A., Lombana-Bermudez, A., Kim, S., & Gasser, U. (2020). Youth and digital citizenship+ (plus): Understanding skills for a digital world. *Berkman Klein Center Research Publication*, (2020-2). https://dx.doi.org/ doi:10.2139/ssrn.3557518

Csikszentmihalyi, M. (1997). Flow and the psychology of discovery and invention. *HarperPerennial, New York, 39*, 1–16.

D'Amico, M., Denov, M., Khan, F., Linds, W., & Akesson, B. (2016). Research as intervention? Exploring the health and well-being of children and youth facing global adversity through participatory visual methods. *Global Public Health: An International Journal for Research, Policy and Practice, 11*(5-6), 528–545. doi:10.1080/17441692.2016.1165719

Davenport, T. H., & Ronanki, R. (2018). Artificial intelligence for the real world. *Harvard Business Review, 96*(1), 108–116.

Davis, K., Katz, S., Santo, R., & James, C. (2010). Fostering Cross-Generational Dialogues About the Ethics of Online Life. *Davis, K., Katz, SL, Santo, R., & James, C.(2010). Fostering cross-generational dialogues about the ethics of online life. The Journal of Media Literacy Education, 2*(2), 124–150. https://ssrn.com/abstract=2979512

De Abreu, B. (2010). Changing Technology= Empowering Students through Media Literacy Education. *New Horizons in Education, 58*(3), 26–33. doi:10.1111/ntwe.12235

Dunckley, V. L. (2014). Gray matters: Too much screen time damages the brain. *Psychology Today*, 27.

Farivar, F., Esmaeelinezhad, O., & Richardson, J. (2022). Digital intrusions or distraction at work and work-Life conflict. *New Technology, Work and Employment, 37*(3), 363–380.

Flanigan, A. E., & Babchuk, W. A. (2022). Digital distraction in the classroom: Exploring instructor perceptions and reactions. *Teaching in Higher Education, 27*(3), 352–370. doi:10.1080/13562517.2020.1724937

Flayelle, M., Maurage, P., & Billieux, J. (2017). Toward a qualitative understanding of binge-watching behaviors: A focus group approach. *Journal of Behavioral Addictions, 6*(4), 457–471. doi:10.1556/2006.6.2017.060

Forzani, E., & Leu, D. J. (2017). Multiple perspectives on literacy as it continuously changes: Reflections on opportunities and challenges when literacy is deictic. *Journal of Education*, *197*(2), 19–24.

Garcia, A. G., McGrew, S., Mirra, N., Tynes, B., & Kahne, J. (2021). Rethinking digital citizenship: Learning about media, literacy, and race in turbulent times. *Educating for civic reasoning and discourse*, 319-352. http://doi.org/ doi:10.31094/2021/2

Giangrande, N., White, R. M., East, M., Jackson, R., Clarke, T., Saloff Coste, M., & Penha-Lopes, G. (2019). A competency framework to assess and activate education for sustainable development: Addressing the UN sustainable development goals 4.7 challenge. *Sustainability*, *11*(10), 2832. doi:10.3390u11102832

Gui, M., Fasoli, M., & Carradore, R. (2017). Digital well-being. Developing a new theoretical tool for media literacy research. *Italian Journal of Sociology of Education*, *9*(1), 155–173. doi:10.14658/pupj-ijse-2017-1-8

Haleem, A., Javaid, M., Qadri, M. A., & Suman, R. (2022). Understanding the role of digital technologies in education: A review. *Sustainable Operations and Computers*, *3*, 275–285. doi:10.1016/j.susoc.2022.05.004

Hämäläinen, E. K., Kiili, C., Marttunen, M., Räikkönen, E., González-Ibáñez, R., & Leppänen, P. H. (2020). Promoting sixth graders' credibility evaluation of Web pages: An intervention study. *Computers in Human Behavior*, *110*, 106372. doi:10.1016/j.chb.2020.106372

Harris, B. N., McCarthy, P. C., Wright, A. M., Schutz, H., Boersma, K. S., Shepherd, S. L., Manning, L. A., Malisch, J. L., & Ellington, R. M. (2020). From panic to pedagogy: Using online active learning to promote inclusive instruction in ecology and evolutionary biology courses and beyond. *Ecology and Evolution*, *10*(22), 12581–12612. doi:10.1002/ece3.6915

Heberle, A. E., Rapa, L. J., & Farago, F. (2020). Critical consciousness in children and adolescents: A systematic review, critical assessment, and recommendations for future research. *Psychological Bulletin*, *146*(6), 525. doi:10.1037/bul0000230

Hobbs, R. (2021). *Media literacy in action: Questioning the media*. Rowman & Littlefield Publishers.

Hobbs, R., Donnelly, K., Friesem, J., & Moen, M. (2013). Learning to engage: How positive attitudes about the news, media literacy, and video production contribute to adolescent civic engagement. *Educational Media International*, *50*(4), 231–246. doi:10.1080/09523987.2013.862364

Höttecke, D., & Allchin, D. (2020). Reconceptualizing nature-of-science education in the age of social media. *Science Education*, *104*(4), 641–666. doi:10.1002ce.21575

Hunt, M. G., Marx, R., Lipson, C., & Young, J. (2018). No more FOMO: Limiting social media decreases loneliness and depression. *Journal of Social and Clinical Psychology*, *37*(10), 751–768. doi:10.1521/jscp.2018.37.10.751

Iordache, C., Mariën, I., & Baelden, D. (2017). Developing digital skills and competences: A quick-scan analysis of 13 digital literacy models. *Italian Journal of Sociology of Education, 9*(Italian Journal of Sociology of Education 9/1), 6-30.

Jemal, A. (2017). Critical consciousness: A critique and critical analysis of the literature. *The Urban Review, 49*, 602–626. doi:10.100711256-017-0411-3

Jenkins, H. (2016). Youth voice, media, and political engagement. *By any media necessary: The new youth activism, 3*(1). https://doi.org/ doi:10.18574/nyu/9781479829712.003.0004

Jenkins, H., Itō, M., & Boyd, D. (2015). *Participatory culture in a networked era: A conversation on youth, learning, commerce, and politics*. John Wiley & Sons.

Jennings, L. B., Parra-Medina, D. M., Hilfinger-Messias, D. K., & McLoughlin, K. (2006). Toward a critical social theory of youth empowerment. *Journal of Community Practice, 14*(1-2), 31–55. doi:10.1300/J125v14n01_03

Jocius, R., O'Byrne, I., Blanton, M., Albert, J., Joshi, D., Robinson, R., & Andrews, A. (2021). Leveraging virtual professional development to build computational thinking literacies in English language arts classrooms. *Contemporary Issues in Technology & Teacher Education, 21*(4), 626–654. https://www.learntechlib.org/primary/p/217717/

Kearney, S., & Maakrun, J. (2020). Let's get engaged: The nexus between digital technologies, engagement and learning. *Education in Science, 10*(12), 357. doi:10.3390/educsci10120357

Khalifa, M. A., Gooden, M. A., & Davis, J. E. (2016). Culturally responsive school leadership: A synthesis of the literature. *Review of Educational Research, 86*(4), 1272–1311. doi:10.3102/0034654316630383

Kozyreva, A., Lewandowsky, S., & Hertwig, R. (2020). Citizens versus the internet: Confronting digital challenges with cognitive tools. *Psychological Science in the Public Interest, 21*(3), 103–156. doi:10.1177/1529100620946707

Kunnath, J. P., & Jackson, A. (2019). Developing student critical consciousness: Twitter as a tool to apply critical literacy in the English classroom. *The Journal of Media Literacy Education, 11*(1), 52–74. doi:10.23860/JMLE-2019-11-1-3

Kutsyuruba, B., Klinger, D. A., & Hussain, A. (2015). Relationships among school climate, school safety, and student achievement and well-being: A review of the literature. *Review of Education, 3*(2), 103–135. doi:10.1002/rev3.3043

Lawless, B., & Chen, Y. W. (2021). *Teaching social justice: Critical tools for the intercultural communication classroom*. Rowman & Littlefield.

Lee, A. Y., & Hancock, J. T. (2023). Developing Digital Resilience: An Educational Intervention Improves Elementary Students' Response to Digital Challenges. *Computers and Education Open, 100144*. Advance online publication. doi:10.1016/j.caeo.2023.100144

Lee, D. (2018). *Design thinking in the classroom: Easy-to-use teaching tools to foster creativity, encourage innovation and unleash potential in every student*. Simon and Schuster.

Leu, D. J., Kinzer, C. K., Coiro, J., Castek, J., & Henry, L. A. (2018). New literacies: A dual-level theory of the changing nature of literacy, instruction, and assessment. In *Theoretical models and processes of literacy* (pp. 319–346). Routledge.

Lewallen, T. C., Hunt, H., Potts-Datema, W., Zaza, S., & Giles, W. (2015). The whole school, whole community, whole child model: A new approach for improving educational attainment and healthy development for students. *The Journal of School Health*, *85*(11), 729–739. doi:10.1111/josh.12310

Liao, C. H., & Wu, J. Y. (2022). Deploying multimodal learning analytics models to explore the impact of digital distraction and peer learning on student performance. *Computers & Education*, *190*, 104599. doi:10.1016/j.compedu.2022.104599

Livingstone, S., Mascheroni, G., & Stoilova, M. (2021). The outcomes of gaining digital skills for young people's lives and wellbeing: A systematic evidence review. *New Media & Society*. Advance online publication. doi:10.1177/14614448211043189

Livingstone, S., & Blum-Ross, A. (2020). *Parenting for a digital future: How hopes and fears about technology shape children's lives*. Oxford University Press.

Livingstone, S., & Sefton-Green, J. (2016). *The class: Living and learning in the digital age.* NYU Press. muse.jhu.edu/book/49200

Maker Castro, E., Wray-Lake, L., & Cohen, A. K. (2022). Critical consciousness and wellbeing in adolescents and young adults: A systematic review. *Adolescent Research Review*, *7*(4), 499–522. doi:10.100740894-022-00188-3

Masi, A., Mendoza Diaz, A., Tully, L., Azim, S. I., Woolfenden, S., Efron, D., & Eapen, V. (2021). Impact of the COVID-19 pandemic on the well-being of children with neurodevelopmental disabilities and their parents. *Journal of Paediatrics and Child Health*, *57*(5), 631–636. doi:10.1111/jpc.15285

May, K. E., & Elder, A. D. (2018). Efficient, helpful, or distracting? A literature review of media multitasking in relation to academic performance. *International Journal of Educational Technology in Higher Education*, *15*(1), 1–17. doi:10.118641239-018-0096-

McDaniel, J. W. (2020). Implementing Personal Devices in Math. Networks. *An Online Journal for Teacher Research*, *22*(3). Advance online publication. doi:10.4148/2470-6353.1314

McGillivray, D., McPherson, G., Jones, J., & McCandlish, A. (2016). Young people, digital media making and critical digital citizenship. *Leisure Studies*, *35*(6), 724–738. doi:10.1080/02614367.2015.1062041

Mirra, N., Garcia, A., & Morrell, E. (2015). *Doing youth participatory action research: Transforming inquiry with researchers, educators, and students*. Routledge.

O'Byrne, W. I. (2018). Empowering students as critical readers and writers in online spaces. In *Best practices in teaching digital literacies* (pp. 233–250). Emerald Publishing Limited., doi:10.1108/S2048-045820180000009018

O'Byrne, W. I., Turner, K., Paciga, K. A., & Stevens, E. (2021). Co-constructing digital futures: Parents and children becoming thoughtful, connected, and critical users of digital technologies. *Algorithmic rights and protections for children*.

OFS Health Care. (2022). *Kids' screen time: How much is too much?* https://www.osfhealthcare.org/blog/kids-screen-time-how-much-is-too-much/

Orlowski, S. K., Lawn, S., Venning, A., Winsall, M., Jones, G. M., Wyld, K., Damarell, R. A., Antezana, G., Schrader, G., Smith, D., Collin, P., & Bidargaddi, N. (2015). Participatory research as one piece of the puzzle: A systematic review of consumer involvement in design of technology-based youth mental health and well-being interventions. *JMIR Human Factors*, *2*(2), e4361. doi:10.2196/humanfactors.4361

Passey, D., Shonfeld, M., Appleby, L., Judge, M., Saito, T., & Smits, A. (2018). Digital agency: Empowering equity in and through education. *Technology. Knowledge and Learning*, *23*, 425–439. doi:10.100710758-018-9384-x

Pérez-Juárez, M. Á., González-Ortega, D., & Aguiar-Pérez, J. M. (2023). Digital Distractions from the Point of View of Higher Education Students. *Sustainability*, *15*(7), 6044. doi:10.3390u15076044

Pinkard, N., Erete, S., Martin, C. K., & McKinney de Royston, M. (2017). Digital youth divas: Exploring narrative-driven curriculum to spark middle school girls' interest in computational activities. *Journal of the Learning Sciences*, *26*(3), 477–516. doi:10.1080/10508406.2017.1307199

Putman, S. M., Hathaway, J. I., Coiro, J., & Quinn, D. J. (2015). Putting on CAPES to search the Internet: A framework for self-regulated learning. *Literacy Learning: The Middle Years*, *23*(3), 38–49.

Redmond, T. (2015). Media literacy is common sense: Bridging Common Core Standards with the media experiences of digital learners: Findings from a case study highlight the benefits of an integrated model of literacy, thereby illustrating the relevance and accessibility of media literacy education. *Middle School Journal*, *46*(3), 10–17. doi:10.1080/00940771.2015.11461910

Reeves, J., & Crowther, T. (2019). Teacher feedback on the use of innovative social media simulations to enhance critical thinking in young people on radicalisation, extremism, sexual exploitation and grooming. *Pastoral Care in Education*, *37*(4), 280–296. doi:10.1080/02643944.2019.1618377

Resnick, M. (2017). *Lifelong kindergarten: Cultivating creativity through projects, passion, peers, and play*. MIT press.

Rideout, V. J., Foehr, U. G., & Roberts, D. F. (2010). *Generation M 2: Media in the Lives of 8-to 18-Year-Olds*. Henry J. Kaiser Family Foundation.

RobbGrieco. M., & Hobbs, R. (2013). A field guide to media literacy education in the United States. Kingston, RI: Media Education Lab, University of Rhode Island.

Rozgonjuk, D., Täht, K., & Vassil, K. (2021). Internet Use at and outside of School in Relation to Low- and High-Stakes Mathematics Test Scores across 3 Years. *International Journal of STEM Education*, *8*. Advance online publication. doi:10.118640594-021-00287-y

Schwartz, S. E., Rhodes, J. E., Liang, B., Sánchez, B., Spencer, R., Kremer, S., & Kanchewa, S. (2014). Mentoring in the digital age: Social media use in adult–youth relationships. *Children and Youth Services Review*, *47*, 205–213. doi:10.1016/j.childyouth.2014.09.004

Seider, S., El-Amin, A., & Kelly, L. L. (2020). The development of critical consciousness. The Oxford handbook of moral development: An interdisciplinary perspective, 203-221.

Serdyukov, P. (2017). Innovation in education: what works, what doesn't, and what to do about it?. *Journal of research in innovative teaching & learning, 10*(1), 4-33. https://doi.org/ doi:10.1108/JRIT-10-2016-0007

Skulmowski, A., & Xu, K. M. (2021). Understanding cognitive load in digital and online learning: A new perspective on extraneous cognitive load. *Educational Psychology Review*, •••, 1–26. doi:10.100710648-021-09624-7

Souers, K., & Hall, P. (2016). *Fostering resilient learners: Strategies for creating a trauma-sensitive classroom*. ASCD.

Spires, H. A., Paul, C. M., & Kerkhoff, S. N. (2019). Digital literacy for the 21st century. In *Advanced methodologies and technologies in library science, information management, and scholarly inquiry* (pp. 12–21). IGI Global.

Street, C., El Asam, A., & Katz, A. (2021). Young people's digital well-being: optimising the potential and minimising the risks. *SOCIAL INCLUSION AND DIGITALISATION*, 105

Stubbs, C. F., & Hullinger, H. (2022). Developing Critical Consciousness in Students for Cultural Transformation. *Journal of the Scholarship of Teaching and Learning for Christians in Higher Education*, *12*(1), 5. doi:10.31380otlched.12.1.9

Türel, Y. K., & Dokumaci, Ö. (2022). Use of media and technology, academic procrastination, and academic achievement in adolescence. *Participatory Educational Research, 9*(2), 481–497. doi:10.17275/per.22.50.9.2

Turner, K. H., Hicks, T., & Zucker, L. (2020). Connected reading: A framework for understanding how adolescents encounter, evaluate, and engage with texts in the digital age. *Reading Research Quarterly*, *55*(2), 291–309. doi:10.1002/rrq.271

UNESCO. (2022). *Reimagining our futures together: A new social contract for education*. UN.

Unsworth, N., & McMillan, B. D. (2017). Attentional disengagements in educational contexts: A diary investigation of everyday mind-wandering and distraction. *Cognitive research: Principles and implications. 2*, 1-20. http://doi.org/ doi:10.1186/s41235-017-0070-7

Valtonen, T., Tedre, M., Mäkitalo, K., & Vartiainen, H. (2019). Media Literacy Education in the Age of Machine Learning. *The Journal of Media Literacy Education*, *11*(2), 20–36. doi:10.23860/JMLE-2019-11-2-2

Watts, R. J., & Hipolito-Delgado, C. P. (2015). Thinking ourselves to liberation?: Advancing sociopolitical action in critical consciousness. *The Urban Review*, *47*, 847–867. doi:10.100711256-015-0341-x

Weinstein, E., & James, C. (2022). *School-Based Initiatives Promoting Digital Citizenship and Healthy Digital Media Use. Handbook of Adolescent Digital Media Use and Mental Health*. Cambridge University Press.

Wineburg, S., Breakstone, J., Mcgrew, S., Smith, M. D., & Ortega, T. (2022). Supplemental Material for Lateral Reading on the Open Internet: A District-Wide Field Study in High School Government Classes. *Journal of Educational Psychology*, *114*(5), 893–909. https://doi.org/10.1037/edu0000740.supp

Wu, J. Y., & Xie, C. (2018). Using time pressure and note-taking to prevent digital distraction behavior and enhance online search performance: Perspectives from the load theory of attention and cognitive control. *Computers in Human Behavior*, *88*, 244–254. doi:10.1016/j.chb.2018.07.008

Chapter 3
Elementary School Teacher Training for Digital Citizenship in the English Language Classroom

Tetiana Holovatenko
https://orcid.org/0000-0002-7545-3253
Borys Grinchenko Kyiv University, Ukraine & University of Minnesota, USA

ABSTRACT

The chapter is a mixed-method study of knowledge, skills, and beliefs of pre-service elementary school teachers on digital citizenship. The quantitative data was gathered through surveying participants. The intervention in the form of a 30-hour module was introduced to participants to identify a change in perceptions. The qualitative data was obtained through the analysis of artefacts in the form of lesson plans. The author grounds the theoretical underpinning into the theory and principles of digital citizenship. The result of the posttest shows a positive change in the knowledge, skills, and beliefs of pre-service elementary school teachers on digital citizenship.

INTRODUCTION

Considerable attention of civic society has been paid to the digitalization of all spheres of life. Attention towards the development of 21st-century skills and growing digitalization have resulted in the creation of digital analogues of those skills. UNESCO during Mobile Learning Week in 2018 noted that digital skills alone cannot adequately respond to challenges imposed by a fast-developing society and economy. The holistic approach suggested by this organization provides for developing digital skills together with numeracy, literacy skills, as well as a set of skills referred to as digital citizenship skills (e.g. intrapersonal, interpersonal skills, critical and innovative thinking etc.) (UNESCO, 2018).

According to Becker (2019), this concept emerged in the 1990s as a response to the challenges of the cyber world and was used to denote the 'hard skills' instrumental in the digital environment. However,

DOI: 10.4018/978-1-6684-8934-5.ch003

later the focus of the concept has changed. For instance, Mossberger et al. (2007) in their definition of digital citizenship focused on the 'soft skills', necessary for digital environments. Despite the fact that the concept was introduced a long time ago, the literature body devoted to it is still growing. For instance, a search of the keyword 'digital citizenship' in English in Google Scholar gives a rough result of 1,370,000 papers. As digital citizenship has been gaining importance on a global scale, there is a need for educational systems to implement this concept and teach it to a new generation of students.

This chapter aims to outline what digital citizenship is and identify the opinions of undergraduate pre-service teachers on digital citizenship, as well as feasible ways of informing them on how to teach elementary school students about digital citizenship.

THEORETICAL BACKGROUND

The author finds it important to state their research positionality. This study is set in one of the Ukrainian universities. With the introduction of lockdown due to Covid-19 and the switch to blended and hy-flex learning in Ukraine due to Russian unprovoked aggression, there is a rapid development of various areas of elementary school students' digital competence. However, digital citizenship is an under-researched notion in Ukrainian scholarly discourse and the author wants to raise awareness of it among pre-service teachers. The search for the Ukrainian equivalent of the notion of digital citizenship ("цифрове громадянство") in Google Scholar returns only around 14,400 results. As a result, the author studies the global experience of teaching pre-service teachers about global citizenship and attempts to bring this knowledge into Ukrainian teacher training discourse. At the same time, the results of this study have replicability potential in any setting and are not restricted to the Ukrainian pre-service teacher system and can be of interest to teacher trainers and pre-service elementary school teachers.

Approaches to Defining the Notion of Digital Citizenship

The notion of digital citizenship is developing together with the development of digital technologies. In this section of the study, the author analyzes the main approaches to defining this concept.

The Applied Educational Systems define digital citizenship as the "responsible use of technology by anyone who uses computers, the Internet, and digital devices to engage with society on any level" (What Is Digital Citizenship & How Do You Teach It?, 2023). A similar approach to defining this concept is suggested by Ribble et al. (2004), who characterizes digital citizenship as norms of anticipated behavior when using technology. Similarly, Isin and Ruppert (2015) claim that individuals' actions in the digital world make their identity as a digital citizen. UNESCO defines digital citizenship as "being able to find, access, use and create information effectively; engage with other users and with content in an active, critical, sensitive and ethical manner; and navigate the online and ICT environment safely and responsibly, being aware of one's own rights" (UNESCO, 2017). Another similar definition is given by Canada's Centre for Digital and Media Literacy, which defines digital citizenship as the ability to navigate digital environments in a way that's safe and responsible and to actively and respectfully engage in these spaces (What is Digital Citizenship?, 2022). These definitions are focused on the importance of ethical engagement with the use of technology.

However, the abovementioned definitions give a general understanding of the notion and focus on the development of one's responsible use of technology and the author of this chapter finds it necessary to analyze this concept through the analysis of its internal structure and the domains it comprises.

Ribble and Bailey in their attempt to outline the scope of the notion of digital citizenship suggested the following structure: Digital access, Digital Commerce, Digital Communication and Collaboration, Digital Etiquette, Digital Fluency, Digital Health and Welfare, Digital Law, Digital Rights and Responsibility, Digital Security and Privacy (Ribble, 2017). Digital access is grounded in the ability of people to freely access technology in any place. Digital commerce is about buying and selling items safely on the Internet. Digital Communication and Collaboration are grounded in the idea of exchanges online. Digital Etiquette is concerned with netiquette, the code of conduct online. Digital Fluency is similar in its context to Digital Literacy because it emphasizes the way how people use and understand technology. However, the structure of digital literacy is not clearly defined in scholarly literature. In an attempt to develop the structure of the notion of digital literacy, Middleton used qualitative grounded theory research. The main pillars of digital literacy according to the author comprise functional skills, safety, finding and selecting relevant information, communication and collaboration, cultural, social, and ethical understanding, and creativity (Middleton, 2022). Ribble understands Digital Health and Welfare as well-being online. Digital Law refers to the responsible use of materials posted on the Internet. Digital Rights and Responsibility are concerned with liberties people can exercise online. Digital Security and Privacy are concerned with the electronic means to assure a person's safety online (Ribble, 2017).

The author finds Ribble's principles to be strongly focused on behaviors of an individual. Ribble's framework is also aligned with the aim of this study, which is centred around the opinions of pre-service teachers about digital citizenship. As a result, the surveying tool to identify pre-service teachers will be structured around perceived knowledge, skills, and beliefs participants have in Ribble's 9 domains of digital citizenship.

In contrast to Ribble's idea, the Council of Europe suggests grounding digital citizenship into contextual (Access, Secure infrastructure, Literacy skills), informational (Reliable information sources, Rights and responsibilities, Participation skills), organizational (Communication channels, Citizenship opportunity, Problem-solving) levels of digital citizenship (Council of Europe, 2019). All contextual principles are basic requirements to enable digital citizenship both in terms of available technology and basic digital literacy skills. All informational level principles are grounded in leveraging students' abilities with knowledge of their rights and responsibilities, ability to choose reliable information sources and ability to participate in society. The organizational level of principles is grounded in the higher order of skills, such as creative thinking and problem-solving, communication and opportunities to exercise citizenship (Council of Europe, 2019).

Ideas of Ribble and the Council of Europe share some common principles, such as the importance of digital access and digital literacy, the importance of rights and responsibilities of online users, and the importance of digital security and communication using digital means. At the same time, the Council of Europe's scope of principles provides more focus on the societal dimension of digital citizenship with a broad perspective on its implementation. Reliance on problem-solving and reliable information sources is a distinguishing feature of the European principles of digital citizenship.

In the European discourse on digital citizenship, there is also an alternative approach to defining digital citizenship. The European Commission in the DigComp 2.2: The Digital Competence Framework for Citizens (2022) identifies digital competence through 5 competence domains: Information and data literacy, Communication and collaboration, Digital content creation, Safety, and Problem solving. Each of these domains is further detailed in the competencies and their descriptors. Engagement in citizenship through digital technologies is a part of the Communication and collaboration domain. According to the paper, digital citizenship is a way to "participate in society through the use of public and private digital services; to seek opportunities for self-empowerment and forcibly citizenship through appropriate digital technologies" (DigComp 2.2: The Digital Competence framework for citizens, 2022). The European Commission categorizes knowledge, skills, and attitudes regarding this competence into four proficiency levels (Table 1).

Table 1. Engagement in citizenship through digital technologies

<table>
<tr><td rowspan="2">FOUNDATION</td><td>At basic level and with guidance, I can:</td><td>• identify simple digital services in order to participate in society.
• recognise simple appropriate digital technologies to empower myself and to participate in society as a citizen.</td></tr>
<tr><td>At basic level and with autonomy and appropriate guidance where needed, I can:</td><td>• identify simple digital services in order to participate in society.
• recognise simple appropriate digital technologies to empower myself and to participate in society as a citizen.</td></tr>
<tr><td rowspan="2">INTERMEDIATE</td><td>On my own and solving straightforward problems, I can:</td><td>• select well-defined and routine digital services in order to participate in society.
indicate well-defined and routine appropriate digital technologies to empower myself and to participate in society as a citizen.</td></tr>
<tr><td>Independently, according to my own needs, and solving well-defined and non-routine problems, I can:</td><td>• select digital services in order to participate in society.
• discuss appropriate digital technologies to empower myself and to participate in society as a citizen.</td></tr>
<tr><td rowspan="2">ADVANCED</td><td>As well as guiding others, I can:</td><td>• propose different digital services to participate in society.
• use appropriate digital technologies to empower myself and to participate in society as a citizen.</td></tr>
<tr><td>At advanced level, according to my own needs and those of others, and in complex contexts, I can:</td><td>• vary the use of the most appropriate digital services in order to participate in society.
• vary the use of the most appropriate digital technologies to empower myself and to participate in society as a citizen.</td></tr>
<tr><td rowspan="2">HIGHLY SPECIALISED</td><td>At highly specialised level, I can:</td><td>• create solutions to complex problems with limited definition that are related to engaging in citizenship through digital technologies.
• integrate my knowledge to contribute to professional practices and knowledge and guide others in engaging in citizenship through digital technologies.</td></tr>
<tr><td>At the most advanced and specialised level, I can:</td><td>• create solutions to solve complex problems with many interacting factors that are related to engaging in citizenship through digital technologies.
• propose new ideas and processes to the field.</td></tr>
</table>

Source: (DigComp 2.2: The Digital Competence framework for citizens, 2022)

The analysis of the European Commission's reference for knowledge, skills, and attitudes in regards to engaging in citizenship via digital technologies shows that one needs to know about different types of digital services on the Internet, secure electronic identification, knowing about AI and one's rights in regards to it, being aware of actionable civil society platforms, being aware of the role of various forms of media in democratic societies. Skills and attitudes are built around the knowledge abovementioned (DigComp 2.2: The Digital Competence framework for citizens, 2022).

The author recognizes the importance of empowering people through using state and private digital services and participating in civic society by digital means, which is at the core of digital citizenship competence. However, in the author's opinion, the European Commission in their definition focuses primarily on how an individual uses public services online. At the same time, it is also important to incorporate the variability of the use of technology when defining the notion of digital citizenship.

An alternative structure of digital citizenship is suggested by Becker. The scholar has developed a model of the digital citizen grounded in digital literacy, digital etiquette, and critical awareness. The model comprises technical, social, environmental, identity-related, and affective dimensions of digital citizenship. These dimensions consist of the various abilities of a person (Becker, 2019). The technical dimension is concerned with the ability of a person to use technology to participate in society. The social dimension of digital citizenship is an umbrella term for socially accepted ways of participation in society. The environmental dimension refers to a person's ability to understand the communication online and act accordingly. Identity-related dimension refers to the ability of a person to perceive personal identity as a constantly revolving notion, which develops in an interaction with the digital environment. The affective dimension is grounded in the belief that a person should be able to accept the uncertainty of the digital world and know how to deal with negative personal experiences online (Becker, 2019).

This model, as well as Ribble's and the Council of Europe's, is deeply rooted in the individual dimension of digital citizenship and the author finds it important to employ it in the analysis of various dimensions of individual digital citizenship that pre-service teachers incorporate in their lesson plans.

The definition of digital citizenship is a complex matter, and its scope should be analyzed through the lens of fostering it from a very young age. All previous frameworks were focused on outlining general approaches to the concept of digital citizenship regardless of the individual characteristics of learners. However, it is also important to see digital citizenship as a constantly revolving idea and identify the beginning of its formation for young learners.

Developing Digital Citizenship Competence From an Early Age

This sub-section is devoted to the analysis of approaches to the development of digital citizenship of young learners. The analysis of the previous frameworks has shown that they do not take into account the psychological characteristics of children. However, the author claims the importance of teaching about digital citizenship from an early age when the basics of a child's perception of the world are fostered. A similar idea is expressed by scholars from abroad (Hollandsworth et al., 2011).

In an attempt to identify the necessary skills to start developing in young learners, the Council of Europe has drafted the spiral progression of competence building from the earliest childhood. According to it, young learners need to have the knowledge and critical understanding of self, the skills of listening and observing, empathy and co-operation skills, responsible and respectful attitudes, and value human dignity and human rights (Council of Europe, 2019). It is important to build up a system of incorporating a digital citizenship framework in teaching young learners and educate teachers on that.

However, in the author's opinion, the Council of Europe framework outlines the sequence of building the components of digital citizenship competence, but it does not provide its scope.

Further analysis of the landscape of approaches towards the development of digital citizenship for children has led to the analysis of UNESCO's Digital Kids Asia-Pacific Framework: Insights into Children's Digital Citizenship. According to this Framework, digital citizenship can be fulfilled through five domains: Digital Literacy, Digital Safety and Resilience, Digital Participation and Agency, Digital Emotional Intelligence, and Digital Creativity and Innovation (UNESCO, 2019).

Digital Literacy is defined as the ability to seek, critically evaluate, and use digital tools and information effectively to make informed decisions. It comprises ICT literacy and information literacy. Safety and resilience is the ability of children to protect themselves and others from harm in the digital space. It comprises the following competencies: an understanding of child rights, personal data, privacy and reputation, promoting and protecting health and well-being and digital resilience. Digital participation and agency is defined as the ability to equitably interact, engage and positively influence society through ICT. IT comprises interacting, sharing and collaborating, civic engagement and netiquette. Digital emotional intelligence is the ability of children to recognize and express emotions in intrapersonal and interpersonal digital interaction. It comprises self-awareness, self-regulation, self-motivation, interpersonal skills and empathy. Digital creativity and innovation is the ability of children to express themselves and explore through the creation of content using ICT tools. It comprises creative literacy and expression in its structure (UNESCO, 2019, pp. 10-11).

The structure of this framework and its domains is research-informed and is grounded in the cognitive, behavioral and social-emotional dimensions of child development. Unlike previous frameworks, this framework was field tested by its authors to prove its validity. Hence, the author thinks it is important to inform pre-service elementary school teachers regardless of the countries where they work about the framework.

This chapter aims at identifying opinions of undergraduate pre-service teachers on digital citizenship and feasible ways of informing them on how to teach elementary school students about digital citizenship.

Opinions and Practices of Teachers About Digital Citizenship

In this section, the author analyzes the **scholarly** body of research on the opinions of teachers about digital citizenship.

Gleason and von Gillern dwelled on how social media can leverage the development of digital citizenship for secondary school students. Scholars claim that the development of digital citizenship requires the use of a curriculum, which provides plenty of opportunities for students to develop an understanding of citizenship and promote civic education, as well as the use of social media to practice skills of active participation in digital society. Gleason and von Gillern have empirically tested a system of three components to develop digital citizenship: creating digital research projects, creating opportunities for students to connect with officials, and practising participation in society through Twitter. The system of these three components taught students how to become engaged in civic activities, lowered barriers to participation in digital citizenship, as well as developed their digital literacies (Gleason & Von Gillern, 2018).

In this project, in the author's opinion, one of the most valuable parts is the practical component of developing digital citizenship competence. It is important to turn the theory of the structure of digital citizenship, which was analyzed in the previous section, into the practice of teaching citizenship. The author of the chapter thinks that by practicing their presence online, students per se develop a range of skills (e.g. communication and collaboration online, digital literacy, etc.) and it is important to keep fostering the range of skills necessary for digital citizenship. In this chapter, the author finds it important to implement the practical component of students practicing teaching about digital citizenship to elementary school students.

Challenges of teaching citizenship issues in English language classrooms have been addressed by Valente (Valente, 2022). The scholar suggests scaffolding in-depth learning based on real-world issues using picturebooks. Some of the activities the author suggests are framing the discussions with students using guiding questions to break students' stereotypes, involving students in all parts of the book, making more meaningful connections to the real world, and enhancing students' collaboration and mutual exchanges (Valente, 2022, pp. 259-283). Activities, suggested by Valente are both highly practical and grounded in the TESOL practices. However, they are tailored to offline teaching. The author thinks that teaching both digital and citizenship skills favorably affects the development of students' core competencies. Hence, further adaptation of these materials to the challenges of teaching digital citizenship teaching is needed.

Moreover, Valente raised an important issue of fostering global citizens in foreign language classrooms. A similar idea is communicated by Jackson, who states that foreign language lessons provide plenty of opportunities for fostering civic identity, citizenship, and cosmopolitanism (Jackson, 2022). At the same time, this study does not provide insights into developing citizenship skills in online settings.

Walters has studied the knowledge, beliefs, and professional practices of elementary school teachers related to digital citizenship. The major findings state that teachers have a relatively high level of knowledge and beliefs on digital citizenship. However, practitioners demonstrated a low level of knowledge about digital law. Based on the correlational analysis, the scholar says that teachers in schools with higher levels of adoption of 1:1 technology tend to implement digital citizenship practices more often (Walters, 2018).

Based on the results of a quantitative study, Webster claims that teachers state the importance of teaching about digital citizenship, yet they 'rather frequently' to 'rather infrequently' implement digital citizenship principles and only a small number of schools have a formal digital citizenship curriculum. According to the findings, digital communication, digital literacy, digital access, and digital etiquette were most often taught at schools. The gender of teachers was a significant factor influencing the digital citizenship implementation practices (Webster, 2018). The scholar suggests developing and implementing comprehensive digital citizenship programs, which focus on educating students about the use of technology (Webster, 2018).

Suppo studied the beliefs and practices of school leaders in Pennsylvania public schools in regard to digital citizenship instruction. The major findings indicate that school leaders perceive digital rights and responsibilities as one of the most important domains to be taught. The majority of school leaders regard digital citizenship as a less important issue after the behavioral issues in schools. There is also no strong correlation between leadership beliefs and curriculum practices implemented. However, contrary to the previous study, in this study, teachers' demographic characteristics, such as age and gender, were not found to be factors that influence digital citizenship and beliefs (Suppo, 2018, pp. 52-83).

Haase studied teacher perceptions and implementation of digital citizenship in elementary classrooms. The scholar found that elementary school teachers with 6 to 10 years of experience demonstrate the highest level of implementation of digital citizenship practices (Haase, 2022). Among this cohort of teachers, those with teaching experience in online settings demonstrated a higher level of preparedness to implement digital citizenship practices. Digital etiquette, digital security and privacy and digital health and welfare were among the top domains taught by teachers (Haase, 2022, pp. 56-106). This result is to some extent aligned with the choices of teachers in the study by Webster (2018).

The analysis of the scholarly research on teachers' perspectives on digital citizenship shows a diverse landscape of practices and opinions. Scholars have used mixed-method studies and various interventions to test the effectiveness of their interventions. These studies show the evolving nature of the notion of digital citizenship and facilitate the discussion of a balanced holistic approach between theory and practice.

METHODOLOGY

This paper contributes to a growing field of research that seeks to explore ways to teach digital citizenship to pre-service elementary school teachers. In this paper, the author shares Ribble's understanding of digital citizenship as the way to follow norms of anticipated behavior when using technology in the digital world. This section describes the experiment set in Borys Grinchenko Kyiv University (Ukraine) for pre-service elementary school teachers on learning their perceptions of digital citizenship and studying the way how they tend to implement the idea of digital citizenship in their lesson plans for elementary school learners of English.

The research is guided by two research questions: (1) What are the knowledge, skills, and attitudes of pre-service elementary school teachers on digital citizenship?; (2) How do pre-service elementary school teachers envision incorporating digital citizenship in elementary school English classrooms?

This paper employs a mixed-method study design. The quantitative methodology includes surveying of pre-service elementary school teachers. The paper is also grounded in the qualitative methodology through the analysis of pre-service elementary school teachers' artefacts in the form of lesson plans to teach English as a Foreign Language by means of digital citizenship.

Participants of the Study

Participants of the study are 13 Year 3 undergraduate students of the specialty Elementary Education of Borys Grinchenko Kyiv University, who take the course Modern Technologies of Teaching Foreign Language in Elementary Schools (12 credits).

Table 2 shows the demographic data of participants. All of them are pre-service elementary school teachers, who are in their first year of teaching internship. A majority of participants undergo the teacher training in the city schools, where 1:1 device distribution is introduced in some grades.

Table 2. Demographic data of participants

	Frequency	Percent
Gender		
Male	0	0
Female	13	100
Prefer not to Answer		
Age		
20-25	13	100
26-30	0	0
31-40	0	0
Including this year, how many years have you been teaching?		
This is my first year	12	92.3%
2-5 years	0	0
6-10 years	0	0
11-15 years	0	0
16-20 years	0	0
21+	0	0
prefer not to answer	1	7.7%
Where is your current school located?		
In the city	10	76.9%
In the rural area	3	23.1%
What is the level of adoption of a 1:1 device program at your current school?		
Not 1:1 at all	1	7.7%
I have a quarter of the number of devices as I have students in my classroom (ex, I have 5 devices and 20 students).	1	7.7%
I have half the number of devices as I have students in my classroom (ex, I have 10 devices and 20 students).	0	0
Some grade levels/classrooms are 1:1 and other grade levels/classrooms are 2:1 or less for device access	0	0
Piloting 1:1 in some classrooms in the school but not mine.	0	0
Piloting 1:1 in some classrooms in the school including mine.	1	7.7%
1:1 at certain grade levels but not mine.	3	23.1%
1:1 at certain grade levels including mine.	2	15.4%
Fully adopted 1:1 at all planned grade levels	0	0
Prefer not to answer	5	38.5

Source: created by Author

Research Design

This is a mixed-method study. Its structure is outlined in Figure 1.

Figure 1. Research design
Source: created by Author

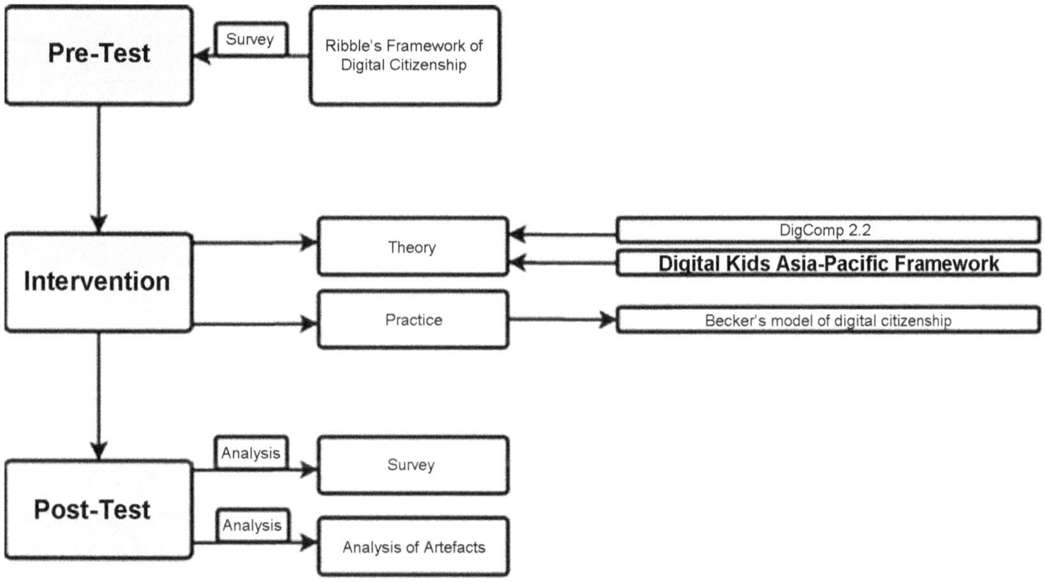

Data Collection and Analysis Plan

The quantitative data in this paper was obtained through the adapted survey developed by Walters (Walters, 2018). The surveying tool was adopted in full. However, the research design was modified to incorporate a pre-test, intervention and post-test. The author has also changed the research questions and qualitative analysis of artefacts.

Participants were asked to share their opinion on digital citizenship using an anonymous survey with Likert Scale and limited open-ended questions. The survey was grounded in Ribble's principles of digital citizenship and structured into the following blocks: Demographic information, Digital literacy, Digital commerce, Digital etiquette, Digital access, Digital communication, Digital law, Digital rights and responsibilities, Digital health and well-being, Digital safety and security, Planning and implementation practices. Participants could choose their answer for 66 statements on a scale from 1 (Not true of me) to 4 (Always true of me). The survey was distributed in Google Forms and data was transferred in Excel, where the data was analyzed. 'Not true for me' was coded as 1, 'Sometimes true of me' was coded as 2; 'Always true of me' was coded as 3, and 'Always true of me' was coded as 4. The author checked the internal reliability of the tool using Alpha Cronbach α=.958, which showed its relatively high consistency (Taber, 2018). The details of the analysis of data are provided in the Results section of this paper.

After the survey, the intervention was organized in the form of a module on language teaching based on digital citizenship. The outline of the module is provided in Table 3. In this table, 2 hours of lectures and practical classes equal 1 teaching class.

Table 3. Outline of the study intervention

Module 4. Gamification in Teaching Foreign Languages to Young Learners				
Topics	**Lectures**	**Practical classes**	**Office hours**	**Self-study work**
Topic 1. The Notion of Digital Citizenship in Teaching Foreign Languages to Elementary School Students	2			
Topic 2. Implementing digital citizenship in teaching Grade 1 and 2 elementary school students.		6		6
Topic 3. Implementing digital citizenship in teaching Grade 3 and 4 elementary school students.		8		6
Test			2	
Total 30 h	2	14	2	12

Source: created by Author

After the intervention, the posttest survey was administered for participants to identify a change in opinions. The author also employed the qualitative methodology by analyzing the artefacts students have developed within the intervention module. All artefacts were uploaded in Nvivo and coded according to Becker's dimensions of digital citizenship (technical, social, environmental, identity-related, and affective dimensions of digital citizenship).

RESULTS

In this section, the author provides the results of the study structured by research questions.

Knowledge and Skills of Pre-Service Elementary School Teachers on Digital Citizenship

The first research question aimed at identifying knowledge, skills, and attitudes of pre-service elementary school teachers on digital citizenship. Students reported they did not have any formal instruction on digital citizenship. Likert scale survey comprised answers 'Not true for me', coded as 1, 'Sometimes true of me', coded as 2; 'Always true of me', coded as 3, and 'Always true of me', coded as 4.

In Table 4 there is an outline of results on the knowledge and skills of participants grouped by blocks according to Ribble. In regards to knowledge and skills, the analysis of the mode shows that most participants agreed the statements are always true about them. However, the analysis of the mean in each category shows a trend to its slight rise in the posttest with a decrease in standard deviation for most of the categories, which means that as a result of the intervention, participants feel they acquired the necessary for digital citizenship knowledge and skills.

Insignificant change is observed in the posttest data on Digital rights and responsibilities compared to the pretest. Relatively low responses were also obtained in the category of Digital safety and security. The author attributes these results to a lack of instruction on these sub-topics during the intervention.

Table 4. The level of pre-service elementary school teacher's knowledge and skills for digital citizenship

	Mean		Median		Mode		SD	
	T1	T2	T1	T2	T1	T2	T1	T2
Digital literacy	3.44	3.74	3.42	4	4	4	.560	.445
Digital commerce	3.33	3.66	3.83	3.83	4	4	.786	.541
Digital etiquette	3.2	3.86	3.2	4	3; 4	4	.722	.34
Digital access	3.23	3.48	3.33	4	4	4	.952	1.03
Digital communication	3.44	3.60	3.71	4	4	4	.707	.706
Digital law	3.19	3.77	3.5	4	4	4	1.00	.520
Digital rights and responsibilities	3.76	3.77	4	4	4	4	.438	.440
Digital health and well-being	3	3.63	3	4	3	4	.968	.778
Digital safety and security	2.75	3.44	2.6	3.8	3	4	1.078	.936

The second part of the first research question was related to the beliefs pre-service elementary school teachers hold about digital citizenship. Table 5 provides an overview of descriptive statistics based on survey. There is a negative change observed in the category Digital literacy: the posttest results demonstrate a decrease in the mean and a rise in standard deviation, which means a higher difference in answers. According to the descriptive statistics, a negative change in beliefs has also happened in the category Digital access. The author explains it with participants being better-informed after the intervention.

Table 5. Beliefs of pre-service elementary school teachers about digital citizenship

	Mean		Median		Mode		SD	
	T1	T2	T1	T2	T1	T2	T1	T2
Digital literacy	3.44	3.29	3.42	3.66	4	4	.560	1.00
Digital commerce	2.85	3.5	2.5	4	3;2	4	1.07	.87
Digital etiquette	3.1	3.62	3.3	4	3	4	.921	1.00
Digital access	3.46	3.17	3.83	3.42	4	4	.652	.534
Digital communication	3.01	3.52	3	4	3	4	.911	1.00
Digital law	3.07	3.66	3	4	3	4	.847	.893
Digital rights and responsibilities	3.35	3.59	4	4	4	4	.976	.787

Main Ways Pre-Service Teachers Utilize to Incorporate Digital Citizenship in Elementary School English Classrooms

The second research question aimed at identifying practices that pre-service elementary school teachers attempt to incorporate in their English lessons after undergoing an intervention module on incorporating digital citizenship issues in language teaching. Participants have designed lesson plans on the topic of their choice and incorporated digital citizenship into language teaching. The author has analyzed and coded the artefacts according to Becker's principles (Table 6).

Most frequently participants focused their attention on instructing elementary students on the use of technology and the social dimension, as well as the affective dimension. Students have noted that it all aligns with a broader agenda of fostering elementary school students to become global citizens.

Table 6. The frequency of implementation of Becker's dimensions of digital citizenship in pre-service elementary school teacher's lesson plans

Becker's dimensions of digital citizenship	Frequency of implementation
Technical dimension	17
Social dimension	15
Environmental dimension	7
Identity-related dimension	4
affective dimension	12

DISCUSSIONS

This section discusses the results of the study based on the research questions, which aimed to identify the knowledge, skills, beliefs, and practices of pre-service elementary school teachers regarding digital citizenship education in English language classrooms.

The first research question is focused on the knowledge, skills, and attitudes of pre-service elementary school teachers regarding digital citizenship. The survey has shown that participants did not receive formal instruction on digital citizenship and have a relatively low level of background knowledge on digital citizenship. This result is contrary to the study by Walters (2018), where participants had relatively high knowledge in all domains of digital citizenship except for digital law. The post-test analysis has shown a slight increase in the mean score and a decrease in standard deviation for most domains of digital citizenship, which means that the intervention has positively impacted participants' perceptions.

The second research question focused on participants' beliefs about digital citizenship. The survey results have demonstrated a slight shift in the participants' beliefs. Such a result can be attributed to raising awareness about digital citizenship after the intervention.

The analysis of the participants' artefacts indicates that they have frequently referred to the technical, social, and affective dimensions of digital citizenship. This result can be attributed to participants having background knowledge on developing young learners' technological skills, knowing the basics of social and emotional learning, and applying it to digital contexts.

This study has shown the positive impact of a designed intervention on pre-service elementary school teachers' knowledge, skills, and beliefs about digital citizenship.

CONCLUSION

The objective of this paper is to outline ways of teaching pre-service elementary school teachers about incorporating digital citizenship in language instruction. The author grounds the theoretical underpinning into the theory and principles of digital citizenship outlined by Ribble and Becker. The paper is a mixed-method study employing quantitative methodology (surveying) and qualitative methodology (artefact analysis). Ribble's framework was implemented in the survey tool adopted from Walters (2018). Becker's approach to structuring digital citizenship was applied to the analysis of students' artefacts. The result of the post-test shows a positive change in the knowledge, skills, and beliefs of pre-service elementary school teachers on digital citizenship. Pre-service elementary school teachers are more confident in developing lesson plans incorporating the digital citizenship domains they are most comfortable with (technical, social, and affective domains).

This chapter is relevant for pre-service and giving service and elementary classroom teachers, teacher trainers and those who provide professional development for them.

ACKNOWLEDGMENT

This study was made possible with the informational support of libraries of the University of Minnesota Twin Cities, where the author was a Visiting Scholar in the 2022-2023 academic year.

REFERENCES

Becker, D. (2019). The digital citizen 2.0. *Arbeiten Aus Anglistik Und Amerikanistik*, *44*(2), 167–194.

Council of Europe. (2019). *Digital Citizenship Education Handbook*. European Commission. https://ec.europa.eu/newsroom/just/items/672450/en

European Commission, Joint Research Centre, Vuorikari, R., Kluzer, S., & Punie, Y. (2022). *DigComp 2.2: The Digital Competence framework for citizens*. Publications Office of the European Union.

Gleason, B., & Von Gillern, S. (2018). Digital citizenship with social media. *Journal of Educational Technology & Society*, *21*(1), 200–212.

Haase, A. (2022). *Teacher Perceptions and Implementation of Digital Citizenship in Elementary Classrooms: A Mixed Methods Study*. ProQuest Dissertations.

Hollandsworth, R., Dowdy, L., & Donovan, J. (2011). Digital citizenship in K-12: It takes a village. *TechTrends*, *55*(4), 37–47. doi:10.100711528-011-0510-z

Isin, E., & Ruppert, E. (2015). Being Digital Citizens. Rowman & Littlefield.

Jackson, L. (2022). Civic identity, citizenship and foreign language education. In C. Lütge, T. Merse, & P. Rauschert (Eds.). Global citizenship in foreign language education: Concepts, practices, connection (pp. 41-61). Routledge. https://doi.org/10.4324/9781003183839

Middleton, S. (2022). *Towards digital citizenship: a digital literacy curriculum to support teachers in the classroom*. Northumbria University.

Mossberger, K., Tolbert, C., & McNeal, R. (2007). *The internet, society, and participation*. The MIT Press.

Ribble, M. (2017). *Nine Themes of Digital Citizenship*. Digital Citizenship. https://www.digitalcitizenship.net/nine-elements.html

Ribble, M., Bailey, G. D., & Ross, T. (2004). Digital citizenship: addressing appropriate technology behavior. Learning and Leading with Technology, 32(1). 6-9. https://eric.ed.gov/?id=EJ695788

Suppo, C. (2018). *Digital citizenship instruction in Pennsylvania public schools: School leaders expressed beliefs and current practices*. ProQuest Dissertations.

Taber, K. (2018). The use of Cronbach's Alpha when developing and reporting research instruments in science education. *Research in Science Education*, *48*(6), 1273–1296. doi:10.100711165-016-9602-2

UNESCO. (2017). *Outcome document of the Conference on Digital Citizenship Education in Asia-Pacific*. UNESCO.

UNESCO. (2018). *Skills for a connected world Programme*. UNESCO.

UNESCO. (2019). *Digital kids Asia-Pacific: insights into children's digital citizenship*. UNESCO. https://unesdoc.unesco.org/ark:/48223/pf0000367985

Valente, D. (2022). Scaffolding in-depth learning: picturebooks for intercultural citizenship in primary English teacher education. In D. Valente, & D. Xerri, Innovative Practices in Early English Language Education (pp. 259-283). Cham: Palgrave Macmillan.

Walters, M. (2018). *Elementary educators' knowledge, beliefs, and planned and implemented practices for digital citizenship*. ProQuest Dissertations.

Webster, J. P. (2018). *Teacher Perceptions and Implementation of Digital Citizenship Curriculum in One-To-One High Schools in Missouri*. ProQuest Dissertations.

What is digital citizenship & how do you teach it? (2023). Applied Educational Systems. https://www.aeseducation.com/blog/what-is-digital-citizenship#:~:text=Digital%20citizenship%20refers%20to%20the,with%20society%20on%20any%20level

What is digital citizenship? (2022). Canada's Centre for Digital and Media Literacy. https://mediasmarts.ca/digital-media-literacy/general-information/digital-media-literacy-fundamentals/what-digital-citizenship

Chapter 4
Empowering Digital Citizenship Through Distance Education:
A Technology-Driven Education Action Plan

Seda Adıgüzel
https://orcid.org/0000-0001-8292-0916
Gazi University, Turkey

Selami Eryılmaz
Gazi University, Turkey

ABSTRACT

The European Union Digital Education Action Plan addresses the imperative of digital citizenship in the landscape of post-pandemic education. This abstract highlights the plan's focus on equipping learners with the skills to navigate the digital sphere responsibly. Emphasizing digital literacy, online safety, and ethical behavior also the plan aims to foster critical engagement with safer technology by promoting inclusivity and equitable access, it seeks to empower individuals to thrive as responsible digital citizens in an interconnected world.

INTRODUCTION

In recent years, the world has experienced a rapid shift towards digitalization, which has changed the way we live, learn, and work. The COVID-19 pandemic has accelerated this trend, leading to a significant increase in the use of digital technologies for education. The Council of Europe Digital Education Action Plan 2021-2027 aims to promote quality digital education for all and highlights the importance of digital citizenship. In this book part, it will have investigated the digital citizenship perceptions of teachers in distance education in the framework of the Council of Europe Digital Education Action Plan 2021-2027.

DOI: 10.4018/978-1-6684-8934-5.ch004

Digital Citizenship

Digital citizenship refers to the responsible use of digital technologies, including the internet, social media, and digital devices, to engage with others, share information, and participate in society (Simsek, Simsek, 2013). It encompasses a range of skills, including digital literacy, online safety, and ethical behaviors. It is particularly crucial for young people who are growing up in an increasingly digital world and need to develop the necessary skills to navigate and participate in this environment. In the context of education, digital citizenship is crucial for students to become responsible and informed digital citizens (Choi, 2016). However, it is also important for teachers to have a good understanding of digital citizenship to guide and support their students. According to Mike Ribble there are nine dimensions of digital citizenship and the nine dimensions of digital citizenship were first proposed by Mike Ribble in his book "Digital Citizenship in Schools" (2015). These dimensions provide a framework for understanding the various aspects of digital citizenship. The nine dimensions of digital citizenship are:

Digital Access: This refers to the ability to access digital technologies and the internet. It includes issues related to the digital divide, such as access to devices and broadband internet. Online learning, collaboration and communication, research and special needs education and career preparation can be educational uses examples for digital access (Hanson, Carlson, 2005).

Digital Commerce: This refers to the safe and responsible use of online financial transactions, such as online shopping and banking. Online learning platforms, e-commerce classes, online marketplaces, fundraising and virtual internships can be some educational uses examples for digital commerce (Bonanno, 2016).

Digital Communication: This element refers to the ability to communicate effectively using digital technologies, including email, social media, and video conferencing. It includes issues related to online etiquette and digital communication skills. Virtual classrooms, online discussion forums, email communication, video conferencing and social media can be some educational uses examples for digital communication (Nguyen, et all, 2020).

Digital Literacy: The element refers to the ability to find, evaluate, and use information effectively and ethically online. It includes issues related to digital research skills and critical thinking. Digital skills training, information literacy, digital citizenship, digital storytelling and online research can be some educational uses examples for digital communication (Buckingham, 2015).

Digital Etiquette: The element refers to the responsible and respectful use of digital technologies, including social media, email, and online chat. It includes issues related to online behavior and etiquette. Online communication, online safety, cyber ethics, online collaboration, netiquette can be some educational uses examples for digital etiquette (Ribble, Bailey, 2004).

Digital Law: The element refers to the understanding of the laws and regulations that govern the use of digital technologies, including issues related to copyright, privacy, and security. Cyber law education, cybersecurity education, intellectual property education, online privacy education and digital rights and responsibilities education can be some educational uses examples for digital law (Tomczyk, 2019).

Digital Rights and Responsibilities: The element refers to the understanding of the rights and responsibilities of digital citizenship, including issues related to online safety, privacy, and security. Online citizenship, digital privacy, online freedom of speech, digital literacy and digital security can be some educational uses examples for digital rights and responsibilities (Pangrazio, Sefton-Green, 2021).

Digital Health and Wellness: This element refers to the physical and emotional health impacts of using digital technologies, including issues related to screen time, cyberbullying, and online addiction.

Digital health education, digital wellness education, digital health monitoring, digital mental health support and digital health research can be some educational uses examples for digital health and wellness (Aungst, Patel, 2020).

Digital Security: The last element refers to the safe and responsible use of digital technologies to protect oneself and others from online threats, including issues related to malware, phishing, and identity theft. Cybersecurity education, digital security policies and procedures, digital security assessments, digital forensics education and digital security research can be some educational uses examples for digital security (Kirlappos, Sasse, 2011).

Digital Citizenship Applications

Digital citizenship applications are tools, methods, roadmap or programs that promote responsible, ethical, and safe behavior online (Mahmoud, 2021). These applications help individuals to develop the necessary skills and knowledge needed to participate effectively and safely in digital environments. For example, digital compass or Interland. Digital Compass is an online game that teaches students in grades 6-9 about digital citizenship skills. The game covers topics such as privacy, cyberbullying, and online safety, and challenges players to make decisions based on real-life scenarios (Allen, Seaman, 2017). Another example is Common Sense Education. Common Sense Education provides a variety of resources, including lesson plans, videos, and interactive tools, to help educators teach digital citizenship skills to students of all ages. The program covers topics such as media literacy, digital safety, and responsible online behavior (Barrow, 2012). Another example is Be Internet Awesome (Interland). Be Internet Awesome is a program created by Google that provides resources for educators, students, and parents to promote safe and responsible online behavior. The program includes interactive games, lesson plans, and resources for parents to help their children stay safe online (Seale, Schoenberger, 2018). Other example is Digital Citizenship Curriculum. The Digital Citizenship Curriculum, created by the International Society for Technology in Education (ISTE), provides a comprehensive framework for teaching digital citizenship skills to students from kindergarten through grade 12. The curriculum covers topics such as digital communication, cyberbullying, and online privacy (Krutka, Carpenter, 2017). Last example is Digital Citizenship Interactive Whiteboard Lessons. It was created by again Common Sense Education, are a series of interactive lessons designed to teach digital citizenship skills to students in grades K-12. The lessons cover topics such as online safety, privacy, and responsible digital behavior. Overall, digital citizenship applications can help individuals develop the necessary skills and knowledge needed to navigate the digital world safely and responsibly. By promoting responsible digital behavior and fostering a culture of digital citizenship, these applications can help create a safer and more positive online environment for everyone.

Education and Digital Citizenship Relation

Education is a critical component of preparing students to be responsible digital citizens. Schools have an important role to play in providing students with the necessary skills and knowledge to navigate the digital world. Digital citizenship education involves teaching students how to use technology safely and ethically, how to protect their privacy and personal information, how to identify and avoid online scams, and how to be good digital citizens (Oxley, 2011). Educational technology can play a critical role in promoting digital citizenship among students. For example, teachers can use educational technologies to teach students about online safety, privacy, and security (Elmali, et all 2020). They can use online resources and tools to help students learn about cyberbullying, identity theft, and online scams. They can also use educational technologies to teach students about digital etiquette, responsible communication, and digital literacy. Educational technologies can also provide opportunities for students to practice digital citizenship skills in a safe and controlled environment. For example, teachers can use online discussion forums and social media platforms to teach students how to communicate respectfully and responsibly online. They can also use educational games and simulations to help students practice critical thinking, problem-solving, and decision-making skills related to digital citizenship. Moreover, educational technologies can help bridge the digital divide by providing access to digital technologies and resources to students who may not have access otherwise. Teachers can use educational technologies to provide online resources and tools to students from underserved communities and help them develop the necessary skills to become active and engaged digital citizens (Richardson & Milovidov, 2019). Beyond the educational technology the relationship between education and digital citizenship is becoming increasingly important as technology continues to shape our daily lives. Digital citizenship refers to the responsible and ethical use of technology, particularly the internet. As technology becomes more integrated into our daily lives, it is essential that we educate students about digital citizenship to ensure they can use technology safely and responsibly.

One of the key aspects of digital citizenship education is teaching students to use technology responsibly. This means helping students understand the potential consequences of their actions online, including cyberbullying, sexting, and online harassment. By teaching students to be responsible digital citizens, we can help create a safer and more respectful online community. Digital citizenship education can also help students develop critical thinking skills. With so much information available online, it is essential that students learn how to evaluate the credibility and reliability of sources. Teaching students to think critically about the information they find online, we can help them become better-informed citizens. Teaching students about online privacy and security is another important aspect. Students need to understand the importance of protecting their personal information online and how to do so effectively. Talking about online privacy and security, teachers can help them avoid identity theft and other online scams. Finally, digital citizenship education can also help promote digital equity. As technology becomes more important in our daily lives, it is essential that all students have access to the necessary tools and resources to succeed in the digital world. By providing students with the necessary skills and knowledge, we can help ensure that everyone has an equal opportunity to succeed in the digital age (Andrews, et al, 2020).

In conclusion, the relationship between education and digital citizenship is becoming increasingly important as technology continues to shape our daily lives. Digital citizenship education is essential for teaching students how to use technology safely and ethically, while education plays a critical role in helping students develop the skills and knowledge necessary to thrive in the digital age. The most important thing is that teachers can help create a safer, more informed, and more equitable digital community with all others support.

Investigation of Digital Citizenship Perceptions of Teachers in Distance Education

As the use of digital technologies in education continues to grow, the concept of digital citizenship has become increasingly important. Digital citizenship refers to the responsible use of digital technologies, including online behavior, digital literacy, and digital safety. In the context of distance education, digital citizenship becomes even more critical, as learners are required to use digital technologies to access educational resources and communicate with their teachers and peers (Baloch, et al, 2020). The investigation of digital citizenship perceptions of teachers in distance education can provide valuable insights into how educators perceive and approach digital citizenship in their teaching. Understanding the perspectives of teachers, educational policymakers and stakeholders may develop effective strategies and policies to promote digital citizenship in distance education. Council of Europe Digital Education Action Plan 2021-2027 is one of the important source which provides a set of priorities and objectives for promoting digital citizenship in education and can serve as a guide for investigating teachers' perceptions.

Some key areas that can be explored in an investigation of digital citizenship perceptions of teachers in distance education include (Ata & Yıldırım, 2019):

- Understanding teachers' views on the importance of digital citizenship in distance education.
- Exploring teachers' understanding of the various dimensions of digital citizenship and how they integrate these dimensions into their teaching.
- Investigating teachers' perceptions of the challenges and opportunities associated with promoting digital citizenship in distance education.
- Identifying the resources and support that teachers need to effectively promote digital citizenship in their teaching.
- Exploring the role of professional development in supporting teachers' digital citizenship practices in distance education.

By investigating these and other areas, managers, families, policymakers and others can gain a better understanding of how to support teachers in promoting digital citizenship in distance education. It can also help to ensure that the learners are equipped with the necessary skills and knowledge to participate responsibly and ethically in the digital world.

How Changed the Teacher's Digital Citizenship Vision With Pandemic

The COVID-19 pandemic has forced many teachers to shift to digital and remote teaching, which has had a significant impact on their views of digital citizenship. There are some ways in which teachers' digital citizenship views may have changed after the pandemic. Firstly, increased emphasis on online safety. The pandemic has led to an increased emphasis on online safety, as teachers and students have spent more time online the teachers may have become more aware of the potential risks and challenges associated with online interactions, and may have implemented more rigorous online safety measures. Secondly, greater awareness of digital etiquette. The teachers and students have had to adapt to new norms and expectations for online interactions. Teachers may have become more attuned to issues such as online bullying, cyber harassment, and inappropriate online behavior, and may have implemented more explicit guidelines for digital etiquette. Thirdly, increased focus on digital literacy. Both teachers and students have had to adapt to new technologies and online learning platforms. Fourthly, greater use of digital communication. The pandemic has led to a greater use of digital communication tools, such as video conferencing, email, and instant messaging therefore the teachers have become more comfortable with these tools and may have integrated them more fully into their teaching practices, leading to a greater emphasis on digital communication skills (Li & Yu, 2022).

On the other hand, with the shift to remote and hybrid learning models, teachers have had to rely more heavily on technology and digital tools to deliver instruction and engage with students (Whalen, 2020). This has brought to the forefront issues related to responsible and ethical use of technology, online safety, and digital literacy. Many teachers have had to adapt quickly to new technologies and digital platforms, and in doing so, have become more aware of the importance of digital citizenship and they have had to consider how to teach their students to be responsible digital citizens, including issues related to cyberbullying, online privacy, and responsible use of social media (Oxley, 2011). Additionally, the pandemic has also highlighted the digital divide and the inequities that exist in access to technology and digital resources. Overall, the pandemic has underscored the importance of digital citizenship and has prompted many teachers to prioritize this aspect of education in their instruction, digital citizenship views, leading to greater awareness of online safety, digital etiquette, digital literacy, and digital communication. As a result, teachers can be more effective at promoting responsible and ethical digital behavior among their students, and may be better equipped to help their students navigate the complex and rapidly evolving digital landscape.

The Importance of Technology Plan

A technology plan is a critical document that outlines the strategic use of technology in an organization (King, South, 2017). Whether it's a business, government agency, or educational institution, a technology plan is essential for ensuring that technology is used effectively to achieve organizational goals and objectives. Clear direction and focus, improved efficiency and productivity, enhanced security and risk management, Improved decision-making, competitive advantage are some reasons explains why the technology plan is important.

Clear direction and focus is important because a technology plan provides a clear direction and focus for an organization's technology efforts. It outlines the organization's goals and objectives for technology, the resources required to achieve those goals, and the timeline for implementation (Umble, et all, 2003). It is important to add the digital citizenship to the technology plans goals and objectives because it should be explained clearly. Improved efficiency and productivity is another important topic because a well-designed technology plan can help an organization improve its efficiency and productivity by identifying opportunities for automation and streamlining processes. Cost savings, increased output, and faster turnaround times, enhanced security and risk management is also important because a technology plan includes measures for enhancing security and managing risk, such as data backup and disaster recovery plans, cybersecurity protocols, and access controls. These can help to protect an organization's valuable assets, including its data and intellectual property, from cyber threats and other risks. It is important to add the digital citizenship to the technology plans goals and objectives because everyone should understand the digital security. Improved decision-making makes it important because a technology plan provides data and insights that can inform decision-making, such as information on technology trends, the cost-benefit analysis of different technology investments, and the impact of technology on business operations (Vessey, 1994) so the organizations make informed decisions about their technology investments and strategies.

As it is mentioned above a technology plan is essential for any organization that wants to use technology effectively to achieve its goals and objectives. It is also important to add the digital citizenship to the technology plans all goals and objectives because it is important for everyone to live in a secure digital world.

USA Technology Plan

The United States has long been a leader in educational technology, and the federal government has played a key role in supporting the integration of technology in education. While there is no centralized national educational technology plan, the U.S. Department of Education has released a number of initiatives and guidelines to support the use of technology in teaching and learning.

One of the key initiatives is the National Education Technology Plan, which was last updated in 2017. The plan provides a roadmap for how technology can transform teaching and learning in the United States and identifies five key goals to guide the integration of technology in education (King & South, 2017).

- The first goal is to personalize learning to improve student engagement and achievement. This goal recognizes that students learn in different ways and at different paces, and that technology can be used to provide more personalized learning experiences. For example, adaptive learning software can provide students with individualized learning paths based on their strengths and weaknesses.
- The second goal is to use technology to assess student learning more effectively. Technology can provide teachers with real-time data on student progress, allowing them to identify areas where students need additional support and adjust instruction accordingly. Additionally, technology can be used to develop more authentic assessments that better reflect the skills and knowledge students will need in the 21st century workforce.
- The third goal is to prepare and support teachers to use technology effectively in the classroom. Teachers need to be equipped with the knowledge and skills to effectively integrate technology into their instruction. The National Education Technology Plan calls for teacher training programs that focus on the use of technology for teaching and learning, as well as ongoing professional development to support teachers as they integrate technology into their practice.
- The fourth goal is to provide schools and communities with the infrastructure needed for effective use of technology. This includes not only access to high-speed internet and other digital tools, but also the policies and procedures needed to ensure that technology is used responsibly and ethically.
- Finally, the fifth goal is to use technology to increase efficiency and productivity in education. This includes using technology to streamline administrative tasks, such as grading and record-keeping, as well as leveraging technology to provide students with access to high-quality educational resources and materials.

In addition to the National Education Technology Plan, the U.S. Department of Education has also launched the Future Ready initiative, which provides resources and support for schools and districts to effectively use technology for teaching and learning (Thomas, 2016). The initiative focuses on seven key areas, including curriculum, instruction, and assessment; professional learning; infrastructure; data and privacy; community partnerships; content and resources; and budget and resources. These areas are also important issues for digital citizenships nine elements.

Council of Europe Digital Education Action Plan 2021-2027

The Council of Europe has recently released its Digital Education Action Plan for 2021-2027, which sets out a vision for how technology can transform education across Europe. The plan recognizes the rapid pace of technological change and the need for education systems to adapt in order to prepare students for the digital world. The Digital Education Action Plan is built around four key themes such as inclusive digital education, digital competence and literacy, quality and efficiency, and innovation and creativity (Caena, Redecker, 2019). Each theme is accompanied by a set of actions and recommendations aimed at supporting the integration of technology in education.

1. Inclusive digital education, focuses on ensuring that all students have equal access to digital resources and that technology is used in a way that promotes equity and inclusion. This includes initiatives to provide access to digital devices and infrastructure for disadvantaged students, as well as efforts to ensure that digital resources and tools are designed with accessibility in mind.
2. Digital competence and literacy, recognizes the importance of ensuring that students have the skills and knowledge needed to thrive in a digital world. Supporting the development of digital literacy and citizenship skills, as well as efforts to integrate digital skills into the curriculum across all subjects is important.
3. Quality and efficiency, focuses on ensuring that technology is used in a way that supports high-quality teaching and learning so the initiatives to promote the use of evidence-based practices and to support teachers in effectively integrating technology into their instruction.
4. Innovation and creativity, recognizes the potential of technology to support new and innovative approaches to teaching and learning because it includes initiatives to promote the development of digital learning resources and tools, as well as efforts to support experimentation and innovation in education (Caena, Redecker, 2019).

Turkey Education Vision 2023

The vision for educational technology in Turkey by 2023 is to create a modern and effective education system that integrates technology into every aspect of teaching and learning. The Turkish government has identified the importance of educational technology in the 21st century and has set ambitious goals for the future (Kirkgöz, 2009). One of the main goals of the Education Vision 2023 in Turkey is to provide equal access to technology and digital resources for all students. This includes ensuring that every student has access to digital devices such as tablets, laptops, and smartphones, as well as high-speed internet connectivity so it will enable students from all regions of the country to have the same opportunities to access digital resources and educational content. Ennhancing the quality of education by using technology to provide personalized and adaptive learning experiences for students is another important element. The use of digital resources such as educational apps, online courses, and interactive learning materials will enable students to learn at their own pace and in a way that suits their individual learning style. This will not only improve student engagement and motivation but also help to improve learning outcomes. The Education Vision 2023 in Turkey also includes a focus on developing the skills and competencies of teachers (Albrahim, 2020) so the teachers with training and professional development opportunities help them integrate technology into their teaching practices effectively. By that all teachers are comfortable and confident in using educational technology to enhance student learning. Other important aspect of the

Education Vision 2023 in Turkey is to promote digital citizenship and online safety. It is important for teaching students about responsible and ethical behavior online, as well as how to protect their personal information and stay safe online. It also includes promoting critical thinking and media literacy skills to help students navigate the vast amount of information available online. It also includes also a focus on promoting innovation and entrepreneurship in the education sector with this element encouraging the development of new educational technology products and services, as well as fostering a culture of innovation in schools and universities getting more important (Gulbahar & Guven, 2008). By doing so, Turkey aims to become a leader in educational technology and provide its students with the best possible education for the future. In summary, the Education Vision 2023 in Turkey by 2023 is to create a modern and effective education system that integrates technology into every aspect of teaching and learning.

Differences and Similarities Between the Council of Europe Digital Education Action Plan 2021-2027, the USA Technology Plan, and Turkey Education Vision 2023 Plans

The Council of Europe Digital Education Action Plan 2021-2027, the USA Technology Plan, and Turkey Education Vision 2023 plans are all aimed at integrating technology in education to improve teaching and learning outcomes. While they share some similarities, there are also some key differences in their approaches and goals.

Similarities:

- All three plans recognize the importance of technology in education and the need to prepare students for the digital world.
- They all aim to support the development of digital literacy and citizenship skills.
- They all emphasize the importance of providing access to digital resources and infrastructure for students and teachers.
- They all recognize the need to support teachers in effectively integrating technology into their instruction.

Differences:

- The Council of Europe plan focuses on promoting inclusion and equity in digital education, whereas the USA Technology Plan emphasizes personalized learning and the Turkey Education Vision 2023 plan emphasizes the development of creative and critical thinking skills.
- The Council of Europe plan emphasizes the importance of promoting evidence-based practices in the use of technology in education, while the USA Technology Plan emphasizes the importance of data-driven decision making.
- The Turkey Education Vision 2023 plan places a greater emphasis on the use of technology to support the development of a knowledge-based economy.
- The Council of Europe plan focuses on promoting innovation in digital education, while the USA Technology Plan and Turkey Education Vision 2023 plan also emphasize the need to provide support for research and development in the field of educational technology.

Overall, while these plans share some common goals and objectives, they also have distinct differences in their approaches and priorities. By learning from each other's experiences and best practices, countries can work together to develop comprehensive and effective strategies for integrating technology in education to improve student outcomes.

CONCLUSION

In today's digital age, digital citizenship has become an essential aspect of education, particularly with the widespread use of technology in teaching and learning. Digital citizenship is a multifaceted concept that encompasses various dimensions, including digital access, literacy, communication, security, and ethics. Education plays a critical role in fostering digital citizenship by equipping students with the skills, knowledge, and attitudes necessary to navigate the digital world responsibly and ethically. The investigation of digital citizenship perceptions of teachers in distance education highlights the importance of understanding the digital citizenship practices of educators and the need for professional development programs to support them. The Council of Europe's Digital Education Action Plan 2021-2027 is a comprehensive framework that aims to promote digital citizenship, among other goals, by providing guidance and support to member states. In Turkey, the Education Vision for 2023 places a strong emphasis on educational technology and digital citizenship. The vision aims to provide equal access to technology and digital resources for all students, enhance the quality of education, develop teacher skills and competencies, promote digital citizenship and online safety, and foster a culture of innovation and entrepreneurship in the education sector.

REFERENCES

Albrahim, F. A. (2020). Online teaching skills and competencies. *Turkish Online Journal of Educational Technology-TOJET*, *19*(1), 9–20.

Allen, I. E., & Seaman, J. (2017). *Digital Compass Learning: Distance Education Enrollment Report 2017*. Babson Survey Research Group.

Andrews, J. C., Walker, K. L., & Kees, J. (2020). Children and online privacy protection: Empowerment from cognitive defense strategies. *Journal of Public Policy & Marketing*, *39*(2), 205–219. doi:10.1177/0743915619883638

Ata, R., & Yıldırım, K. (2019). Turkish pre-service teachers' perceptions of digital citizenship in education programs. *Journal of Information Technology Education*, *18*, 419–436. doi:10.28945/4392

Aungst, T. D., & Patel, R. (2020). Integrating digital health into the curriculum—Considerations on the current landscape and future developments. *Journal of Medical Education and Curricular Development*, *7*, 2382120519901275. doi:10.1177/2382120519901275 PMID:32010795

Baloch, I., Kaye, T., Koomar, S., & McBurnie, C. (2020). Pakistan topic brief: providing distance learning to hard-to-reach children in urban slums and remote areas. (EdTech Hub Helpdesk Response No 17) DOI:. 3885606. doi:10.5281/zenodo

Barrow, R. (2012). *Common sense and the curriculum* (Vol. 13). Routledge. doi:10.4324/9780203815694

Bonanno, K. (2016). Let's talk software: Digital commerce, legalities and copyright awareness, and online wellbeing. *Education Technology Solutions*, 71, 56–58.

Buckingham, D. (2015). Defining digital literacy-What do young people need to know about digital media? *Nordic journal of digital literacy, 10* (Jubileumsnummer), 21-35.

Caena, F., & Redecker, C. (2019). Aligning teacher competence frameworks to 21st century challenges: The case for the European Digital Competence Framework for Educators (Digcompedu). *European Journal of Education, 54*(3), 356–369. doi:10.1111/ejed.12345

Choi, M. (2016). A concept analysis of digital citizenship for democratic citizenship education in the internet age. *Theory and Research in Social Education, 44*(4), 565–607. doi:10.1080/00933104.2016.1210549

Elmali, F., Tekin, A., & Polat, E. (2020). A study on digital citizenship: Preschool teacher candidates vs. computer education and instructional technology teacher candidates. *Turkish Online Journal of Distance Education, 21*(4), 251–269. doi:10.17718/tojde.803423

Gulbahar, Y., & Guven, I. (2008). A survey on ICT usage and the perceptions of social studies teachers in Turkey. *Journal of Educational Technology & Society, 11*(3), 37–51.

Hanson, K., & Carlson, B. (2005). *Effective Access: Teachers' Use of Digital Resources in STEM Teaching*. Education Development Center, Inc.

Ketchen, D. J. Jr, Snow, C. C., & Street, V. L. (2004). Improving firm performance by matching strategic decision-making processes to competitive dynamics. *The Academy of Management Perspectives, 18*(4), 29–43. doi:10.5465/ame.2004.15268671

King, J., & South, J. (2017). *Reimagining the role of technology in higher education: A supplement to the national education technology plan*. US Department of Education, Office of Educational Technology.

King, J., & South, J. (2017). *Reimagining the role of technology in higher education: A supplement to the national education technology plan*. US Department of Education, Office of Educational Technology.

Kirkgöz, Y. (2009). Globalization and English language policy in Turkey. *Educational Policy, 23*(5), 663–684. doi:10.1177/0895904808316319

Kirlappos, I., & Sasse, M. A. (2011). Security education against phishing: A modest proposal for a major rethink. *IEEE Security and Privacy, 10*(2), 24–32. doi:10.1109/MSP.2011.179

Krutka, D. G., & Carpenter, J. P. (2017). Digital Citizenship in the Curriculum. *Educational Leadership, 75*(3), 50–55.

Li, M., & Yu, Z. (2022). Teachers' satisfaction, role, and digital literacy during the COVID-19 pandemic. *Sustainability (Basel), 14*(3), 1121. doi:10.3390u14031121

Mahmoud, A. H. (2021). *Cultivating Digital Citizenship By Creating An Effective Roadmap For New Generation Of Distance Learning*.

Nguyen, M. H., Gruber, J., Fuchs, J., Marler, W., Hunsaker, A., & Hargittai, E. (2020). <?covid19?> Changes in Digital Communication During the COVID-19 Global Pandemic: Implications for Digital Inequality and Future Research. *Social Media + Society*, *6*(3), 2056305120948255. doi:10.1177/2056305120948255 PMID:34192039

Oxley, C. (2011). Digital citizenship: Developing an ethical and responsible online culture. *Access*, *25*(3), 5–9.

Oxley, C. (2011). Digital citizenship: Developing an ethical and responsible online culture. *Access*, *25*(3), 5–9.

Pangrazio, L., & Sefton-Green, J. (2021). Digital rights, digital citizenship and digital literacy: What's the difference? NAER. *Journal of New Approaches in Educational Research*, *10*(1), 15–27. doi:10.7821/naer.2021.1.616

Ribble, M. S., & Bailey, G. D. (2004). Digital citizenship focus questions for implementation. *Learning and Leading with Technology*, *32*(2), 12–15.

Ribble, M. (2015). *Digital citizenship in schools: Nine elements all students should know*. International Society for Technology in Education.

Richardson, J., & Milovidov, E. (2019). *Digital citizenship education handbook: Being online, well-being online, and rights online*. Council of Europe.

Seale, J., & Schoenberger, N. (2018). Be internet awesome: A critical analysis of google's child-focused internet safety program. *Emerging Library & Information Perspectives*, *1*, 34–58. doi:10.5206/elip.v1i1.366

Simsek, E., & Simsek, A. (2013). New literacies for digital citizenship. *Contemporary Educational Technology*, *4*(2), 126–137. doi:10.30935/cedtech/6097

Thomas, S. (2016). *Future Ready Learning: Reimagining the Role of Technology in Education. 2016 National Education Technology Plan*. Office of Educational Technology, US Department of Education.

Tomczyk, Ł. (2019). What do teachers know about digital safety? *Computers in the Schools*, *36*(3), 167–187. doi:10.1080/07380569.2019.1642728

Umble, E. J., Haft, R. R., & Umble, M. M. (2003). Enterprise resource planning: Implementation procedurs and critical success factors. *European Journal of Operational Research*, *146*(2), 241–257. doi:10.1016/S0377-2217(02)00547-7

Vessey, I. (1994). The effect of information presentation on decision making: A cost-benefit analysis. *Information & Management*, *27*(2), 103–119. doi:10.1016/0378-7206(94)90010-8

Whalen, J. (2020). Should teachers be trained in emergency remote teaching? Lessons learned from the COVID-19 pandemic. *Journal of Technology and Teacher Education*, *28*(2), 189–199.

Chapter 5
Challenging Digital Citizenship Discourse Through the Lens of (Dis)ability

Mary Frances Rice
University of New Mexico, USA

Shernette Dunn
The United States Air Force Academy Preparatory School, USA

ABSTRACT

Digital citizenship is a term that has been used globally. However, many schools have also traditionally operated on a perceived need to behaviorally control children's activity. These control mechanisms have been presented in communities using terms such as citizenship–where the children and community are led to believe that strict obedience is for the common good. In this chapter, the authors share the experiences of Green, a child identified with multiple disabilities. The authors address laws in the United States and inclusive ethics and demonstrate the potential harm that is done when digital citizenship is defined around a narrow set of compliance behaviors that are too rigid for many students without support. The authors propose an alternative framing of digital citizenship that accounts for inclusive ethics and that encircles adults and institutions in collateral obligations regarding digital technologies' ethical use. The authors end with suggestions for classroom practices that support these definitions and offer an alternative ending for Green.

INTRODUCTION

Green Lantern, an eighth grader who had been identified as having multiple disabilities, including attention deficit hyperactivity disorder (ADHD), autism spectrum disorder (ASD), anxiety disorder, and a sensory disorder, walked into his reading class. He was 25 minutes late because he had been using the restroom. Green took out his school-provided chrome book and began to make his way to a preferred website. His teacher observed Green and attempted to redirect him.

DOI: 10.4018/978-1-6684-8934-5.ch005

"Green!" she said in a loud voice. "That is NOT an approved site. Go to page 10 in the story we are reading and do your work."

Green lowered his head. He reached into his pocket and pulled out some fidget magnets. His breathing became more rapid. He stared into the screen and continued to view the preferred site. His teacher noticed.

"Go to Ms. Winnuka," she ordered.

Green slammed his book and walked out. Ms. Winnuka was his special education teacher, specifically tasked with supporting him with behavior and emotional skills.

When Ms. Winnuka saw Green open the door, she started shouting.

"What are you doing here? Get out your chrome book! Do your work!"

Green sat down at an empty desk. He opened his Chromebook and went to his preferred site.
A few moments later, Ms. Winnuka passed by Green's desk.

"Just do your work. Do your work. If you don't do your work, I will make sure that your electives are taken off your IEP. DO YOUR WORK."

"My head hurts," Green replied.

"Go to the office," Ms. Winnuka ordered.

Green headed to the office. The office assistant was frustrated as well.

"Well, if you aren't going to do your work, then give me your Chromebook and you can just do puzzles."

"Okay," Green said, surrendering the Chromebook and sat at a table in the office waiting room and pulled out a puzzle of a farm scene. He worked on the puzzle, putting the pieces together and taking them apart for three hours until the end of the day. When the dismissal bell rang, Green retrieved his Chromebook, visited his locker to pick up his backpack, and went out to the car to meet his brothers and mother.

"How is the birthday boy!?!" his mother exclaimed.

Green burst into tears.

"I hate everyone! I hate school!"

Green's experience with his Chromebook, his use or perceived misuse of it by school officials, and the subsequent confiscation illustrate several critical issues in digital citizenship. First, Green was trying to use his chromebook during unsanctioned time. Second, he was visiting unapproved sites. Third, he was not completing assignments that were expected of him. In traditional thinking about digital citizenship, this is an open and shut case: Green is exhibiting poor citizenship. According to Ribble (2015), digital citizenship is described as a set of learned behaviors.

What are the appropriate behaviors in a digital society? How can an individual learn what is appropriate and what isn't? These are core questions ... The goal of digital citizenship is to provide a consistent message to students and educational professionals so that they can become productive and responsible users of digital technologies. (p. 10)

The terms *productive* and *responsible* have special application here since at the end of the day, Green was perceived to be neither of these things. He was on the computer when he was not supposed to be and visiting sites that were deemed unworthy. It was also considered irresponsible of him to be on the chromebook at that time. In every sense, Green was using his device to act in ways that were beyond the control of the teacher and which were deemed particularly disruptive and annoying to her. Moreover, Green continued to be passed around as a human being from room to room and person to person until he landed in the office doing a puzzle. At that point, it was somehow no longer important to ensure that Green was doing work that was grade and skill level appropriate like it had been when Green wanted to be on the chromebook earlier. Without the device and under the control of an adult, doing some task just to take up time, did not seem to bother anyone.

When Green's mother called the school and they had a meeting about Green, the school came to her even more forcefully with assertions that Green was not being responsible or productive and that he perhaps needed to lose the opportunity to use digital devices at school altogether. Green's mother was understandably frustrated. She did not like having to monitor Green's device use at home either. The surveillance systems installed on the computers by the school often flagged Green's machine for looking at unapproved sites and engaging in activities like playing games. It was embarrassing for her as a mother to be constantly receiving promptings from the school that she needed to go and intervene, not to mention intrusive into her personal time and family life in general.

Yet, Green's mother also did not see how yanking the use of the device altogether would prepare Green for the expectations of high school. If he did not learn to use a device, where exactly would Green learn to be a *good digital citizen*? She also wondered whether the teacher was attempting to place Green under undue exclusion considering laws like the Individuals with Disabilities in Education Act (2004). If others were using digital devices in school to do work and communicate, could Green be excluded under school policies without an evaluation and team decision making? Thus, Green's experience opens space for interrogating the discourses of digital citizenship alongside understandings about disabilities and obligations to serve all children in schools.

The topic of disability, digital technology, and accessibility has grown slowly across practice, policy, and research over at least a thirty-year period, and is finally gaining critical mass and influence — and many people, especially those engaged in digital technology would have a sense of this. What has received much less attention is the radical way in which disability stands to recast our understanding of digital technologies, society, power, and participation. (Goggin, 2016; p. 2)

The rest of this proposed chapter will highlight common ideas in digital citizenship discourse around productivity and responsibility and re-narrate these as cudgel terms that obscure the interconnected obligations of adults and children to one another in school settings, particularly about digital technologies.

In this chapter, we will also address laws in the United States and inclusive ethics and demonstrate the potential harm that is done when digital citizenship is defined around a narrow set of compliance behaviors that are too rigid for many students without support. Finally, we will propose an alternative framing of digital citizenship that accounts for inclusive ethics and that encircles adults and institutions in collateral obligations regarding digital technologies' ethical use. We will end with some suggestions for classroom practices that support these definitions and offer an alternative ending for Green and his mother.

DEFINING DIGITAL CITIZENSHIP

Digital citizenship is a term that has been used globally and has been around for decades. Due to the widespread use of internet and digital technologies in the workplace and society, there is also an increased demand for digital learning in schools (Milenkova & Lendzhova, 2021, p 2). As a result of this socially centered need to build a society around internet technologies, digital citizenship is centered in the field of education and teaching children about how to use technologies (Prasetiyo et al., 2023). However, U.S. schools have also traditionally operated on a perceived need to behaviorally control children's activity–ostensibly to ensure that all the children learn equally (Sharp et. al, 2017). These control mechanisms have been presented in communities using terms such as citizenship–where the children and community are led to believe that strict obedience is for the common good (Gutuk, 2022). To be sure, there is also a tradition in schools of using the communal educational experience to become aware of one another's needs (Ravitch, 2016). With the increased use of social media and the varying demographics of users who engage in online spaces, that discourse of collective agreement to obey for the common good as part of being in school has had space to merge with the need for those who use digital tools and applications to be prosocial and oriented away toward harming other people with the things they do and say online. Although some educators may not have given the confluence of obedience discourse in schools, the practical need for civility in shared spaces–physical and virtual, and the very real danger of harms online as impetuses for understanding digital citizenship, there seems to be a expression of hope that in being educated as digital citizens, there will be a reduction in cyberbullying and other unsavory forms of online and social media usage among a variety of users, including users with disabilities.

To support the need for more digital citizens education, Dunn (2023) cited that having the knowledge that cultural factors and global differences with usage of technology presents challenges for varied online users, "educators, administrators, and governments could begin to introduce more media literacy and digital citizenry classes in the K-12 school systems around the world" (p.12). She also posited that "these classes could also be introduced in a variety of adult education training to include colleges and universities" (p. 13). While digital citizenship has taken on many definitions over the years, one definition that is currently used is that of Milenkova & Lendzhova, 2021 who noted that "Digital citizenship refers to awareness, as well as the ability of sifting out the fake news, and to be critical of social life and what is happening in general. Digital citizenship also means activity and to take a position towards specific events in the life trajectory. In the modern world, society is facing various crises which require a specific approach for managing, along with behavioral strategies" (p 2). Being that online learning is an important part of teaching and learning in K-12 schools there have been measures to ensure that educators have the necessary tools and resources necessary to teach digital citizenship. In the United States, two senators have facilitated a bill that totals over 20 million dollars to provide funding for digital literacy, information literacy, and media literacy skills. These skills are being taught through the International Society for Technology in Education (ISTE) standards especially standard 2 that is titled digital citizen and has four subcomponents and is listed below.

"2--Digital Citizen--Students recognize the rights, responsibilities, and opportunities of living, learning and working in an interconnected digital world, and they act and model in ways that are safe, legal and ethical.

2a--Students cultivate and manage their digital identity and reputation and are aware of the permanence of their actions in the digital world.

2b--Students engage in positive, safe, legal and ethical behavior when using technology, including social interactions online or when using networked devices.

2c--Students demonstrate an understanding of and respect for the rights and obligations of using and sharing intellectual property.

2d--Students manage their personal data to maintain digital privacy and security and are aware of data-collection technology used to track their navigation online. (ISTE 2016)" (Dawkins, 2020).

Some of the elements associated with digital citizenship include cyberbullying, digital footprint, digital privacy, digital netiquette, and digital identity. Which specifically focuses on online behaviors. At the granular level the term digital citizenship is a noun and describes the responsible use of the Internet by individuals to perform a variety of functions. According to the Cambridge Dictionary, digital citizenship is defined as "the state of being skilled in using the internet in order to communicate with others, buy and sell things, and take part in politics, and understanding how to do this in a safe and responsible way" (Cambridge University Press, n.d.). In the same vein, scholars often examine who are the digital citizens that form these communities that encompass citizenship. The Cambridge Dictionary also defines digital citizens as "someone who is skilled in using the internet in order to communicate with others, buy and sell things, and take part in politics, and who understands how to do this in a safe and responsible way" (Cambridge University Press, n.d.). With both definitions focusing on using the Internet to conduct business or communicate in an effective manner, it is imperative that digit citizens have the skills necessary to communicate effectively in these online spaces. However, these skills are not innate and have to be taught whether explicitly or implicitly. While literacy skills are explicitly taught in schools, a new form of literacy that has gained traction in recent years and has become the focus in the educational sector is digital literacy. While digital citizenship may not be a subject on its own, with the closeness of digital literacy, students learn skills from their digital literacy instructions that enable them to become good citizens. While becoming "good" digital citizens and being digitally literate is expected, not all learners are benefiting from the digital literacy curriculum to become good digital citizens. While children's voices, and experiences are not always represented in the debates and policies about digital policy and laws, children with disabilities fear even worse in these spheres because their needs and rights and digital citizens' needs are sometimes overlooked (Alper & Googings, 2017). This is also supported by Fisher et al. (2022), who noted that individuals with individuals with intellectual and developmental disabilities are at a greater risk as it pertains to digital literacy because these individuals have less access and less participation across four digital citizenship elements which puts them at a greater risk for digital social isolation and community disengagement.

In the field of education technology and teacher education, there are several scholars who focus their research on digital citizenry and have tried to coin a definition for this term. One such scholar is Choi (2016), who concluded that digital citizenship was multifaceted and complicated. Based on the context and location, the definition has evolved as various national and global laws come into play. With the variations in the definition of digital citizenry and the impact this definition and usage is having on the larger community, there is a call for more emphasis to be placed on teaching digital citizenship in K-12 schools (Prasetiyo et. al, 2023). However, for this to be effectively taught in the schools, pre-service teachers must also learn effective measures to teach and implement digital citizenship practices. To support preservice teachers becoming more prepared with digital citizenship curriculum and practices Prasetiyo, 2023, cited (Choi et al., 2018), and noted that, "[t]eacher education should plan to offer a course that helps students develop their digital citizenship competencies. They see this as a necessary step in teacher education programs because teachers are expected to effectively support all learners in the use of technology. However, these authors did not make a specific allowance for students with disabilities.

In other studies, Sakorn (2022), supports the need for teachers to be competent digital citizens who are effective in teaching the skills needed to effectively use technology and the internet. He noted that teachers should feel competent to teach the skills that a digital citizen needs and drew on the definitions of digital citizen as was cited by several authors. For example, Sakorn (2022) posited that teachers need to be effective digital citizens because, apart from the ability to manage classes, computer self-efficiency also indicates teachers' competency as digital citizens" (p.514). Sakorn also cited Schuler (2002) definition of digital citizen and noted that "digital citizens as members of a digital environment who are competent in utilizing the internet regularly and effectively" (p.524).

As time progresses, the definition of digital citizenship has revolved. In the late 2000s, Ribble and Bailey (2007), asserted that digital citizens have a positive attitude toward using technology in supporting collaboration, learning, and productivity and accept personal responsibility for lifelong learning. In addition, Ribble (2008) suggested that the definition encompasses the understanding of cultural and societal issues related to technology. Ribble (2008) also emphasized the need for legal and ethical issues to be considered as contributing factors to the responsible use of information technology. This definition has become a foundation for the definition and has been used by several other scholars in the field. To illustrate, Emali et al., 2020 cited Ribble's definition and posited that "[a]lthough digital citizenship was initially described as online access, it has recently been re-defined as using digital technologies in a safe and responsible manner, or in other words, behavioral norms in digital environments" (p. 251). Table one compares these definitions.

Table 1. Definitions of digital citizenship, key components, and challenges

Definition of Digital Citizenship	Key Components	Challenges to Key Components
"the state of being skilled in using the internet in order to communicate with others, buy and sell things, and take part in politics, and understanding how to do this in a safe and responsible way" (Cambridge English Dictionary, n.d)	● Communication skills ● Political participation ● Sufficient awareness of norms around safety and responsibility	● Individuals are not all equally able to communicate in the same language with speech/hearing and to the same cultural expectations. ● Individuals do not have the same access to political participation such as organizing and voting, and they are not all equally safe in their communities. ● Neurological diversity and community norms will shift individual awareness of safety and responsibility.
"Digital citizenship refers to awareness, as well as the ability of sifting out the fake news, and to be critical of social life and what is happening in general. Digital citizenship also means activity and to take a position towards specific events in the life trajectory. In the modern world, society is facing various crises which require a specific approach for managing, along with behavioral strategies" (Milenkova & Lendzhova, 2021, p 2).	● Advocacy ● Awareness ● Shared responsibility	● Individuals with learning differences may not be at a cognitive level commensurate with their biological age to navigate certain digital spaces effectively to decipher facts from "fake" information. ● Measures and personnel will be needed to provide explicit instruction on appropriate usage of the internet and other ICTs.
"Although digital citizenship was initially described as online access, it has recently been re-defined as using digital technologies in a safe and responsible manner, or in other words, behavioral norms in digital environments" (Emali et al., 2020, p.251).	● Social safety ● responsibility	● Individuals may not have the knowledge and skills needed to know what acceptable online behaviors or digital norms are.
Digital citizenship itself is a useful framework widely accepted as a foundation for acting responsibly when interacting with technology. Prasetiyo et al. 2023, (p. 390).	● Social responsibility using the internet and other communication technologies	Neurological and cognitive differences may present challenges to act responsibility when interacting with technology.
Digital citizenship goes beyond conversations about personal responsibility. It's about being active citizens who see possibilities instead of problems, and opportunities instead of risks as they curate a positive and effective digital footprint (International Society for Technology Education Standards (ISTE) (n.d).	● Responsibility ● Opportunities ● Digital footprints	● Not all individuals with disabilities are able to take advantage of the opportunities presented with internet access. ● Neurological differences may present challenges in creating a positive and effective digital footprint. ● Neurological, cognitive, and physical differences may prevent students from transferring knowledge learned in one context to another.

As we have attempted to define digital citizenship, it is important to note that the term may have different meaning based on context. Fisher et al. (2021), supported this context based definition by stating that "Digital citizenship definitions vary depending on the target group (e.g., adult citizens, youth, school administrators, technologists, families)" (p. 264). They also promote a focus on an individual's skills and the practices and supports that help individuals use technology to better their communities as they use technologies to educate themselves and for online civic engagement where they participate in their communities and contribute to the greater society through justice oriented citizenry (Fisher et al.,

2021). Again, key to the elements of context is the idea that a student like Green may not be able to do many of most of the prerequisite skills that would be expected of him to perform digital citizenship. He would certainly need explicit instruction, planning, and practice. Further, as also stated, depending on Green's own unique individual circumstances there may be some aspects of digital citizenship that he can never do. Does this mean that Green can *never* use digital devices to learn? Not so fast. There are critical legalities and ethics that should be brought to bear on this situation as well.

UNDERSTANDING INCLUSIVE LEGALITIES AND ETHICS

Green's mother is correct in asking questions about whether it is legal and ethical to simply become frustrated, give up, and confiscate Green's technological device under the banner of so-called breaches of digital citizenship. After all, Green has been identified as displaying restrictive, repetitive behaviors and does not respond to social cues at the same rate and in the same manner as his peers (Chen et al., 2009). With this understanding, Green's behavior is predictable and in line with what we know about him as a person with Autism. Punishing Green for having Autism is unacceptable under federal law (IDEA, 2004).

Beyond simply avoiding coercion and punishing students is not enough. The Individuals with Disabilities Education Act (IDEA, 2004) requires that qualified individuals between ages 3 and 21 receive a free and appropriate public education (FAPE). When schools are providing FAPE, that means that they are offering educational services that meet the needs of the students at no cost to them or to their families (Education Law Center, 2020). The services that are provided as part of FAPE are supervised by a school leader. This leader acts as the representative of what is referred to as the *local educational agency* (LEA). The LEA representatives are supposed to participate with caregivers/legal guardians, teachers, and other experts to design what are called *individualized educational programs* (IEPs). The purposes of these IEPs are to (1) establish learning goals for the students and (2) name and describe the services that the students will receive (Education Law Center, 2020).

Students are also supposed to receive access to instruction using general education curriculum and opportunities to interact with their peers with and without disabilities "to the maximum extent appropriate" (IDEA, 20 U.S.C. § 1412(a)(5); Mead and Green, 2012).

According to Rice and Payzey (2022)

Historically, such access to instruction and to peers has been regarded to be in the least restrictive environment (LRE) when instruction is provided in a traditional general education classroom. It was tacit that a general education classroom was in a physical building (Education Law Center, 2020). As fully online learning through online charter schools gained momentum in the 2010s, insights about serving students with disabilities in online spaces were emerging (Swenson & Ryder, 2016). In these online charter schools, school leaders grappled with how to guide the placement of students with disabilities into and within online settings to provide the best educational experiences possible. Students with disabilities continue to be entitled to FAPE within the full range of digital and online learning programs, as it is their right to exercise educational choices commensurate with the choices available to students without disabilities (ADAA, 2008; IDEA, 2004; Section 504 of the Rehabilitation Act of 1973).

Now that there are more opportunities to participate in online and digital learning opportunities, there may seem to be some gray areas about the LRE. However, revisions to federal laws such as Section 504 of the Rehabilitation Act of 1973 are underway and recently the Office of Civil Rights and the Department of Justice released a joint *Dear Colleague* letter (Clarke & Lhamon, 2023). In this letter, the DOJ and OCR affirm that individuals cannot be denied access to digital learning opportunities available to peers because of their disability status. In what has been released so far by way of attempts to update federal policies, it seems clear that the federal guidance is trending towards continuing to require that Green have access to and be supported in having access to digital learning opportunities.

REINTERPRETING DIGITAL CITIZENSHIP DEFINITIONS THROUGH THE LENS OF INCLUSIVE ETHICS

The alternative to these definitions that emphasize capitalist production and performativity, and personal responsibility are inclusive ethical principles. Costanza-Chock (2020) proposed a set of inclusive design principles that attempt to loosen the grip of historical narratives around power in institutional settings. The ethical principles that we propose here are based on Costanza-Chock's (2020) work.

1. Digital citizenship policies and practices in schools should sustain, heal and and empower students.

In the aftermath of these incidents, Green and his family need someone to reach out to them to offer help, hope, and healing. They do not need any additional blame cast their way and they certainly do not need any more threats. Schools cannot claim that they care about 'citizenship' in any transformative way when their interactions with young people and their families suggests authoritarian domination are their primary strategies.

2. Digital citizenship policies and practices in schools should center the needs of the youth who are the most vulnerable but aim to protect all in the school community.

Green is obviously among the most vulnerable. He has been positioned in the school as having defects and deficiencies even outside of technology use. This deficit positioning was exacerbated when devices came into classrooms because of the ways in which digital programs and applications provide experiences for him in school that are more positive and attractive than the ones he is having with people. Also, the nature of his brain biology makes him susceptible to routines and even though he does not like confrontation with the teacher, it has become a habit. These factors all make him vulnerable, and it is his experience that should be protected by policies and practices.

3. Digital citizenship policies and practices in schools should consider the effects of the policies ahead of their intentions.

It is understandable that those who wrote the original policies around device use or who agreed to surveillance software did not realize all the implications of what might happen. However, at this point, much more is known about how problematic blanket policies can be for certain students. Moreover, there has not been much scrutiny into how the surveillance has been used. For example, some teachers reported flagging questionable content for any site they just did want children going to (Rice, 2022). Not meaning any harm is not enough. We must do no harm when it comes to how we administrate technologies (Constanza-Shock, 2020).

4. Digital citizenship policies and practices in schools should aim to be accountable, accessible collaborations, rather than one-and-done processes

The usual process in schools is for upper level administrators to send home a computer use agreement and parents read it over and sign it without much ado. There is really no opportunity for choice or comment on the matter. In such documents, the school district does a better job telling parents what children may not do and protecting itself against liability than it does telling parents what the district will do in reciprocation to project and serve children (Rice & Ortiz, 2021). These types of digital citizenship policies developments and deployments must change.

5. Digital citizenship policies and practices in schools should involve administrators and legal entities as facilitators rather than primary members or ultimate experts.

When new policies are made, it is important to determine who needs to be involved in the process of designing these policies and procedures and how. Most of this chapter, we have been quite critical of the teachers in this story, but in truth, they have also been left out of much decision making when it comes to determining what policies are appropriate for them and for students (Rice, 2021). Certainly, there will need to be administrators and lawyers and those usual folks, but to make digital citizenship policies and practices that are meaningful in Green's school community, more decision makers are needed to participate.

6. Digital citizenship policies and practices in schools should consider the lived experiences of the youth and families, as well as teachers and others that have to be subject to these policies–as such they are positioned to make important contributions to policy making and implementation processes.

Hearing and airing these painful experiences is often quite unpleasant for school leaders and it can be embarrassing for teachers and others. It also can be unpleasant to relive these moments for students. Green and his mother certainly were rattled while sharing what happened to them. Even so, gathering lived experiences and thinking about how individuals in these digital learning environments encounter, engage with, and address issues is important. The Online Learning Consortium (Shellgren & Gunder, 2023) recently issued a statement on this topic calling for more narrative work. Rice (in press) has also detailed how narratives specifically support equity work for students who have been positioned as having disabilities in schools.

7. Digital citizenship policies and practices in schools should result in shared tools and ideas that are made available within communities.

During the pandemic, teachers had already established and developed practices and routines for supporting students. These included posting schedules, using timers, singing tunes to initiate transitions, and more. Yet, they did not share these with parents–many of whom were at home struggling at kitchen tables and on sofas to help children settle and do work (Rice & Deschaine, 2021). Digital citizenship policies and procedures should not be another instance where tools and ideas are developed and then clutched tightly to the vest by schools. When teachers find ideas that support Green, they should share that with his mother. Likewise, they should be willing to listen to what Green's mother says helps him use the device in the most positive ways to the most positive ends at home.

8. Digital citizenship policies and practices in schools should work toward outcomes that are sustainable for young people and families.

When school officials become frustrated with some aspect of digital learning and involve parents, it is often to pressure the parents to do something highly labor-intensive to compensate for the lack of support the school is prepared to provide (Ortiz et al., 2021). In Green's case, they have summoned his mother to try to convince her to see their point-of-view–that Green should be subject to some considerable punishment that is beyond their legal rights to administer and/or Green should lose the opportunity to have a device and Green's mother should compensate in some way. This might include additional supervision to ensure homework is accessed through a family computer, additional supervision of paper copies of homework, making space for and keeping track of textbooks, finding alternative means to communicate with teachers about assignments, and so forth. and procedures must consider whose labor is being used? in what ways? for how long? to achieve what goals? under what consent?

9. Digital citizenship policies and practices in schools should be non-exploitative to all–including entities such as the planet Earth.

In addition to the additional work potentially foisted on Green's mother to manage his device use without further support from the school or do additional work since he might have no device, there are potential environmental impacts from making paper copies, additional driving of vehicles, and other consequences. What is crucial is that all the effects are considered during decision making processes.

10. Digital citizenship policies and practices in schools should consider perspectives of traditional indigenous, and local knowledge communities before seeking solutions outside of the community.

Often decisions about policies and practices are pulled from spaces other than local knowledges and communities. What is typical is for some school leaders to devise a policy that may or may not be grounded in their own community needs and then that policy is shared through channels like staff meetings, news sources, social media, school websites or places. Other leaders take up the policy or practice without even thinking very much about it (Rice & Bailon, 2022). What is better is to take the time to gather information from various perspectives and make decisions that are sound given the circumstances, with some expectation that some adjustments and monitoring will always be needed.

In Green's case, the school officials seemed to be making his mother feel that her son was a lone bad actor and all the other adolescents in that school were navigating device use without incident. However, common sense would suggest that many adolescents would be having trouble determining when, how, and why to use their devices in school. Instead of an indignation framing, it would be wiser to learn more about what local wisdom reveals about how to monitor children, how to allow them agency while protecting them from harm, and what community tolerance really was for what the school had framed as problematic.

RENARRATING GREEN'S STORY AS SHARED RESPONSIBILITY AND UNITED ADVOCACY

Instead of the very bad day that Green was subject to, it would be important for all the adults who were going to work with Green to understand his needs and plan appropriate support. Moreover, this support should draw on his strengths. Scott (2017) outlined several strategies by which technological devices could be leveraged to support students with Autism in being successful in general education settings. For example, Scott (2017) highlighted applications that helped students use the multimodal capabilities of devices to communicate their needs as well as identify soothing sounds and images. Green could have used some of these applications for soothing himself when he started to feel anxious so that he would not have to be using the restroom so much. There might also be ways to help Green develop soothing strategies that are offline, but digital strategies can be part of the plan since Green has shown that he likes his devices and finds it comforting.

Scott (2017) also identified uses for applications on devices where students used video or other strategies to model and practice communication. Maybe Green would feel more comfortable communicating with his teacher using digital communication strategies even when they are in the same room, perhaps especially when he is feeling anxious or upset or he can tell that she is. Recall that the teacher also needs to develop appropriate response strategies so that she can develop a working relationship with Green. In this way we see the problem with framing digital citizenship as an individual endeavor that is the responsibility of the student. Green is clearly demonstrating that at school part of his device use stems from efforts to soothe himself when he is in confrontation with his teacher. Finding strategies to avoid this where the teacher is involved should be a priority. It would also be worth investigating more about Green's online behavior and whether his behaviors with his device are that different from other children his age and in what ways and making plans for support based on what is normal and appropriate at home within the context of the community.

Finally, it is vital that Green's right to FAPE is understood in the context of decisions and planning that are made for helping him be successful at school. Green sitting and doing a puzzle in the office until school is out, is not FAPE. Taking the device away from Green also does not give him any meaningful opportunity to achieve any of the other positive goals of so-called digital citizenship. Thus, while he cannot behave in ways that have been deemed irresponsible, he also cannot leave a 'positive digital footprint' which is important, according to organizations like ISTE (n.d.). Overall, it is probably a stronger course for all students to focus on what students *should be* doing as part of digital citizenship rather than what schools usually do, which is look at what they *should not* be.

CONCLUSION

In this chapter, we met Green, a child that had been identified with multiple disabilities and we saw what happened to him when school officials clung doggedly to principles of digital citizenship as a compliance project rather than framing it in the context of students' rights to FAPE and their needs as humans with varying needs and strengths. We recommend that those in a position to develop, evaluate, and enact policies about digital literacies do so with a careful eye for student rights and with care that citizenship is more than a discourse for asserting control over children and devices.

REFERENCES

Alper, M., & Goggin, G. (2017). Digital technology and rights in the lives of children with disabilities. *New Media & Society, 19*(5), 726–740. doi:10.1177/1461444816686323

Americans with Disabilities Amendments Act of 2008, Public Law 110–325, 122 Stat. 3553 (2008).

Cambridge University Press. (n.d.). Digital Citizenship. In *Cambridge dictionary*. https://dictionary.cambridge.org/us/dictionary/english/digital-citizenship

Cambridge University Press. (n.d.). Digital Citizens. In *Cambridge dictionary*. https://dictionary.cambridge.org/us/dictionary/english/digital-citizenship

Chen, Y., Rodgers, J., & Mcconachie, H. (2009). Restricted and repetitive behaviours, sensory processing and cognitive style in children with autism spectrum disorders. *Journal of Autism and Developmental Disorders, 39*(4), 635–642. doi:10.100710803-008-0663-6 PMID:19015969

Choi, M. (2016). A Concept Analysis of Digital Citizenship for Democratic Citizenship Education in the Internet Age. *Theory and Research in Social Education, 44*(4), 565–607. doi:10.1080/00933104.2016.1210549

Clarke, K., & Lhamon, C. E. (2023). *Dear Colleague: Online Accessibility.* Washington DC: United States Department of Justice Civil Rights Division. United States Department of Education Office of Civil Rights. https://www2.ed.gov/about/offices/list/ocr/docs/postsec-online-access-051923.pdf

Dawkins, A. (2020). Digital Citizenship, Digital Legacy, and School Librarians. *Library Technology Reports, 56*(5), 17+. https://link.gale.com/apps/doc/A632141257/ITOF?u=gale15691&sid=bookmark-ITOF&xid=753c1ab5

Dunn, S. (2023). Cultural connections based upon technology availability. In V. Brown (Ed.), *Global societal impact of digital literacy*. Kendall Hunt.

Education Law Center. (2020, August). *Fact sheet: Resolving special education disagreements.* Education Law Center. https://www.elc-pa.org/wp-content/uploads/2019/08/Resolving-Special-Ed-Disagreements-Rev-Aug-2019.pdf

Elmali, F, Tekin, A., & Polat, E. (2020). A study on digital citizenship: Preschool Teacher Candidates vs. Computer Education and instructional technology teacher candidates. *Turkish Online Journal of Distance Education, 21*(4)-15

Fisher, K., Williamson, H., Guerra, N., & Kupferman, S. (2021). Digital citizenship: Technology access and use for youth with and without intellectual and developmental disabilities. *Inclusion (Washington, D.C.), 9*(4), 263–275. doi:10.1352/2326-6988-9.4.263

Goggin, G. (2016). Reimagining digital citizenship via disability. In A. McCosker, S. Vivienne, & A. Johns (Eds.), *Negotiating digital citizenship: Control, contest, culture* (pp. 61–80). Rowman and Littlefield.

Gutek, G. L. (2022). *A history of the western educational experience*. Waveland Press.

Individuals with Disabilities Education Act (IDEA). 20 U.S.C. § 1400 (2004). International Society for Technology in Education. Digital Citizens in Education. https://www.iste.org/areas-of-focus/digital-citizenship

Mead, J. F., & Green, P. C. (2012). Special education and school choice: a special leadership challenge. In J. Crockett, B. Billingsley, & M. L. Boscardin (Eds.), *Handbook of leadership in special education* (pp. 115–128). Routledge.

Milenkova, V., & Lendzhova, V. (2021). *Digital Citizenship and Digital Literacy in the Conditions of Social Crisis. Computers, 10(4), 40.* MDPI AG., doi:10.3390/computers10040040

Ortiz, K., Rice, M., Curry, T., Mellard, D., & Kennedy, K. (2021). Parent perceptions of virtual school support for their children with disabilities. *American Journal of Distance Education*, *35*(4), 276–292. doi:10.1080/08923647.2021.1979343

Prasetiyo, W. H., Sumardjoko, B., Muhibbin, A., Naidu, N. B. M., & Muthali'in, A. (2023). Promoting Digital Citizenship among Student-Teachers: The Role of Project-Based Learning in Improving Appropriate Online Behaviors. [PER]. *Participatory Educational Research*, *10*(1), 389–407. doi:10.17275/per.23.21.10.1

Ravitch, D. (2016). *The death and life of the great American school system: How testing and choice are undermining education*. Basic books.

Rehabilitation Act of 1973. Section 504 Regulations, 34 C.F.R. § 104.1 et seq.

Ribble, M. (2008). Passport to Digital Citizenship: Journey toward Appropriate Technology Use at School and Home. *Learning and Leading with Technology*, *36*, 14–17.

Ribble, M. (2015). *Digital citizenship in schools: Nine elements all students should know*. International Society for Technology in Education.

Ribble, M., & Bailey, G. (2007). *Digital Citizenship in Schools*. ISTE.

Rice, M. (2021). Reconceptualizing teacher professional learning about technology integration as intra-active entanglements. *Professional Development in Education*, *47*(3), 524–537. doi:10.1080/19415257.2021.1891953

Rice, M. (2022). Special education teachers' use of technologies during the COVID-19 era (Spring 2020—Fall 2021). *TechTrends*, *66*(2), 310–326. doi:10.100711528-022-00700-5 PMID:35098256

Rice, M. (in press). Using Narrative Inquiry research methodology in online and distance education environments. *Educational Technology Research and Development.*

Rice, M., & Bailon, M. (2023). A qualitative analysis of publicly available standards and guidance about digital literacies in U.S. states. *Education and Information Technologies*, *28*(6), 6927–6948. doi:10.100710639-022-11482-x PMID:36465420

Rice, M., & Deschaine, M. (2020). Orienting towards teacher education in online environments for all students. *The Educational Forum*, *84*(2), 114–125. doi:10.1080/00131725.2020.1702747

Rice, M., & Ortiz, K. (2022). Reframing parental involvement in online learning as shared work. *Journal of Research on Technology in Education*. doi:10.1080/15391523.2022.2030269

Sakorn, W. (2022). Assessing the digital citizenship self-efficacy of pre-service teachers. *Journal of Educational Issues.*, *8*(2), 513. doi:10.5296/jei.v8i2.20246

Scott, L. (2017). Use of technology for children with Autism: Academic success in the Least Restrictive Environment. *Journal of Educational Leadership in Action*, *5*(1), 5. https://digitalcommons.lindenwood.edu/cgi/viewcontent.cgi?article=1051&context=ela

Sharp, R., Green, A., & Lewis, J. (2017). *Education and social control: A study in progressive primary education* (Vol. 49). Routledge. doi:10.4324/9781315210179

Shellgren, M., & Gunder, A. (2022). *Advancing universal access to quality digital learning through global coalitions and narrative practices*. Online Learning Consortium.

Swenson, S., & Ryder, R. E. (2016). *Dear Colleague: Virtual Schools*. Washington DC: Department of Education Office of Special Education and Rehabilitative Services. https://sites.ed.gov/idea/files/dcl-virtual-schools-08-05-2016.pdf

Chapter 6
Digital Image Ethics:
Social Practices and Technological Mediations

Paulo Barroso
https://orcid.org/0000-0001-7638-5064
NOVA Institute of Communication, NOVA University of Lisbon, Portugal

ABSTRACT

The digital transforms and influences the social practices, the mediations, and the quality of the information in the new digital sphere of modern visual cultures. Digital information is as permanent as it is transitory, real and ephemeral, current and obsolete. Therefore, ethical concerns must be able to keep up with the rapid development of the digital. This chapter is based on exploratory research on the digital image used daily in mass media and social media. The relevance of this anchoring in the digital image is due to the increasingly visual information world, raising fundamental ethical implications. There are ethical criteria (e.g., truth, facts, authenticity) that guide good information practices, whether in the mass media or in social media. They are central standards for responsible citizenship. Following a theoretical-conceptual strategy, the purpose is to argue the ethical use of digital images in daily social practices and technological mediations.

1. INTRODUCTION

In the interdisciplinary scope of media sociology and communication ethics, the connexion between the digital image and ethics of social practice and technological mediation are increasingly pressing and changing. Of particular concern is the digitization of the image of the mass media and social media. The recent digitalization of photography and, in a broader sense, of everyday life is a determining factor for research, as media sociology and ethics of communication are now obliged to approach the digital or miss investigating and theorizing an extensive and significant cultural activity. The ways of life, collective uses and customs, cultural patterns, everyday lifestyles, social practices and mediated social relations, for example, are in permanent transformation and are now digital and due to the new technological devices.

DOI: 10.4018/978-1-6684-8934-5.ch006

Ethics is important at any time or situation. In a current world of misinformation, ethics is even more important in the context of the use and effects of new digital technologies like digital images. Ethics is eminently a practical field concerned with what we should do in any given situation. According to Elliott and Spence in *Ethics for a digital era*, as Aristotle would say, ethical learning, in this case, digital ethics as it applies to the media, comes through individual and group reflective practice and not just through the reading of articles and books on the topic (Elliott & Spence, 2018).

Therefore, digital ethics is relevant for digital images in real life. Digital ethics play a crucial role in the modern world, particularly when it comes to digital images and the rampant spread of misinformation and fake content. As Beever, McDaniel and Stanlick point out in *Understanding Digital Ethics: Cases and Contexts*, digital ethics is a way of thinking about the ethical implications of the digital, or digital technologies and the cultures they co-constitute. Digital ethics is akin to other emergent fields of applied ethics. The ease of manipulating images with advanced software tools has made it increasingly difficult to discern between authentic and doctored visuals. This has serious implications for society, as misinformation and fake images can lead to the spread of false narratives, the erosion of trust, and even real-world consequences. Some real-life cases highlight the practical importance of digital ethics for digital images in the context of misinformation and fakes. For instance, the "deepfake" phenomenon that involves the use of artificial intelligence to manipulate audio and video content, often resulting in highly realistic but entirely fabricated portrayals. In 2019, a deepfake video of Facebook CEO Mark Zuckerberg was created, where he appeared to confess to controlling people's stolen data. This video raises awareness about the potential misuse of deepfakes and highlights the need for ethical concerns in creating and sharing digital content, as these technologies can be exploited to deceive and manipulate audiences.

Another real-life case highlighting the practical importance of digital ethics for digital images in the context of misinformation and fakes is the warped reality through Photoshop and other image editing tools. These technologies can be used to create images that depict events in a way that didn't happen. During the 2014 Ukrainian crisis, a photograph of two Ukrainian fighter jets was manipulated to appear as if they were attacking Russian forces. Another problem arose when this image went viral, fuelling misinformation and potentially escalating the conflict. This case underscores the ethical responsibility of content creators and consumers to verify the authenticity of images before accepting them as truth.

An important issue related to this case has to do with another real-life case regarding the authenticity of journalism. Misleading or altered images can make their way into news stories, undermining the credibility of journalism. When the news media publish digitally manipulated images, it is shown the ethical obligations of media outlets to verify the authenticity of images before publishing them, as well as the public's need to critically assess the visuals they encounter. Digital ethics in the realm of digital images is of paramount importance in combating the spread of misinformation and fake content. The real-life cases mentioned above emphasize the potential consequences of unchecked manipulation and deception. By upholding ethical standards when creating, sharing, and consuming digital images, individuals, news media, public schools (where digital literacy and media literacy are taught and social values and ethical principles are instilled in the curricula), and society at large can contribute to a more truthful, trustworthy, and informed digital landscape. Digital and media literacies refer to the ability of someone to evaluate and produce media in a digital format. Someone who is media literate can understand media, that are now predominantly digital, without being too influenced by it (Rogers, 2016).

Are digital technologies leading to a transition from traditional analogue photography to modern digital photography? Will this transition be a technological paradigm shift? A rupture or evolutionary continuity? If digital photography is a transformation of photochemical photography, is digital photography still photography? Is it legal and ethical to modify press photography? To what extent is post-photographic computer intervention tolerable and acceptable? Can the digital imply and encompass the ethical?

Following a theoretical-conceptual strategy, this chapter is based on theoretical references (and not on a literature review) in which the meanings of each key concept (viz. digital, ethics and image) for the investigation are identified. The chapter intents to understand how these concepts are related to each other in line with the research and the classical normative and ethical theory. The chapter is not oriented towards practical application despite it presents some examples to understand the ethical implications of the use of digital images. The purpose is to conceptualize the digital image ethics and to argue the ethical use of digital images and its implications in the daily social practices and technological mediations.

Therefore, considering the growing secularization and globalization of the world, on the one hand, and the extraordinary development of technology, on the other, two fundamental domains of life become critical: information/representation and truth. Given the technological advances and the emergence of programs for easy scanning and image editing, information/representation and truth will be in deficit in visual societies and cultures, but not the image since cultures are increasingly visual and digital. Today, the presence of images in general (especially photographs) suggest a pictorial turn, as Mitchell (1994, p. 9) called it. "As digital images have become increasingly important items of exchange in the worldwide electronic information economy and as traditional conceptions of image truth, authenticity, and originality have consequently been challenged, ethical and legal dilemmas have emerged.", says Mitchell (1992, p. 52). Therefore, "many of the traditions, standards, and laws developed in the predigital era seem inadequate when they are extended to the new situations created by the new technology" (Mitchell, 1992, p. 52).

In *The reconfigured eye: Visual truth in the post-photographic era*, Mitchell (1992, p. 55) mentions that, for photojournalists, ethical issues are dramatically presented as a form of creative control, individual and institutional responsibility for image content and the formulation of codes of conduct.

In modern visual cultures, everything that is given to the eye and satisfies the *skopéô*, Greek term to define the scopic, act of contemplating, the look, observation, surveillance, i.e. the visual desire through the profusion of images corresponds to the multiform transformations (mainly social, cultural and communicational) driven by technique. As Stiegler points out in the introduction to *Technics and time 1: The Fault of Epimetheu*, "the modern age is essentially that of modern technics" (Stiegler, 1998, p. 7). Stiegler refers to technique as production (*poiesis*), as a "way of revealing… Like *poiesis*, it brings into being what is not." (Stiegler, 1998, p. 9). In this perspective, Stiegler recovers the *Nicomachean Ethics*, where Aristotle states that all art (*tekhne*) is concerned with bringing something into existence and seeks the technical and theoretical means of producing something that belongs to the category of possibility and whose cause is in the producer and not what is produced.

The traditional credibility of the image carrying and showing its referent has ended. There is a problem of lack of credibility of the image which is not attributable to the image or to the transition from the analogue to the digital; it is in human action, i.e. it's the way we use/modify the image. The ethical problem is about the free use of the image and the subjacent intention that the lie of the image serves, it is under the control of the individual to impose an ethical direction on his lie. In this perspective, the image (photography) has neither ethical sustainability nor credibility, as it would always be a fiction from which one could not expect reliable documentary proof of reality and events. This is how the digital image is found in an era of manipulation, misinformation, and false news either in the consumption of information in the news media or in the access to information in social networks. In the digital era of photography and, above all, of equally digital and visual cultures, the literal and direct connections between photography (the image) and reality (the things) are compromised. Without these traditional rigid and unquestionable connections, ethical questions arise about the authenticity of photography, as doubts arise regarding the truth revealed and the ability to distinguish facts from opinions, news from fake news, events from simulacra.

The reality is distinct from the representation of reality, even if this representation is literal. As a sign, an image can never be what it represents and replaces. When they are digitally modified, even for reasons of technical correction of the image, press photographs represent "another reality", situations that do not exist. These images create a virtual reality, as they add, eliminate, or modify visual elements present in the image. The digital modification of photographs is increasingly improved, due to the perfection of image techniques and practices based on computer programs. Thus, it is increasingly difficult to realize that a given photograph has been digitally modified.

Digital technology implies ethical questions, especially when digital technology is used to adulterate the truth of the photographic language. The technique is amoral, but the use of the technique can be moral or immoral. As a tool, procedure, practice or set of processes of an art to accomplish a task, the technique is not the same as the use of the technique that underlies an intention. Therefore, this practice of image configures a form of censorship. It would not be acceptable in terms of trust, credibility, and information accuracy for a means of communication to falsify information by manipulating the photograph through *Photoshop*, for example. Given its social functions (e.g. to inform in the public interest), a media cannot modify the content of the photographs published. For ethical reasons (the practice of substituting facts for simulacra in an illusory media artifice) and for reasons linked to the truth (the original photograph), the editorial line and its audience, modifying the content of published images is building a different narrative. Even when they are correction retouching or image quality adjustment, digital retouching in photojournalism is an intervention that affects the presented information (the truth). Therefore, these image practices and photography post-production procedures have ethical implications independent of either the intention of the media or the consequence.

The ethics of the digital practice of the image in photojournalism underlie the correct or incorrect action of the moral agent, i.e. the use of the image, as well as the intention, motivation and purpose of the action. The digital technological image in the mass media and social media both provides deception and illusion as well as creativity and freedom. Following this duality of opportunities, digitizing photographs brings advantages in image editing (e.g. technical corrections), but also disadvantages (e.g. imperceptible manipulations). However, there is a duty of trust on the part of the public, as there is an expectation to find out about an event when looking for a media for this purpose. Furthermore, the public is credulous and have no *a priori* reason to doubt the information received. Journalists are seen as privileged intermediaries and competent to treat and selection what happens and transform it to news.

Indeed, digitalization of photography has had a strong impact, to the point of being considered an irreversible technological revolution. Digital photography editing techniques have evolved rapidly and have enabled a wealth of new effects and changes in images. Digitization is the process of converting physical to digital. The digitization process is a dynamic method, a system or mode of activity involving a set of actions and manipulations to obtain a result or effect. The digitization process is, therefore, the action or effect of converting into a digital format, what can be manipulated by a computer, or any digital medium such as a smartphone.

When the image is a press photograph that has a referential and informative function, the composition and editing of the image in the post-production phase raises critical and deontological questions. When moving from analogue to digital, the image becomes technological. Despite the digitalization as a technological change without direct human intervention, the image is intervened, and the action is subject to ethical scrutiny. The action and intervention may distort (with or without intention) the referential relationship between things and the image, with effects both on the perception of the image and on the understanding of the information conveyed by the photograph's content, misleading the public.

If the pre-production or post-production of photography already demonstrate subjective human intervention in the content and form of the photograph, the digitization process photographers acquire more power of intervention over the image and become less recorders or observers impartial of reality. Photography is not a simple, transparent, and exact reproduction or reflection of reality, of things and states of affairs, of events. Digitally intervened or transformed photography is more critical. Digital photography is characterized by immediacy (speed), flexibility and mobility in its production, editing, transmission and use in general. These characteristics do not imply or invite the manipulation of digital photography any more than the characteristics of analogue photography. The eventual adulteration of the image depends on the action in the photograph, which can oscillate between respecting the integrity and authenticity of the visual content or not. Analogue or digital, photography varies in the dialectic between reality and fiction, i.e. it oscillates between reproducing things and events as they really are and creating something new, different, imagined. The same happens with other registers, such as painting, cinema, or literature, which can be real or fictional, documental or expressive. Thus, although digital photography has a technological wealth of possibilities and transformative interventions, it does not necessarily have to be more susceptible to manipulation and falsehood, i.e. to the loss of informative quality and ability to rule the truth. Mainly in journalism, with the fulfilment of the image's informative, referential, and testimonial function, as this activity is governed by principles and values and has a brake on technological possibilities and interventions: deontological codes.

It would not be fair to approach ethics without mentioning its precursor: Aristotle. Despite the twenty-five centuries that separate us from Aristotle and his first and original considerations about what ethics is and what it serves, Aristotle's considerations remain current, whose heritage in relation to the importance of virtue and values appears to be if perennial and absolute. In this perspective, the reference to Aristotle, in his Nicomachean Ethics, serves as a motto and clarification of intentions on ethics, both as a domain and cultivation of virtues and as a field of study. If we should not seek to know what ethics is to acquire knowledge, we should seek to apply it and thus give it meaning. In the *Nicomachean ethics*, Aristotle states: "The branch of philosophy we are dealing with at present is not purely theoretical like the others, because it is not in order to acquire knowledge that we are considering what virtue is, but to become good people" (Aristotle, 2004, 1103b). Accordingly, "otherwise there would be no point in it," concludes Aristotle (2004, 1103b).

This excerpt from Aristotle's *Nicomachean ethics* presupposes three significant and elementary aspects: a) ethical concerns are timeless and concern everything we do or say to others; b) the fundamentals of human action are of interest to knowledge and, above all, to social practice, because ethics does not determine ideal behaviour, does not analyse or make value judgments, does not establish the best decisions, but it is applied, i.e. it has to do with the concrete conduct of life, as it allows you to live life well; c) ethics resides in an essential element, virtue (from the Latin *vir*, moral quality of doing good), which is the means to collective happiness and for Aristotle is the ultimate end to be achieved.

According to Schellekens, in Aesthetics & morality: "on the one hand, the relation between aesthetic and moral value targets one of the most important fields open to philosophical investigation in virtue of involving a subject that holds a very significant place in our relations" (Schellekens, 2007, p. 3). The relation between aesthetic and moral value concerns about what kind of person one wants to be, what goals one should aspire to, and how one should treat other persons, states Schellekens. "All form part of

Moral Philosophy, and any relation it may bear to another branch of learning that affects or is affected by it deserves our serious attention" (Schellekens, 2007, p. 3).

In turn, Spinoza's *Ethics* points out that "to act absolutely in obedience to virtue is in us the same thing as to act, to live, or to preserve one's being" (Spinoza, 2009, p. 280). Spinoza clarifies that these three terms (to act, to live, or to preserve) are identical in meaning. For Spinoza, to act in obedience to virtue is the same thing as to act, to live, or to preserve one's being "in accordance with the dictates of reason on the basis of seeking what is useful to one's self" (Spinoza, 2009, p. 280). To act by virtue is to act according to the laws of nature. If virtue is, for Spinoza, acting according to the laws of nature itself and that no one strives to conserve his being, then "happiness consists in man's power of preserving his own being" and "virtue is to be desired for its own sake" (Spinoza, 2009, p. 274).

The subject dealt with by ethics (understood as moral philosophy) is specifically character (*ethos*), which defines socially established ideal models of good conduct. According to Deigh, the subject matter of ethics is the morality of action: "the main problems of ethics arise, as our example of your finding a lost purse containing a huge wad of cash illustrates, from reflection on situations in life that involve matters of morality" (Deigh, 2010, p. 7). For Deigh, ethics is the philosophical study of morality. "It is a study of what are good and bad ends to pursue in life and what it is right and wrong to do in the conduct of life" (Deigh, 2010, p. 7). Ethics is therefore, above all, a practical discipline; its primary aim is to determine how one ought to live and what actions one ought to do in the conduct of one's life (Deigh, 2010).

The subject dealt with by ethics is the notion of morality, comprising standards of correct and wise conduct whose authority in practical thinking is determined by reason and not by custom, according to Aristotle (2004, 1131b) and Deigh (2010). The standards of correct and wise conduct apply to every field or aspects of life. In the communication field, ethical communication presupposes important principles and values, such as transparency, truth and authenticity of the transmitted information; principles and values that, in all situations, should not allow means to overlap with ends, i.e. the ends, however important they may be, cannot justify the means adopted to achieve them at all costs and overriding others or the truth. Like Kantian human ethics, points out Klikauer, "ethical communication treats communication not as a means", for example to achieve personal or organisational goals, profits and the like, but as an end in itself. "This establishes communication as a fundamentally ethical issue that regulates the way in which humans relate to each other" (Klikauer, 2008, p. 227). "Such ethical communication is about what to do and as such always includes an element of what ought to be, transcending the objective reality of work affairs – what is – through communicative action" (Klikauer, 2008, p. 230). Therefore, ethical communication provides a "somewhat built-in procedure that ensures participants' awareness of ethics," a somewhat universal-ethical that demands necessarily always encompasses an element of Kantian what ought to be.

2.THE DIGITAL AGE

The digital age, also known as the information age, is characterized by the widespread use and reliance on digital technologies, particularly computers, the internet, and digital communication systems. The use and influence of digital technologies are two of the several key characteristics that define the digital age since this digital age is marked by the pervasive presence of digital technologies, including computers, smartphones, tablets, and other internet-connected devices. These technologies enable the processing, storage, and transmission of vast amounts of digital information.

There are other key characteristics that define the digital age: connectivity and global communication, easy and immediate access to information, several digital transformations in different aspects of daily and collective life, data and information explosion, social media and online communities and mediations, digital economy and e-commerce in a global and unique virtual market, technological innovation, and the digital divide. These key characteristics are briefly described in the following paragraphs.

The digital age is characterized by the interconnectedness of individuals, organizations, and societies through the internet and digital communication networks. This circumstance facilitates instant global communication, transcending geographical boundaries and enabling real-time information exchange.

With the advent of the internet and digital platforms, access to information has become democratized. People can readily access a vast amount of knowledge, data, and resources online, transforming the way of seeking and consuming information.

The digital age has brought about significant transformations across various industries and sectors. It has revolutionized business processes, education, healthcare, entertainment, and many other domains, leading to increased efficiency, innovation, and new opportunities. However, ethical frameworks such as deontology may not account for the rapid transformations and mutations of digital media (Gallagher, 2020).

In the digital age, an enormous volume of data, which assumes the excessive level and beyond the necessary as a plethora, is generated and collected. This data serves as a valuable resource for analysis, decision-making, and innovation. Big data analytics and artificial intelligence have emerged as crucial tools for deriving insights and creating value from this data.

The rise of social media platforms and online communities has transformed the way people connect, communicate, and share information. Social media has become the new public sphere, a central platform for social interactions and mediations, news dissemination, and online activism and citizenship participation.

The digital age has facilitated the growth of the digital economy and e-commerce. Online marketplaces, digital payment systems, and digital platforms have reshaped the way goods and services are produced, distributed, and consumed.

The digital age is also characterized by rapid technological advancements and innovation. Emerging technologies such as artificial intelligence, blockchain, virtual reality, and the Internet of Things continue to shape the digital landscape and drive societal and economic changes.

Finally, while the digital age has brought immense opportunities, it has also highlighted the digital divide, i.e. the gap between those who have access to digital technologies and those who do not. Socioeconomic factors, geographic location, and infrastructure limitations contribute to disparities in digital access and skills. The digital divide refers to the unequal access to digital technologies and the internet. Ethical considerations arise when individuals or communities are excluded from the benefits of digital technologies, potentially exacerbating existing social and economic inequalities.

Overall, the digital age is defined by the transformative power of digital technologies, connectivity, access to information, and the profound impact they have on various aspects of society, economy, and human interaction.

2.1. Technological Mediation by Digital Image and Digital Technologies

A digital image is a visual representation of information that is stored and displayed in a digital format. It is a two-dimensional array of pixels (picture elements) that collectively form an image. Each pixel represents a specific colour or grayscale value, and when combined, they create the visual content of the image. A digital image is typically created using various devices such as digital cameras, scanners, or generated digitally through software applications. They may be easily stored, edited, and shared as binary data in several file formats containing the necessary information to reproduce the image accurately. Digital images have numerous advantages over traditional analogue images, including the ability to be easily manipulated, edited, stored, and shared across different platforms and devices. They have become an integral part of modern communication, art, design, entertainment, and various other fields. However, digital images also have numerous disadvantages, implying ethical concerns about its authenticity. Digital images are made technologically, using digital technologies, devices that reduce or remove the human creative intervention in image production.

A photograph is always a photograph of something. The relationship between the photograph and what the photograph shows is causal (Mitchell, 1992). Thus, if a photograph is a photograph of a particular individual, it follows that this individual exists or existed, i.e. there is a particular individual of whom this is the photograph. However, the same degree of confidence cannot be assumed in digital photographic images, which we produce, transmit and receive every day through social media.

According to Mitchell (1992, p. 4), we might choose to regard the digitally encoded, computer-processable image as simply a new, nonchemical form of photograph or as single frame video, just as the automobile was initially seen as a horseless carriage and radio as wireless telegraphy. "Indeed the terms 'electronic photography,' 'still video,' and 'digital camera' have rapidly gained currency. But such metaphors obscure the importance of this new information format and its far-reaching consequences for our visual culture" (Mitchell, 1992, p. 4). "Although a digital image may look just like a photograph when it is published in a newspaper", states Mitchell, "it actually differs as profoundly from a traditional photograph as does a photograph from a painting" (Mitchell, 1992, p. 4). "The difference is grounded in fundamental physical characteristics that have logical and cultural consequences" (Mitchell, 1992, p. 4).

In *Technics and Time 2: Disorientation*, Stiegler highlights the importance of the referent in photography, as Barthes had already pointed out: the past is present in photography, which implies the return of the dead in the structure and phenomenon of the image. In photography, the past is re-presented. According to Stiegler (2009, p. 14), the intentionality of photography is the reference, as a certainty, that the photographed object was. The referent is necessarily the real object that was placed before the lens, without which there would be no photography, contrary to fictional production and the practice of digital imaging, which can pretend or create reality without having seen it.

Digital technologies refer to a wide range of technologies that utilize digital information and processes to perform tasks, store and transmit data, and enable various digital applications. These technologies include digital imaging and photography, computers, internet and networking, digital communication (email, instant messaging, video conferencing, etc. allowing real-time or asynchronous communication and information exchange over digital networks), digital media and entertainment, mobile and smart devices, artificial intelligence and machine learning, cloud computing, or the internet of things. "In the digital age – rooted in computers, the Internet, satellites, and the World Wide Web – ideally all types of persons will use all types of media services for all types of audiences." (Christians, 2011, p. 195). These are just a few examples of digital technologies, and the field continues to evolve rapidly. Digital technologies transform various industries, revolutionizing communication processes and information access and sharing, and open new possibilities in several areas, such as automation, data analysis, and connectivity. "The emergence of new digital technologies – including computers and computer networks as communication media, as well as Internet-enabled mobile phones – has evoked an extensive range of ethical issues and challenges" (Ess, 2011, p. 204).

"We live today with a digital mediatization", considering that "the digital and virtual dimensions of communicative interaction and the digital devices produce signs (specially images) that generate hyperreality with effects on our perception and sensation" (Barroso, 2022, p. 9). Cultures are converted into a cyberculture or, in a more radical perspective, into another world, a virtual and global sphere without borders. "The revolution in information technologies and the impact of computerization and virtualization of social actions lead to structural transformations in societies and cultures, which are now digital" (Barroso, 2022, p. 10). The diverse and daily use of digital images and digital technologies transforms social interactions, providing technological mediations. Technological mediation implies the influence and impact of technology on human activities, interactions, and the overall shaping of society. It involves the role of technology as an intermediary or facilitator in mediating human experiences, relationships, and processes.

In the context of digital image ethics and social practice, technological mediation refers to the ways in which technology influences and shapes the creation, sharing, and consumption of digital images, as well as the ethical considerations associated with these activities. It encompasses the effects of technology on image manipulation, privacy and consent, self-presentation, visual storytelling, algorithmic bias, deepfakes, and misinformation.

Technological mediation can manifest in various forms, including the development of digital image editing software, social media platforms, image recognition algorithms, artificial intelligence applications, and other digital tools and technologies that enable the creation, dissemination, and interpretation of digital images.

Through technological mediation, digital images are transformed and mediated by technology, which can introduce new possibilities, challenges, and ethical implications. It is important to critically examine and understand how technology shapes our interactions with digital images, as well as its impact on social dynamics, power structures, and ethical considerations.

Regarding the digital imaging and photography, digital imaging technologies involve the capture, storage, and manipulation of images using digital cameras, scanners, and image editing software. Digital photography replaces traditional film-based photography, allowing for immediate viewing, editing, and sharing of images.

Digital technologies have indeed led to a transition from traditional analogue photography to modern digital photography. Digital photography has become increasingly popular and widespread, largely due to advancements in digital cameras, image sensors, and image processing technology.

There are some key factors driving the transition from traditional analogue photography to modern digital photography, such as:

Table 1. Key factors driving the transition from traditional analogue photography to modern digital photography

Ease of use	Digital cameras are generally more user-friendly compared to traditional film cameras.
Cost efficiency	Digital photography does not need the purchasing and developing film rolls and photographers can take a virtually unlimited number of photos.
Immediate results + post-processing	It is possible to view instantly the captured images, which can be transferred to a computer for further editing and creative possibilities in post-processing.
Image storage and preservation	Digital photography is easier to store and preserve on various media, reducing the physical space required for storing prints or negatives.
Sharing and accessibility	Digital photography enables instant sharing and dissemination of images through various online platforms, social media, and email.

While digital photography has become the dominant form of photography in many areas, it's important to note that analogue photography still holds its value and continues to have a dedicated following. Some photographers and enthusiasts appreciate the unique qualities, aesthetic, and the process associated with shooting on film. Ultimately, the transition from analogue to digital photography represents a shift in technology and workflow, offering new possibilities and convenience for photographers in the digital age.

The relationship between ethics and the news media is permanent and demanding, raising several ethical dilemmas. According to Campbell, Martin and Fabos (2014), a profound ethical dilemma that journalists occasionally face, especially in the aftermath of 9/11, is: "when is it right to protect government secrets, and when should those secrets be revealed to the public? How must editors weigh such decisions when national security bumps up against citizens' need for information?".

For Campbell, Martin and Fabos (2014, p. 18), back in 2006, President George W. Bush criticized the news media for not showing enough "good news" about U.S. efforts to bring democracy to Iraq. Bush's criticisms raised ethical questions about the complex relationship between political-government and the news media during times of war (Campbell, Martin & Fabos, 2014).

In the sociological and ethical context of journalistic documentary photography referring to events and situations of war, for example, the digitization of the press image, which is a recent process and a phenomenon, takes on a particular milestone. The recentness of the digitization of photography and, in a broader sense, of everyday life is, in itself, a determining factor in the investigation, as sociology is now forced to accept the digital in its investigations so as not to omit or stop theorizing entire tracks and significant aspects of cultural activities, as recognized by Lyon: "Sociology is now obliged to come to terms with the digital, or miss investigating and theorizing whole swathes of significant cultural activity" (Bauman & Lyon, 2013, p. 35). The modern lifestyles, collective uses and customs, cultural patterns, everyday behaviours, social practices and mediated social relations, for example, are in permanent transformation and are now digital, through technological devices. If the object of study is in this dynamic process of digital and technological transformation, the disciplines themselves must also be attentive to the phenomena they study and are transformed; they themselves must transform, i.e. adapt and be equipped to be able to understand, describe or explain the phenomena they study.

Digital technology implies ethical questions, especially when digital technology is used to adulterate the truth of the photographic language in the journalism discursive space. The technique is amoral, but the use made of the technique can be moral or immoral. As a tool, procedure, practice or set of processes of an art to accomplish a task, the technique is not the same as the use of the technique that underlies an intention. Therefore, this practice of image configures a form of censorship, as it would not be acceptable in terms of trust, credibility and information accuracy for a means of communication, mainly of which you are a regular reader, to falsify information by manipulating the photograph through Photoshop, for example, and failing to present the original information to give their readers other adulterated information (Fontcuberta, 2014).

Given its social functions, namely to inform in the public interest, timeliness, relevance and objectivity, a media outlet cannot modify the content of the photographs it publishes. For ethical reasons (the practice of substituting facts for simulacra in an illusory media artifice) and for reasons linked to the truth (the original photograph), the editorial line and its audience, modifying the content of published images is building a narrative of its own. Even when they are correction retouching or image quality adjustment, digital retouching in photojournalism is an intervention that affects the presented truth. Therefore, these image practices and photography post-production procedures have ethical implications independent of either the intention or purpose of the media or the consequence. Digital retouching simply comes across as a cheating and illegal procedure; on the other hand, we consider the mystification of the double take perfectly licit – two photographs taken by the photographer with a small difference (Fontcuberta, 2014).

The ethics of the digital practice of the image in photojournalism underlies the correct or incorrect action of the moral agent, i.e. the use made of the image, as well as the intention, motivation and purpose of the action or use of the image. As Wheeler refers in *Phototruth or photofiction? Ethics and Media Imagery in the Digital Age*, the ethics of an altered image implies the nature of the alteration, but also the use of the photograph: "the ethics of an altered image entail not only the nature of the alteration but also the photo's use, that is, its environment" (Wheeler, 2002, p. 101). According to Wheeler, ethical journalistic photographs reflect reality as readers expect it, so it only makes sense to recognize that expectations change dramatically with changes in context.

The most recent technological and scientific developments have profoundly transformed the devices and instruments we use daily and, consequently, their social effects. New technological devices bring with them new uses and social practices of these devices. The instrumental and socio-cultural transformations are so pronounced that some authors refer to it as a "technological revolution", such as Drucker (1961) in "The technological revolution: Notes on the relationship of technology, science and culture" who already warned of the implications of digital for more than 60 years. For Drucker, what has brought about the explosive change in the human condition is a fundamental change in the concept of technology.

According to McLuhan, the media are reshaping and restructuring patterns of social interdependence and all aspects of the human live. For example, the use of smartphones, social networks, and the consumption of digital information through these technological means. These digital uses are making us numb, deaf, blind and dumb (McLuhan, 1994, p. 17). McLuhan emphasizes the technological fact as a determinant for human transformations. This technological revolution is also understood as a "multimedia revolution", according to Sartori (2011, pp. 7-9) in *Homo videns: Televisione e post-pensiero*; or "digital revolution" for Byung-Chul Han, in *In the swarm: Digital prospects* (Han, 2017, p. 10). Whatever the name for the revolution, it is undeniably a technological paradigm shift between traditional analogue photography and modern digital photography.

2.2. Digital Photography: Is It Still Photography?

If photography is the product of a process, digital photography refers to the process of capturing, storing, and manipulating images using digital technology. It involves using digital cameras or other devices equipped with image sensors to capture photographs in the form of digital files.

In digital photography, light enters the camera through the lens and is captured by an electronic image sensor. The image sensor converts the incoming light into digital data, representing the image as an array of pixels. These pixels contain information about the colour and intensity of light captured at each specific point in the image.

The digital image data is then processed and stored in a digital format. Unlike traditional film photography, which uses a chemical process to develop and print images, digital photography eliminates the need for film and allows for immediate viewing and manipulation of the captured images.

Digital photography offers several advantages over film photography: a) immediate feedback: with digital cameras, photographers can instantly review the images they have captured on the camera's LCD screen (this immediate feedback allows for on-the-spot adjustments, ensuring that the desired results are achieved); b) image storage and preservation: digital images can be easily stored, duplicated, and backed up on various digital media, such as hard drives, memory cards, or cloud storage (this makes it convenient to organize and preserve large collections of photographs without the need for physical prints or negatives); c) image editing and manipulation: digital images can be easily edited, enhanced, and manipulated using photo editing software (photographers have greater control over adjusting aspects like brightness, contrast, color balance, and cropping, allowing for creative possibilities and fine-tuning of the final image); d) sharing and distribution: digital photography enables instant sharing and distribution of images through various online platforms, social media, and email (photographers can easily showcase their work to a global audience and engage with others interested in their photography); e) cost efficiency: unlike film photography, which requires purchasing and developing film rolls, digital photography eliminates the ongoing costs associated with film and processing (photographers can take a virtually unlimited number of photos without worrying about the expense of film or printing).

Digital photography has become the standard in photography due to its convenience, flexibility, and the continuous advancements in digital camera technology. It has democratized photography, allowing enthusiasts and professionals alike to capture, manipulate, and share images with greater ease and creative control.

If digital photography is a transformation of photochemical photography, is digital photography still photography? Effectively, digital photography is still considered a form of photography, but it represents a transformation from traditional photochemical photography and therefore the core principles of capturing and preserving images don't remain the same. While both digital and analogue photography involve the creation of visual representations through the use of light, the former reduces or eliminates human participation in the creative process of designing the image, which is produced automatically, while the latter gives primacy to the mastery of photographic technique by the person who records the image. The digital image shook the public's faith in photography (Schwartz, 2003, p.28).

Digital photography represents a transformation in the tools and techniques used, while analogue photography represents an art. The art of analogue photography refers to the creative practice of capturing and producing photographs using traditional, film-based methods. Analogue photography involves using film rolls, chemical processes, and darkroom techniques to develop and print images. It embraces the unique characteristics and aesthetic qualities associated with traditional photographic processes.

Analogue photography requires a deeper understanding of various elements, including exposure, composition, lighting, and the technical aspects of using film cameras. It involves a more deliberate and considered approach, as photographers have limited exposures on a roll of film and need to carefully compose each shot.

One of the distinctive aspects of analogue photography is the anticipation and suspense that comes with waiting for the film to be developed. Unlike digital photography, where images can be viewed instantly, analogue photographers often experience a sense of excitement and curiosity to see the results after the film is processed (cf. Table 2-1 above).

Analogue photography also offers a unique aesthetic appeal. Film-based images can exhibit qualities such as grain, subtle colour shifts, and the distinct tonal range that can create a nostalgic or artistic atmosphere. Additionally, the hands-on nature of working with film, developing chemicals, and enlarging prints in the darkroom allows photographers to have a more tactile and immersive experience in the creation process.

The art of analogue photography extends beyond just capturing the image. It involves the skilful manipulation of exposure, development techniques, and printing methods to enhance the result and convey the photographer's artistic vision. Darkroom processes, such as dodging, burning, and toning, offer additional creative opportunities to shape the final appearance of the photograph.

Analogue photography continues to be practiced and appreciated by photographers, artists, and enthusiasts who value the unique qualities, craftsmanship, and artistic expression that it offers. While digital photography has become more prevalent, the art of analogue photography remains a respected and cherished medium in the world of photography.

Regarding the journalistic context, is it legal and ethical to modify press photography? The legality and ethics of modifying press photography can vary depending on the specific context and purpose of the modifications. It is important to consider both legal and ethical considerations when discussing the modification of press photographs.

Some key points to consider regarding legal considerations are copyright and defamation. Modifying a press photograph without permission from the copyright holder may infringe on their rights. The copyright laws protect the original expression of an image, including modifications. It is essential to respect copyright laws and obtain proper authorization before modifying press photographs. Modifying a press photograph in a way that falsely portrays individuals or events, leading to reputational harm, may potentially lead to legal consequences, such as defamation claims.

In turn, some key points to consider regarding ethical considerations are truth and accuracy, journalistic standards, and consent and respect. Press photography is expected to present an accurate and true representation of events and subjects. Modifying press photographs in a way that distorts reality or misleads the audience can be considered ethically problematic, as it compromises the integrity of the image and undermines trust in journalism. Journalistic ethics and standards typically discourage the modification of press photographs in ways that alter the essential context or the meaning of the image. Journalists are expected to report and present events honestly and without manipulation. If modifications involve individuals depicted in the press photograph, their rights to privacy and dignity should be respected. Modifying press photographs in a way that violates their consent or portrays them in a negative or misleading light can be ethically questionable.

With digital technologies, the relationship between photography and reality has changed, it has become more flexible, volatile, closer to what Baudrillard (1997) calls simulation than the realism inherent in traditional photography, i.e. than the tendency to represent reality in its real aspect (Barroso, 2023).

According to Fontcuberta, the image is born from the projection of a scene on the photosensitive surface in the principle of reality inherent in traditional photography, the camera appears at the beginning of the 19th century linked to the notions of objectivity, truth, identity, colour, document, archive (Fontcuberta, 2014). But the trans-substantiality of digital photography frees itself from memory and replaces the question of the representation of reality with the question of the construction of meaning. Assigning the function of an organ of social memory to the image, according to Agamben (1998), makes photographs in the press iconic cultural objects, as historic as the events themselves.

According to Fontcuberta, the digital image no longer shares the essential functions of photography aimed at authenticating the experience, since its transgressive effect is similar to that of the Trojan horse: infiltrating behind the walls of credibility to launch the definitive blow (Fontcuberta, 2014). Digital photographs no longer authenticate the world; on the contrary, they produce and reproduce fictions that credulous, unwary, unconscious or illiterate audiences take as authentic, as they are simulacra that build simulations, according to Baudrillard (1997).

It is important to note that there can be instances where minimal and necessary modifications, such as cropping or adjusting brightness/contrast, may be considered acceptable within journalistic practices to improve visual clarity or meet specific publishing requirements. However, significant alterations that change the factual content or context of the image should be approached with caution and carefully considered in terms of their legal and ethical implications.

Ultimately, both legal and ethical frameworks play a crucial role in guiding the appropriate modification of press photography. It is advisable to consult professional guidelines, codes of ethics, and legal experts to ensure compliance and ethical conduct when working with press photographs.

The indexicality of photography does not exclude or contradict its iconicity, as both are linked. Photography can give indexical evidence of facts. As indexical signs, photographs come close to images that express the truth and in the sense in which photographs are acceptable as documents of facts, it can be said that they affirm what they represent. However, the photograph cannot be registered to tell what it shows, nor can it say if it has been manipulated or not, as we need additional evidence about the circumstances in which it was recorded, developed or digitally processed. This is why even photographs cannot state or lie explicitly, but can only be used to lie (Nöth, 2011, p. 310).

In *Journalism and new media*, Pavlik points out that traditional newsworthiness criteria, such as objectivity, characterize responsible journalism that seeks to avoid partiality and sensationalism and, instead, promote ethical values such as equity when you present accurate, complete and balanced news coverage that reflects all sides of an issue. According to Pavlik (2001, p. 93), "the issue of who is a journalist in the digital age also raises what is perhaps the most vexing ethical question facing journalism today". For Pavlik, the fundamental standards that define responsible journalism are typical of an ethical press, whether online or offline, which must be guided by the search for the truth.

Therefore, considering Pavlik's ideas, what is important to point out is the evolving landscape of information in the digital age, focusing on the impact of new media technologies on the field of journalism. The transformation of news information in the context of new media involves several aspects, such as:

Table 2. Implications of news information transformations

Digital disruption and transformation	New media technologies have disrupted traditional journalism practices. The rise of the internet, social media, and digital platforms has challenged traditional news organizations, leading to a rethinking of journalistic practices.
Multimedia and convergence	The convergence of various media formats (e.g. text, audio, video, and interactive elements) in the digital realm. Journalists have embraced multimedia formats to engage audiences and enhance their understanding of news stories.
Audience engagement and participation	New media technologies allow for more interactive and participatory forms of journalism, enabling readers to comment, share, and even contribute to news content in the digital era.
Mobile journalism	The rise of mobile devices as platforms for news consumption and distribution. Mobile technology has changed the way news is accessed and consumed (namely through images), as well as its implications for content delivery and understanding.
Credibility and verification	In the age of digital misinformation, the challenges of maintaining information credibility are crucial. Strategies for verifying information in a rapidly changing and unreliable digital landscape are necessary.
Global impact	New media technologies have transformed information on a global scale, allowing for greater cross-border reporting, collaboration, and the dissemination of diverse perspectives.
Ethical considerations	Ethical challenges arise with new media technologies, such as privacy concerns, the spread of fake news, and the need for transparency in journalistic practices.

Regarding the last implication, the news media must be ethically adapted and evolved in response to the digital revolution. The transformation of news information in the digital age has brought about several ethical implications that impact journalism, media organizations, and society at large. These implications revolve around issues of accuracy and truthfulness, credibility, privacy, transparency, and the role of journalism in a rapidly changing information landscape of misinformation. The digital era has made it easier for misinformation to spread quickly. The ethical challenge lies in discerning between honest errors and deliberate attempts to deceive. Citizens and journalists must be vigilant in verifying sources and information to prevent the amplification of false narratives.

3.DIGITAL ETHICS AND DIGITAL IMAGE ETHICS

Digital ethics refers to the moral principles and guidelines that govern individuals' behaviour and interactions in the digital realm. It encompasses the ethical considerations and responsibilities associated with the use of digital technologies, online platforms, and the internet.

Digital ethics involves applying established ethical principles and values to the digital domain, considering the unique challenges and complexities posed by technology. It encompasses a wide range of topics, including privacy (respecting individuals' rights to control their personal information and safeguarding their privacy in the digital age), security (protecting digital systems, networks, and personal data from unauthorized access, cyberattacks, and breaches), intellectual property (recognizing and respecting the rights of creators and owners of digital content, such as images, videos, software, and written works), online behaviour (promoting ethical conduct, respect, and civility in online interactions, including addressing cyberbullying, harassment, hate speech, and trolling), digital rights (advocating for individuals' rights to access and participate in the digital world, including freedom of expression, access to information, and digital inclusion), and social responsibility (transparency and accountability, i.e. holding technology companies, platforms, and governments accountable for their actions, policies, and the ethical implications of their digital practices).

Digital ethics also contemplates the social impact of the digital. Considering the broader social, cultural, and environmental impact of digital technologies, including issues of inequality, digital divide, and the responsible use of emerging technologies.

Digital ethics provides a framework for individuals, organizations, policymakers, and technology developers to navigate the ethical challenges and dilemmas arising from the use and impact of digital technologies. It aims to promote responsible, ethical behaviour, and to ensure that technology is developed and used in a manner that aligns with fundamental ethical principles and values.

One of the most challenging aspects about developing a sense of digital ethics, or ethics that influence our use of digital technologies is that information in digital form can be duplicated without prior authorization, while simultaneously having the characteristic of disappearing with the push of a button. Digital information is both permanent, and transitory. It is real, and it is ephemeral. It can be here today, and the computer one uses can be obsolete tomorrow, making it difficult, if not impossible to retrieve stored digital data. Additionally, in the context of news, information that binds us together may also tear us apart (Hanson, 2011, p. 819).

From what has been exposed throughout this chapter, the emergence of an ethics of the digital image is advocated. Given the current global context of the profusion of digital images, these are easily edited and altered because they are digital, on the one hand, and are more appealing and misleading than texts, on the other hand, as they condense all the information in an easy-to-understand language and that dispenses with decoding in a given language, such as words. The profusion of images characterizes modern cultures, which are increasingly visual, technological and virtual, but also effectively influence generalized public opinion, especially when images are manipulated and there is not enough and necessary digital literacy and media literacy in citizens to combat this plethora and scourge of adulteration of truth through image. In *Understanding Digital Ethics: Cases and Contexts*, Beever, McDaniel and Stanlick state that literacy in digital ethics is often sadly lacking in both contemporary discourse and the design of technologies that support our everyday conversations and interactions in digital spaces.

What can citizens do for themselves to minimize the effects and consequences of the excess of misinformation in contemporary visual cultures and to fight the adulteration of truth in the image, mainly? There are several basic and interconnected procedures to be followed so as not to be deceived by false or out of context information. For example, verify: i) the source evaluation; ii) the expertise of the author; iii) the supporting evidence; iv) the biases and conflicts of interest.

Table 3. Basic procedures to be followed so as not to be deceived by false or out of context information

Factors to evaluate the source of information	Factors to evaluate the expertise of author	Aspects to evaluate the supporting evidence	Aspects to evaluate the biases and conflicts of interest
Authorship; publication date; publisher or source origin; bias and objectivity; evidence and documentation; consistency with other sources; peer review; reputation and reviews; purpose and audience.	Educational background; professional affiliations; work experience; publication record; citations and references; reputation and recognition; online presence and credentials.	Credibility of sources; relevance; currency; sample size and methodology; consistency with other sources.	Bias: political; cultural; commercial; sensationalism; gender. Conflicts of interest: financial interests; personal relationships; professional advancement; ideological affiliations.

The growth of digital media and communication and the influence of the digital shaping social global life "largely corresponds with the rise of literacy" (Flew & McElhinney, 2006, p. 288). The new digital media and communication technologies influence the social interactions. As Elliott and Spence (2018, p. 78)., E. H. (2018). *Ethics for a digital era*, "digital mass communication allows for people to easily connect across time and space to share talents, skills, and perspectives so that they can come together to create change." The growth of the new digital media technologies must be accompanied by a similar growth of digital and media literacies based on levels, i.e. stages of skills or user's competences for digital devices and instruments:

Table 4. The growth of digital and media literacies based on levels

Level 1: basic digital and media literacies	Level 2: intermediate digital and media literacies	Level 3: complex digital and media literacies
Use of simple digital tools to access information through media images. E.g. access to news media site on the World Wide Web.	Understand digital content of media images and recognize the veracity of the information. E.g. identify images and who produces them, understand the reasons why the images are shown and how they should be used (what they show/inform).	Use and master digital tools and carry out fact-checking. E.g. use of image editing/manipulation software or information verification platforms.

Digital image ethics refers to the moral principles and responsibilities associated with the use, manipulation, and dissemination of digital images in the digital age. With the advancement of technology and the proliferation of digital media, digital image ethics has become a relevant field of study, addressing specific ethical and legal issues related to the creation, sharing, and use of digital images.

Some key aspects of digital image ethics include the opposition between authenticity vs. inauthenticity of the images, the question regarding the privacy and consent, copyright and intellectual property problems, context and representation issues, social responsibility implication, and transparency.

The authenticity of digital images is a critical aspect of ethics. This involves ensuring that images are not falsified or manipulated in a misleading manner to distort reality (Barroso, 2023, p. 145). Image manipulation can have negative ethical and social consequences, such as spreading false information, distorting the truth, or deceiving the public.

Regarding the privacy and consent, respect for privacy and consent is crucial in digital media ethics (Ess, 2011). Unauthorized publication of personal images, invasion of privacy, and misuse of other people's images without their consent are considered ethical violations. Privacy and security are two important aspects the way digital technologies are used imply and have an impact on individuals and society. The digital realm involves the collection, storage, and processing of vast amounts of personal data. Ethical concerns arise regarding how this data is handled, protected, and used by organizations. Issues such as data breaches, unauthorized access, and the misuse of personal information highlight the importance of ethical considerations in digital contexts.

Concerning the copyright and intellectual property, digital image ethics also address issues of copyright and intellectual property (Ess, 2011). It is important to respect the rights of creators of digital images and obtain appropriate permission when using or sharing images that are protected by copyright.

In the context and representation, digital image ethics considers the appropriate context and representation of images. It is important to avoid using images in a misleading or out-of-context manner, especially when it comes to sensitive issues such as race, gender, religion, and culture.

About the social responsibility, digital image ethics includes the social responsibility of considering the impact of images on individuals and society. This involves avoiding the dissemination of images that may cause harm or offend specific individuals or groups.

Finally, regarding the transparency, which is an important principle in digital image ethics, it includes providing clear information about the origin of images, authorship, any manipulations performed (if any), and any intention behind the image.

Digital image ethics is an evolving field as technology advances. It is essential for users of digital media, such as photographers, journalists, influencers, and the general public, to consider these ethical aspects when creating, sharing, and consuming digital images.

Concerning the question previously posed, if can the digital imply and encompass the ethical, the answer must be categorically affirmative. The digital realm can imply and encompass ethical considerations. While digital technologies are neutral tools, the way they are used and the impact they have on individuals and society raise ethical questions, such as digital manipulation and misinformation. The ease of digital manipulation, which is a new moral problem (Christians, 2011), and the rapid dissemination of information in the digital realm raise ethical concerns related to misinformation, fake news, and the potential for the manipulation of public opinion. Ethical considerations involve the responsible use of digital platforms and the promotion of accurate and trustworthy information.

Linked to the digital manipulation and misinformation, we find the algorithmic bias and discrimination as another impact on individuals and society raised by the uses of the digital technologies. This is another example of a fundamental ethical question. Algorithms and machine learning systems used in digital contexts can perpetuate biases and discrimination, leading to unfair outcomes. Ethical considerations revolve around addressing algorithmic biases, ensuring fairness, and promoting transparency and accountability in algorithmic decision-making.

As mentioned before, digital rights and freedom of expression are also implied by the way digital technologies are used and the social impact they have, which also raise ethical questions. Ethical considerations encompass the protection of digital rights and freedoms, including freedom of expression, privacy, and access to information. Balancing these rights with other societal values, such as security and public interest, is a complex ethical challenge.

Concurrently with this aspect, there are concerns regarding the digital well-being and mental health, on the one hand, and the environmental impact, on the other hand. The digital realm has implications for individual well-being and mental health as well as for the environmental footprint, including energy consumption, electronic waste, and the carbon emissions associated with digital infrastructure. Ethical considerations include addressing issues such as digital addiction, online harassment, and the responsible design of digital interfaces to promote positive user experiences, but also minimizing the negative environmental impact of digital technologies and promoting sustainability.

These are just a few examples of how the digital realm intersects with ethical considerations. It is important for digital technologies' users to navigate these issues with a sense of responsibility, awareness, and adherence to ethical principles to ensure that the digital realm promotes the well-being and values of individuals and society.

Ethics has to do with the quality of what is ethical. Thus, the digital practice of the image in journalism implies ethical questions because the digital practice dematerializes the photograph as a document and indicates the vulnerability of the quality of its informative content. Digital practice aggregates a diverse set of possibilities and interventions on photography, modifying it in terms of content or forms and moving away from the belief that photography is a mirror of reality. If a photograph is published in the press after being digitally intervened, doubts arise as to the integrity and authenticity of its visual elements and as to its ability to fulfil the function of informing in the field of journalism. Considering the specificity (referentiality and objectivity) of journalistic documentary photography as proof and a regime of truth, the digital practice of the image in photojournalism raises ethical and deontological questions and concerns and will have implications for the credibility and trust of the public in case of manipulation.

As digital technologies proliferate, the disintermediation: "becomes the point at which trust would be determined... In traditional media industries, the trust the audience placed in the industry for factual information, professional standards, and quality news – was monitored at the author stage, and at the middle-stage of the message as it was prepared for transmission" (Hanson, 2011, p. 815)

Ethics has to do with the fundamental principles and values that guide human action and behaviour. Ethics means the correction of habits and social customs, on the one hand, and the quality of character, nature, on the other. It is a comprehensive term, implying both the action of a moral agent (what a person does and what he should do) and what a person is and should be in psychological terms (his principles, values, virtues). The nature of ethics is formative and disciplinary. It constitutes a moral guide for social action. Ethics cultivates character and virtue, on the one hand, and leads to good, correct and just behaviour, action, habit and custom, on the other, as prescribed by the Kantian deontological ethics with corrective action and the imperative of the universal norm or the Aristotelian virtue ethics with the direct relationship between character, action and habit (Schwartz, 2023, pp. 41-47; Gallagher, 2020, p. 70). Virtue ethics finds extensive application in digital media ethics (Ess, 2011, p. 213). Therefore, ethical concerns should not be understood as doctrines or obstacles to freedom of behaviour and action. On the contrary, ethical concerns act as incentives for a person to be a better citizen, in individual and social terms, and to be a better professional, in terms of ethics applied to the exercise of a trade.

4. CONCLUSION

The rapid advancement of technology has revolutionized the way we create, share, and consume digital images. With the increasing prevalence of social media platforms and digital photography, it is crucial to address the ethical considerations surrounding digital image creation, social practices related to image sharing, and the role of technology in mediating these processes. This discussion aims to explore the complex relationship between digital image ethics, social practice, and technological mediation.

Digital image ethics involves the authenticity and manipulation issue. The ease of digital manipulation raises questions about the authenticity and integrity of digital images. Ethical considerations arise when images are altered or manipulated to deceive viewers or misrepresent reality. Therefore, it is imperative to discuss the ethical implications of image manipulation in different contexts, such as journalism (especial photojournalism), advertising, and social media.

Digital image ethics also involves the privacy and consent issue. The widespread sharing of digital images raises concerns about privacy and consent, namely using images. This issue addresses the ethical challenges associated with sharing images without the explicit consent of the individuals depicted. Thus, it is necessary to discuss the importance of respecting privacy rights in the digital age and the potential consequences of violating them.

Human beings are gregarious and everything they do or say implies or affects others. Human beings are also ethical animals, since "we grade and evaluate, and compare and admire, and claim and justify" (Blackburn, 2003, p. 4). Therefore, ethically guiding everything we do or say is important, especially in a global, virtual era mediated by new communication and information devices. Thus, the question of social practices ethically regulated and guided by fundamental moral values and principles arises. Social practice involves the self-representation and both the individual and collective identities. In this context, it is necessary to explore how digital images influence self-representation and the construction of identity, to discuss the impact of social media platforms on shaping individuals' self-image and the pressures to present an idealized version of oneself, and to examine the ethical implications of self-presentation and the potential consequences of perpetuating unrealistic beauty standards.

Social practice also involves the visual storytelling and empathy because we must discuss the power of digital images in storytelling and fostering empathy, explore the ethical responsibilities of photographers and image creators in representing sensitive or vulnerable subjects, and address the potential for digital images to shape public opinion, influence policy decisions, and drive social change.

In turn, technological mediation is based on algorithmic bias and representation and for this reason it is necessary to examine the role of algorithms in curating and disseminating digital images on social media platforms, to discuss the ethical concerns regarding algorithmic bias and its impact on image representation, visibility, and diversity, and to address the responsibility of technology companies in mitigating bias and promoting fair representation in image algorithms.

Technological mediation implies the serious social phenomena of deepfakes and misinformation, which justifies the need of an ethical challenge posed by the emergence of new technology devices. There are effective and potential consequences of deepfakes in every area of social and cultural life. Citizens must have media and digital literacy and be capable to explore the ethical considerations related to detecting, debunking, and preventing the spread of digital image misinformation.

In short, the ethical dimensions of digital image creation, social practices, and technological mediation are complex and multifaceted. It is essential for citizens or digital media users to critically engage with these issues and establish guidelines and practices that prioritize authenticity, privacy, consent, and fair representation. By fostering ethical awareness and responsible use of digital images, we can navigate the evolving landscape of technology and social media while promoting a more inclusive, empathetic, and ethically conscious society.

REFERENCES

Agamben, G. (1998). *Image et mémoire*. Éditions Hoëbeke.

Aristotle. (2004). *Nicomachean ethics*. Cambridge University Press.

Barroso, P. M. (2022). From reality to the hyperreality of the simulation. *Texto Livre – Linguagem e Tecnologia ,15*, 1-11.

Barroso, P. M. (2023). Baudrillard on the symbolic violence of the image. *Versus - Quaderni di studi semiotici, 136,* 135-154.

Baudrillard, J. (1997). *Simulacra and simulation*. The University of Michigan Press.

Bauman, Z., & Lyon, D. (2013). *Liquid surveillance*. Polity Press.

Beever, J., McDaniel, R., & Stanlick, N. (2020). *Understanding digital ethics: Cases and contexts*. Routledge.

Blackburn, S. (2003). *Ethics: A very short introduction*. Oxford University Press. doi:10.1093/actrade/9780192804426.001.0001

Campbell, R., Martin, C. R., & Fabos, B. (2014). *Media & culture: Mass communication in a digital age*. Bedford / St. Martin's.

Christians, C. (2011). Journalism ethics in theory and practice. In G. Cheney, S. May, & D. Munshi (Eds.), *The Handbook of Communication Ethics* (pp. 190–203). Routledge. doi:10.4324/9780203890400.ch12

Deigh, J. (2010). *An introduction to ethics*. Cambridge University Press. doi:10.1017/CBO9780511750519

Drucker, P. F. (1961). The technological revolution: Notes on the relationship of technology, science and culture. *Technology and Culture*, *2*(4), 342–351. doi:10.2307/3100889

Elliott, D., & Spence, E. H. (2018). *Ethics for a digital era*. Wiley Blackwell.

Ess, C. (2011). Ethical dimensions of new technology/media. In G. Cheney, S. May, & D. Munshi (Eds.), *The Handbook of Communication Ethics* (pp. 204–220). Routledge. doi:10.4324/9780203890400.ch13

Flew, T., & McElhinney, S. (2006). Globalization and the structure of new media industries. In L. Lievrouw & S. Livingstone (Eds.), *The Handbook of New Media* (pp. 287–306). Sage.

Fontcuberta, J. (2014). *A câmera de Pandora: A fotografí@ depois da fotografia* [translation: *Pandora's Camera: Photography After Photography*]. Gustavo Gili.

Gallagher, J. R. (2020). A pedagogy of ethical interface production based on virtue ethics. In J. Reyman & E. Sparby (Eds.), *Digital ethics: Rhetoric and responsibility in online aggression* (pp. 69–84). Routledge.

Han, B.-C. (2017). *In the swarm: Digital prospects*. Massachusetts Institute of Technology. doi:10.7551/mitpress/11055.001.0001

Hanson, J. (2011). Authors, authority, ownership, and ethics in digital media and news. In R. S. Fortner & P. M. Fackler (Eds.), *The Handbook of Global Communication and Media Ethics* (pp. 803–822). Wiley-Blackwell. doi:10.1002/9781444390629.ch41

Klikauer, T. (2008). *Management communication: Communicative ethics and action*. Palgrave MacMillan. doi:10.1057/9780230583238

McLuhan, M. (1994). *Understanding media: The extensions of man*. The MIT Press.

Mitchell, W. J. T. (1992). *The reconfigured eye: Visual truth in the post-photographic era*. The MIT Press.

Mitchell, W. J. T. (1994). *Picture theory: Essays on verbal and visual representation*. Chicago University Press.

Nöth, W. (2011). Visual semiotics: Key features and an application to picture ads. In E. Margolis & L. Pauwels (Eds.), *The SAGE Handbook of Visual Research Methods* (pp. 298–316). Sage. doi:10.4135/9781446268278.n16

Pavlik, J. V. (2001). *Journalism and new media*. Columbia University Press. doi:10.7312/pavl11482

Rogers, R. (2016). The harm of video games: The ethics behind regulating minors' access to violent video games in light of the supreme court ruling. In A. Davisson & P. Booth (Eds.), *Controversies in digital ethics* (pp. 216–230). Bloomsbury.

Sartori, G. (2011). *Homo videns: Televisione e post-pensiero*. Editori Laterza.

Schellekens, E. (2007). Aesthetics & morality. *Continuum*.

Schwartz, D. (2003). Professional oversight: Policing the credibility of photojournalism. In L. Gross, J. S. Katz, & J. Ruby (Eds.), *Image ethics in the digital age* (pp. 27–51). University of Minnesota Press.

Schwartz, M. (2023). *Adaptive ethics for digital transformation: A new approach for enterprise leadrs*. IT Revolution Press.

Spinoza, B. (2009). *Ethics*. The Floating Press.

Stiegler, B. (1998). *Technics and time 1: The Fault of Epimetheu*. Stanford University Press. doi:10.1515/9781503616738

Stiegler, B. (2009). *Technics and time 2: Disorientation*. Stanford University Press.

Wheeler, T. H. (2002). *Phototruth or photofiction? Ethics and Media Imagery in the Digital Age*. Lawrence Erlbaum Associates.

KEY TERMS AND DEFINITIONS

Digital Ethics: The moral principles and guidelines that govern individuals' behaviour and interactions in the digital realm, including ethical considerations and responsibilities associated with the use of digital technologies, online platforms, and the internet.

Digital Image: A visual representation of information that is stored and displayed in a digital format, a two-dimensional array of pixels that collectively form an image.

Digital Image Ethics: The moral principles and responsibilities associated with the use, manipulation, and dissemination of digital images in the digital age, addressing specific ethical and legal issues related to the creation, sharing, and use of digital images.

Digital Photography: The process of capturing, storing, and manipulating images using digital technology, e.g. digital cameras or other devices equipped with image sensors to capture photographs in the form of digital files.

Digital Technologies: Devices that utilize digital information and processes to perform tasks, store and transmit data, and enable various digital applications.

Ethics: A branch of philosophy that deals with moral principles, values, and rules that guide human behaviour; it examines questions about what is right or wrong, good or bad, and how we should behave, providing frameworks for making decisions and evaluating actions based on principles such as fairness, justice, honesty, and respect for others.

Technological Mediation: Ways in which technology influences and shapes the creation, sharing, and consumption of digital images; it encompasses the effects of technology on image manipulation, privacy and consent, self-presentation, visual storytelling, algorithmic bias, deepfakes, and misinformation.

Chapter 7
Developing Writers in the New Digital Age:
Ethical Stances for Writers and Teachers

Kristen Turner
Drew University, USA

Gary Pankiewicz
Montclair State University, USA

ABSTRACT

The teaching of writing is an ethical endeavor on many fronts. This chapter focuses on aspects of teaching ethical practices that have evolved over the last three decades as teachers of writing have adapted instruction to account for new tools that impact how a writer composes and the nature of the final product created. Many of these aspects are again at a turning point with the explosion in AI technologies in recent years, and it is essential that writing teachers see this moment with "historical consciousness" (Bruce, 2003, p. 13) and help writers to understand the ethical implications of being a writer in the age of AI. Teachers must recognize the history of writing development, view writing as a form of inquiry, and understand the complexities of authorship and attribution from a sociocultural and digital perspective so that they can help young writers develop ethical stances in this new era of writing.

For the last thirty years, teachers of writing have adapted instruction to account for new tools that impact how a writer composes and the nature of the final product created. Research in writing instruction evolved alongside technological innovation, and though schools have been slow to adopt wide-scale practices that fully embrace inquiry-based writing classrooms that promote digital, multimodal composition, real-world writing practices have changed quickly and continue to adapt at lightning pace. Given the context of an ever-evolving digital world, educators should not have been surprised when large language models (LLM), commonly referred to as artificial intelligence (AI), became publicly accessible, and students at all levels began using them for school-based writing tasks. However, the seemingly overnight explosion of AI use caught many off guard.

DOI: 10.4018/978-1-6684-8934-5.ch007

For example, Ms. G., a high school English teacher, was surprised to read one of her student's final research papers connecting Arthur Miller's (2000) *The Crucible* to the historical era of McCarthyism. Ms G. had engaged her class in an exploration of the text that included dramatic performance, close reading, text annotation, shared discussion, and journal writing. She designed inquiry-based activities that scaffolded students' examination into the historical context of Salem in 1692 and its thematic connections to the McCarthy hearings of the 1950s. She modeled her own inquiry and writing process as students developed questions, researched print and non-print resources, and composed in a workshop model. Having read some of the students' thinking as they developed their ideas and drafted their papers, she was excited to see the final products.

However, Ms. G. was shocked when she received Perry's final paper. The paper incorporated ideas and characters that had not been represented in his previous drafts, and the voice did not reflect Perry's previous writing. After looking at the revision history in his document, she discovered big blocks of text that appeared to be copied and pasted into Perry's essay. Ms. G began to suspect that Perry had used an AI tool, an act she considered to be cheating. When Ms. G. pulled Perry aside to address him, he appeared proud of his latest draft. After weeks of struggling to write, he was delighted to share how his computer support was no longer limited to grammar and spelling help. He started by using a cloud-based platform to revise and edit some of his sentences, but Perry soon discovered that he could use ChatGPT to generate writing that was much different than his own, and he liked what he was producing with it.

Ms. G. had heard of platforms like ChatGPT, but she assumed that her close work with students as they developed their essays would avoid any issue of cheating or plagiarism that these tools might inspire. After reading Perry's essay and talking with him about it, she began to suspect that more of her students were using AI in their drafts and revisions, and she did not know how to proceed. She wanted their writing to be original, ideas crafted by them. Perry, on the other hand, had assumed that the AI tool was intended to help him compose, and he and his peers embraced the new technology. They did not, necessarily, see it as cheating like their teacher did.

Ms. G was not alone in her concerns, as many teachers wondered how to prevent students from using AI for their writing after tools like ChatGPT became publicly accessible. Responses ranged from eliminating computer-based writing altogether - going back to paper and pencil drafting - to embracing the technology as a tool to help writers at all stages of their processes. The conversation around use of AI has sparked professional organizations to investigate the impact of AI on writing, and, in some cases, to offer citation guides for use of AI. The Modern Language Association-Conference on College Composition & Communication (MLA-CCCC) Task Force on Writing and AI (2023) outlined some of the risks to students, teachers, and programs. They noted the immediate problems of students submitting responses that were generated through AI as their own, teachers' lack of support and accurate training to understand how LLM relate to their disciplines, and difficulty in updating policies that align with the fast-moving changes in technology.

The MLA-CCCC Task Force (2023) recognized the benefits of AI too. For example, they noted that for language acquisition, students can use AI to support linguistic choices and translation, as well as to answer questions about texts in another language. In literacy studies, students can use LLM for creative wordplay, to ask questions about literary texts, and, with more expertise, generate literature and digital writing. Student writers can use LLM to provoke thought and develop and revise drafts as well as to integrate still images, sound, and moving images in multimodal compositions. The task force also recognized opportunities in the areas of language, literacy, and writing instruction to provide resources for instructional planning and differentiation for various students. Central to the MLA-CCCC recommendations for teacher support, guiding resources, and revised policy is a focus on ethical conduct for writers and the writing teacher's responsibility in helping students to develop ethical stances. For a teacher to engage in this work, they must also take an ethical stance, one that is rooted in an understanding of what writing is and how it is mediated by technology.

REFRAMING THE QUESTION: WHAT IS WRITING?

In exploring the ethics of digital literacy, Turner (2019b) posed several questions:

Underlying questions about how technologies have changed the ways individuals read, write, and interact are questions about the ethics of participation in a digital world. As users consume and create seemingly infinite content, what are the moral guidelines that must be considered? How do we teach students to be responsible, ethical citizens in a digital world? (p. 2)

This connection between digital literacy and ethics invited conversations that crossed researcher and teacher practice, inquiries that suggested five dimensions of teaching the ethics of digital literacy: ethical contexts, ethical communities, ethical selves, ethical stances, and ethical practices. The work presented came together under the assumption that digital literacy included "knowledge and skills that allow for critical consumption, creation, interaction, and communication using digital technologies" (p.1). This definition required an understanding of *writing* that moved beyond "thought on paper" (*The Neglected R*, 2003, p. 13).

In fact, the definition of writing has evolved over time, and pushback about the value and authenticity of various forms of writing has existed alongside this evolution. Those who are worried that the writing produced by AI is not writing at all may merely be the next generation of traditionalists who fear the unknown. As Ong (2002) stated:

Most persons are surprised, and many distressed, to learn that essentially the same objections commonly urged today against computers were urged by Plato in the Phaedrus (274–7) and in the Seventh Letter against writing. Writing, Plato has Socrates say in the Phaedrus, is inhuman, pretending to establish outside the mind what in reality can be only in the mind. It is a thing, a manufactured product. (p. 78)

From the time of Socrates and Plato, both of whom suggested that writing was an inferior form of communication to orality (Ong, 2002), humans have questioned advancements in technology that change the ways in which people interact, communicate, and think. The introduction of print through the printing press, screen-based writing through computers, networked writing through social media, and computer-generated writing through AI are all examples of moments when literacy practices changed, new norms needed to be developed, and humanity evolved.

As Ong (2002) stated, "Writing and print and the computer are all ways of technologizing the word" (p. 79). These technologies usher in new Discourses, or ways of being (Gee, 1996), that require an understanding of ethics within contexts of participation so that "technology, properly interiorized, does not degrade human life but on the contrary enhances it" (Ong, 2002, p. 81). In order to understand the ethics associated with writing in a digital age, defining writing is a primary task. Is writing equivalent to the types of script developed around 3500 BC (Ong, 2002)? Is it the pictures or other "recording devices"

(Ong, 2002, p. 81) used by humans across cultures for much longer? Is writing the product created to express meaning, the cognitive process that happens inside the mind (Flower & Hayes, 1981), or some other socially constructed interaction (Gee, 1996)?

Ong (2002) argued that "writing (and especially alphabetic writing) is a technology, calling for the use of tools and other equipment: styli or brushes or pens, carefully prepared surfaces such as paper, animal skins, strips of wood, as well as inks or paints, and much more" (p. 80). And while "Plato was thinking of writing as an external, alien technology" (p. 81), viewing writing, and all the tools associated with it, as something that evolves, is perhaps a more ethical stance for writers and teachers of writing to take. In fact, it might be an essential one.

According to the National Council of Teachers of English (2019), literacy in a digital age is "interconnected, dynamic, and malleable" and involves "communicative and sociocultural practices shared among communities." Writing in this context includes print-based, linear texts, as well as digital, multimodal, networked compositions. Technologies shape writing and the ethical practices associated with both the processes of composition and the products created. The teaching of writing, then, is an ethical endeavor on many fronts. This chapter will focus on aspects of teaching ethical practices that have evolved over the last three decades. Many of these aspects are again at a turning point with the explosion in AI technologies in recent years, and it is essential that writing teachers see this moment with "historical consciousness" (Bruce, 2003, p. 13), as a point on a path of evolution of literacy practices, and help writers to understand the ethical implications of being a writer in the age of AI. Teachers must recognize the history of writing development, view writing as a form of inquiry, and understand the complexities of authorship and attribution from a sociocultural and digital perspective so that they can help young writers develop ethical stances in this new era of writing.

HISTORICAL CONTEXT

Much has shifted in the nature of writing in the last thirty years. In reflecting on the task assigned by Ms. G. described in the opening paragraphs of this chapter, thirty years ago most students asked to complete this kind of task only had access to a print-based library to conduct research. Teachers may have encouraged the use of word-processing programs to finalize written reports, and as tools developed, the use of spelling and grammar checking to help edit and revise work. When accessible, computers supported formatting, including updating footnotes and bibliographies to appropriate style guidelines. As an advancement beyond the typewriter, early word processing also supported writers in organizing with the ability to move, add, or delete large sections of text. Of course, these traditional research tasks were almost entirely focused on alphabetic text, and for the most part, they were product, not process-driven.

With increased research into the nature of writing (e.g., Elbow, 1973; Emig, 1968; Flower & Hayes, 1981; Heath, 1983; Murray, 1968), a push to focus on writing as a process began to shift instructional practice. In 2003 the National Writing Project published *Because Writing Matters,* which posed core questions about writing and writing instruction:

- Why does writing matter?
- What does research say about the teaching of writing?
- What do we mean by "writing processes"?
- What are some features of an effective writing classroom?

- How can writing be used to develop critical thinking?
- How does writing fit into learning across disciplines?
- What kind of professional development prepares teachers to teach and use writing?
- What does a school-wide writing program look like?
- What are fair ways to assess writing? (National Writing Project & Nagin, 2003, p. 4)

These questions are worth revisiting in an age of AI. If tasks like the one described above can lead to relatively effective products that are constructed entirely by AI, then the nature of writing, why it matters, and how to teach it must all evolve.

This is not the first moment in time when the impact of technologies on writing has been questioned. Bruce (2003) presented a trajectory of "literacy transformations" (p. 15) that included the following:

- primitive symbol systems
- complex oral language
- early writing
- manuscript writing
- print literacy
- video literacy
- digital/multimedia/hypertext literacy
- virtual reality (p. 15)

Since the time of Bruce's publication, social media and AI have proven to be important influences on the trajectory. At each point, Bruce contended, "We change along with the world around us, and who we become is determined by how we react to change. When a change occurs, it is often disruptive" (p. 30). Twenty years ago, the disruptive change in literacy practices was investigated by researchers associated with the New London Group, who ultimately articulated a theory of New Literacies (Leu, et al., 2004). Researchers like Lankshear and Knobel (2003) argued "that the future is already here, that we are literally staring at it, but that within school education we are not really engaging with it" (p. 19).

This work was advanced by the National Writing Project, which published *Because Digital Writing Matters* (2010) less than a decade after its flagship book. The authors of this text argued:

Writing today is pervasively and generally digital: composed with digital tools; created out of word, image, sound, and motion; circulated in digital environments; and consumed across a wide range of digital platforms.... Digital writing matters because we live in a networked world and there's no going back. Because, quite simply, digital is [emphasis original]. (p. ix)

Despite the push by researchers and theorists for schools to change practice, shifting mindsets and systems was difficult. In a 2008 Pew Internet and American Life survey, most teenagers did not view the writing they did in digital spaces as "***real*** writing [emphasis original]" (Lenhart, et. al, 2008, p. i), and Hicks and Turner (2011) documented similar tensions in writing teachers' perceptions of the nature of digital writing. The "deep grammar of school" (Lankshear & Knobel, 2003, p. 30) conflicted with the reality of human practices and literacy evolution.

The disconnect was likely a difference in what the National Writing Project (2010) called "the ecology of writing and communication and, indeed, what it means to write - to create and compose and share"

(p. 4). Whereas writers in their real-world Discourses (Gee, 1996) had embraced "dramatic changes" (p. 4), schools resisted. Selfe (2004) believed this resistance stemmed from a divide between those "raised on print texts and invested in the cultural systems of print literacy" (p. 34) and those who were coming of age in an era of new media. Turner wrote about this kind of generational divide in relation to adolescents' *digitalk*. Spurred by technologies that blended talking and writing, digital communities of practice adopted language conventions that broke from Standard Written English, and adolescents embraced the shorthand and playfulness of the discourse. Turner (2013) questioned whether their digitalk was "error or strength" (p. 114), and Turner, et al. (2014) articulated the "conventions of digitalk" (p. 157) that demonstrated how adolescent, digital communities helped language structures to evolve.

This line of research pushed against the idea that elements of digitalk that entered students' academic writing were representative of a decline in writing practices and ability. Instead, digitalk was born of the affordances of the technologies being used, which in turn influenced writing practices. This stance embraced the "opportunity" (Jungwirth, 2003, p. 90) presented by the technologies, rather than the "threat" that they posed, to reshaping writing practices. The question of "threat or opportunity" (p.90) is one that can be asked of any technology in relation to writing. Ethically, writers and teachers must ask the question honestly and not assume that tech is equal only to threat or to opportunity based on a preconceived mindset.

The National Writing Project, et al. (2010) stated, "The writer in us knows that tools matter, that tools shape the user as the user shapes the tools" (p. vii). If AI is the next tool to drastically impact the ways that writers create, then many of the arguments made twenty years ago are just as relevant today. As the National Writing Project, et al. (2010) proposed, "Supporting students in the process of writing, studying the craft of writing, and helping students analyze and understand the rhetorical situation for their writing thus remain hallmarks of an effective writing curriculum that aims to create reflective, flexible, self-aware writers" (p. 43). Figuring out how to do that in the context of a world with AI that is readily accessible to all writers is the ethical path for teachers to follow.

Even so, "something feels significantly different at this *particular* moment" (National Writing Project et al., 2010, p. 142), and teachers of writing will need to not only move beyond limited conceptions of writing as print-based, linear text, but they will also need to rethink what it means to research, to co-author, and to seek support during the writing process. They will need to identify the ethical gray areas where writers will need to develop new literacies that involve the ethics of participation in a community that very well may already embrace AI as a valid tool for generating ideas and developing content. As Selfe (2004) alluded to two decades ago, the tools allow for new forms of text that "are changing our understanding of what it means to be literate in the twenty-first century and help us understand our own role in relation to change" (p. 44). This evolution calls for the view of writing as inquiry and a re-imagining of ethical practices in citation and attribution.

WRITING AS INQUIRY

In arguing that writing instruction is essential, The National Writing Project & Nagin (2003) stated,

The very difficulty of writing is its virtue: it requires that students move beyond rote learning and simply reproducing information, facts, dates, and formulae. Students must also learn how to question their own assumptions and reflect critically on an alternative or an opposing viewpoint. (p. 22-23)

Hillocks (1995) contended that the necessary component of writing and writing instruction in order to develop young writers was inquiry. Without inquiry, writing would be "trivial and inconsequential" (Hillocks, 1982, p. 659).

The connection between writing and inquiry became stronger with the "digital turn" (Mills, 2010, p. 246) and the growing focus on how literacy practices were mediated by various technologies. As the nature of writing changed and publication to a broad audience by all writers who were connected to the Internet became possible, inquiry expanded beyond the walls of the school library. Students who were assigned research connected to *The Crucible*, for example, were able to access new kinds of sources, including, perhaps, video interviews of individuals who had lived during the McCarthy era. In the early 2000s, Bishop (2003) argued,

They [students] need to work with multiple sources and media - not just a textbook. They need to become active learners, to collaborate, and to understand the perspectives of others. In short they need to learn how to learn, and they must ask (find problems), investigate (multiple sources / media), create (engage actively in learning), discuss (collaborate and debate), and reflect to do that. (p. 210)

Like Hillocks, Bishop lamented that schools needed to change their curricula to better engage students in authentic inquiry work.

Ironically, this kind of practice was developing outside of school in the digital spaces where adolescents could be found "hanging out, messing around, and geeking out" (Ito, et al., 2013). Through ongoing research into the digital practices of teens beginning in 2005, the Connected Learning Alliance (n.d.) documented the integration of "personal interests, supportive relationships, and access to opportunities" that define inquiries of learning for young people who explore digital media online. By pursuing passions, students can find mentors in digital spaces who help them learn and develop opportunities to connect that learning to future pathways. Online communities become a big part of a connected learner's out-of-school digital engagement.

Clearly, technologies mediate the Discourses (Gee, 1996) to which students belong, impacting their writing practices. Unfortunately, school-based writing often does not engage students in real-world practice. Inquiry occurs when students drive the lessons in the classroom; even today, in many contexts, it is teachers that are in control of what, when, and how topics are studied, or, as Lankshear & Knobel (2003) stated, school assumes that the teacher "is to *direct* learning" (p. 30). These authors made the case two decades ago that schools need to "become knowledge producers for their communities" (p. 106), putting inquiry and research at the heart of what schools do. This evolution in instructional practices has not yet occurred writ large.

Barron & Darling-Hammond (2008), along with other prominent education researchers, recommended inquiry-based learning to meet the needs of twenty-first-century skills (i.e., managing the demands of changing information, technologies, jobs, and social conditions). Their body of research found that students were most successful "when they are taught how to learn as well as what to learn" and "students learn more deeply when they can apply classroom-gathered knowledge to real-world problems, and when they take part in projects that require sustained engagement and collaboration" (p. 8). Writing teachers and writing teacher educators took up this line of inquiry, documenting shifts in practice that can occur when inquiry is connected to digital literacy. For example, Marchetti and O'Dell (2018) complicated the notion of what it means to assign research: "When we assign students to write about a text we've chosen and ideas we have preselected and in a structure we have determined for them, we are essentially giving student a Get Out of Planning Free card" (p. 23). They argued that teacher-created topics, questions, and resources do not help students to do the difficult work of developing inquiry questions and exploring them using all the tools available to them. Their research into practice shared ways that students can engage in alternate forms of writing and inquiry to develop original thinking.

Similarly, Turner and Hicks (2017) advanced the notion of digital arguments, pointing to the key skill of making connections between claims and evidence in the context of opportunities to communicate to a more global audience. This notion of digital argumentation not only expanded and complicated research-based writing expectations but it prompted writing teachers to contemplate their ethical stance and positioning. Turner and Hicks suggested, "Crafting arguments in a digital world could be one of our greatest opportunities to improve dialogue across cultures and continents, or it could contribute to creating or continuing bitter divides" (p. 7). In other words, the structure of research tasks might have an impact on a student's writing and thinking skills in addition to providing students opportunities to make a difference in the world through more authentic and socially relevant inquiries and expression.

This trajectory of research into practice that connected inquiry and writing in a digital age has influenced - and been influenced by - the stance taken by professional organizations. Recently, the National Association for Media Literacy Education (NAMLE) published "Core Principles of Media Literacy Education Implications of Practice" (2023), which explained their position on media literacy education with an emphasis on "the complex dynamics between individuals, media, and the systems and structures that shape our world." In particular, one of these core principles points to critical inquiry around the media industry's role in society as it relates to the specific issues of equity, inclusion, social justice and sustainability (NAMLE, 2023, Core Principle 9).

Connecting digital practice to inquiry and action is the underlying focus of Youth Participatory Action Research (YPAR), where the goal is to provide pedagogical strategies that promote transformational resistance (Cammarota & Fine, 2008). Cammarota (2017) explained the pedagogy of transformational resistance derived from a YPAR program implemented in Tucson Unified School District, where youth were engaged in inquiries about socio-economic issues that existed in their schools and communities. Examples of self-selected themes included immigration and border policies and the unfair treatment of Latinas in schools, workplaces, and governments. Students presented their analyses to family, school leaders, and public officials in written reports, presentations, and video documentation. According to Cammarota, the insights gained through documentation and analysis "lead to resistances that have the potential to transform young people's subjectivities while allowing them to envision ways of learning to counteract oppressive and reproductive schooling" (p. 189-190). Similarly, Garcia and Mirra (2021) noted the importance of linking the composing that students do in school settings to civic learning, and Bertrand and Lozenski (2023) looked at the far-reaching potential for YPAR findings to lead to shifts in policy and practice, where the weight of YPAR might be found in "youth organizing and institutionally sanctioned decision-making, as mobilizers of power bases" (456).

YPAR helps to bring the tenets of Connected Learning into the classroom, and multimodality becomes important on two fronts: (1) the multimodal resources that students might use to research and (2) the different modes of expression that could be used to communicate their inquiries. Decades ago, the National Council of Teachers of English (NCTE) provided a position statement on Multimodal Literacies (2005) stating, "The techniques of acquiring, organizing, and creatively using multimodal information should become an increasingly important component of the English/Language Arts classroom." More recently, NCTE's position statement "Definition of Literacy in a Digital Age" (2019) advocated critical and thoughtful exploration and engagement across a wide variety of inclusive texts and tools/modalities–noting how varied formats, genres, and mediums provide new insight and perspectives. The field has, quite clearly, connected inquiry and digital writing, yet the concerns about AI seem to negate this important connection, focusing instead on the product the students bring back to the teacher, rather than the process through which they come to understand what they know or believe. As NCTE (2019) contended,

As empowered learners engage in literacy practices, they [students] need opportunities to move from consumers to producers of content. More specifically, learners need to move from content consumers to content curators to content creators. These stages do not have to operate in a sequence, nor should they be mutually exclusive as learners fully utilize the reader/writer nature of digital texts.

As AI launches writing into a new digital era, students, as empowered learners, need to grapple with their agency - and any agency they relinquish to the technologies - within the reader/writer/inquirer nature of digital texts. Writing teachers must remember the deep connection between writing and inquiry and consider the role AI might have in the inquiry process itself as students develop their ethical stances. The kind of tasks and expectations of how writers engage in inquiry might need to shift in response to the technologies that are now available.

COMPLEXITIES OF AUTHORSHIP AND ATTRIBUTION

Citations, paraphrasing, plagiarism. These concepts have been centered in writing instruction as a means of acknowledging that published ideas belong to an author. For students in Ms. G.'s class, citing sources in 2023 likely looked very different than it did for their counterparts thirty years ago. A quick look back at the history of citation as it relates to composition classrooms reveals a pattern of technology innovation coupled with the unique challenges posed to protect authorship and integrate multimodal literacies. The MLA, for one, updated its style guide to respond to the ways in which writers conduct inquiry and create products–including the use of AI. These updates will inevitably need to change alongside the evolution of writing.

The history of intellectual property law as it relates to classroom practice in the composition classroom is ripe with complexity. Logie (1998) explained that copyrights and patents are rooted in Article 1. Section 8 of the Constitution with language "to frame arguments that public rights should take precedence over creators' rights" (p. 202). Arguments that expanded the range and duration of copyright became more common in the 1990s. As one example, the reproducibility of print texts through photocopiers foreshadowed the necessity for copyright protections in other electronic media, such as audio tape and video recordings. While educators could rely on US law to protect scholarly activity within United States classrooms in the past, Logie pointed to the change with the use of Internet-capable computers and the need to address "appropriate practices common in a networked environment" (p. 210). He said,

As we move into an increasingly networked, increasingly collaborative creative context, the challenge is to make sure the composing strategies we teach to students balance our sense of what is right against the developing global definitions of what is legal–and to insure that when the gap between the two is too large, we take action and use our close proximity to the composing process to inform the development of future policies. (Logie, 1998, p. 211)

It appears that this statement from decades ago could just as likely be said today.

MLA shows a consistent responsiveness to support newer technologies in writing. In 1999, when Internet-based information became accessible to the masses, MLA revised the fifth edition of their style guide to account for Internet source attribution. As noted in a timeline posted on the MLA website:

In 1999, facing the sudden information overload introduced by the Internet, the MLA "revamped" its fifth edition "for the Internet age," as noted on the back cover of the handbook (Gibaldi, MLA Handbook [1999]). The handbook cautioned that "[a]ssessing Internet resources is a particular challenge. Whereas the print publications . . . are generally issued by reputable publishers . . . relatively few electronic publications currently have comparable authority" (26). It also provided tips on using scholarly metapages, gateways, search engines, chat rooms, and Boolean operators (20–22). In discussing works found through online subscription services such as America Online or Lexis-Nexis, the handbook recommended providing in the citation the keyword or "sequence of topics"used to locate the material (195). (MLA, 2023b)

MLA has updated their style guide many times to account for changes in technology that impacted the way writers composed. From word processors to CD-ROMs to databases to the Internet, the organization has been responsive, providing guidance to help authors meet ethical standards of attribution. Most recently, MLA addressed the issue of AI and citation on their website. They suggested that authors,

- cite a generative AI tool whenever you paraphrase, quote, or incorporate into your own work any content (whether text, image, data, or other) that was created by it
- acknowledge all functional uses of the tool (like editing your prose or translating words) in a note, your text, or another suitable location
- take care to vet the secondary sources it cites (MLA, 2023a)

Style guides provide writing teachers an opportunity to "teach citation as a community practice" (Switaj, 2015, p. 223), but they also attempt to simplify attribution into a clear set of rules. In reality, authorship and attribution are complex in a digital era where sociocultural perspectives of writing suggest that composing does not occur in isolation (Rhodes & Alexander, 2014). As articulated by National Writing Project et al. (2010),

The focus on the single author and his or her solitary genius is, in academia, of the utmost importance in many areas, for example in literary studies. It is also evident in academia's focus on citation practices, which continually reconstitute the idea of original authorship. However, much digital writing is done... collaboratively, across time and space as well as across documents, and with ... "remixing" as a key practice for invention and composing - that is, writing by appropriation: taking bits, pieces, and ideas and compiling and remixing them in new and innovative ways. (p. 80)

McGrail, et. al (2021) acknowledged this complexity in their framework for assessing digital, multimodal composition. In defining "originality" as a domain of multimodal writing, they argued that it "does not require a reliance on producing an absolutely original product but rather creating a product that is informed by that which already exists but is transformed in a new, fresh way" (p. 291). In fact, *transformativeness* is a concept related to copyright that Hobbs (2010) argued was important for educators to understand. The "Code of Best Practices in Fair Use for Media Literacy Education" (National Association for Media Literacy Education et al., 2008/2018) provided a "foundation to understanding the principles of fair use" for literacy teachers to guide their students.

Originality is something that often gets lost in "schoolish" (Whitney, 2011, p. 55) writing when the audience is limited to the teacher and modes are limited to standardized essays. The research paper is a genre that often becomes regurgitation of knowledge that already exists, rather than the product of student inquiry. It is often an inauthentic endeavor that does not invite transformation. Outside of school, research *must* be original, transformative in some way, or it will likely not be published. Inside of school, however, research is often deemed original as long as students cite their sources according to the accepted style guide. Furthermore, in a digital world, authentic inquiry often results in multimodal creations that "may or may not have components never before presented in a text" (McGrail, et. al, 2021, p. 291). Authors remix and attribute their work in order to show transformativeness, resulting in compositions that look very different than a traditional research essay.

In digital writing, remixing is a legitimate composing practice (Palmeri, 2012; Garcia, 2016), yet fair use does not eliminate the need for attribution, and Hobbs (2019) stressed that when curated resource lists or "search engines decontextualize creative content, removing it from its original context, learners may not fully appreciate the rights of authors of copyrighted content" (p. 155). She argued that students must be able "to explain where the media came from and who created it… [and] how the meaning and value of the image was changed through the particular context of the use" (p. 156). In an era where remixing is considered ethical, attribution of ideas continues to be important. This complexity raises questions for the use of AI in writing.

It is true that LLM are inherently a compilation of uncited work, and in fact authors are currently questioning the use of their copyrighted materials to train the models (O'Brien, 2023). Citing AI as a source or listing it as co-author does not solve the problem of hidden authorship that Hobbs (2019) described. The questions asked about the tool and its impact on writing may be different in many ways; the questions about ethical practice remain similar. How do we attribute ideas, words, phrases, and use of media in the new digital age? It seems that technology, once again, "has increased our sense of both vulnerability and accountability" (Vicinus & Eisner, 2008, p. 163). While it is imperative for teachers of writing and their students to grapple with these questions, the field itself is at a juncture where the ethical stance must be to consider how the tools are changing the nature of writing and human inquiry and interaction. As Turner (2019a) stated, "This learning is not always straightforward nor easy. It is, however, foundational for ethical practices in a digital world" (p. 163).

THE NEXT ERA

…the distinction Jeff Bezos draws between first- and second-phase automation. First-phase automation involves using new technologies to do old processes more quickly and efficiently, whereas second-phase automation involves fundamentally changing the underlying process of an existing practice and, instead, doing things in a new way. (Lankshear & Knobel, 2003, p. 81)

Two decades ago Lankshear & Knobel (2003) warned that educators needed to change the way they viewed new technologies. Rather than "adapting them to familiar routines (p. 31) or "doing old things in new ways" (p. 106), schools needed to fundamentally change. Unfortunately, as Johnson-Eilola (2004) stated back then, we *still* "tend, despite all of our sophisticated theorizing, to teach writing much as we have long taught it: the creative production of original words in linear streams that some reader receives and understands" (p. 200). With the introduction of AI into students' writing practice, teachers need to do things in a new way, adopting a second-phase automation mindset.

This stance is ethical, drawing from a long history of the evolution of writing practices and an understanding of the role of writing to develop thinking and to share meaning. In this mindset, just as calculators and computers did not "weaken the mind" (Ong, 2002, p.78), AI exists as the next tool to mediate literacy practices, not simply destroy them. In fact, the MLA-CCCC Task Force (2023) contends that "critical AI literacy is now part of digital literacy, and students and teachers should be made aware of bias and inaccuracy in model outputs and the particular vulnerability of students who may not yet have sufficient expertise to critically evaluate language model outputs, including seeing them as sentient." In fact, awareness is just the first step of new literacy learning that both teachers and students must make. In addition to using AI critically in a variety of ways, writers must also challenge the underlying system that creates, promotes, and distributes the technology. Companies and software developers need to be held accountable for the tools.

Therefore, while embracing the tool, teaching the ethical use of the tool in an inquiry-based context, and grappling with how to attribute authorship in products created using the tool must become part of the writing classroom so too must advocacy and activism to ensure that AI is designed ethically and promotes ethical practice. Questions about what this ethical use looks like are just now being investigated. Most would likely agree that solely using AI-generated text and submitting it as one's own is unethical. But, if AI is used as a resource to generate and refine ideas that can be remixed in tandem with research, revision, and writing through the critical thinking and writing process of a living author, then the ethical use of AI becomes more complicated.

Writing teachers have begun to explore the opportunities that AI presents. For example, O'Bryan (2023) suggested using an AI platform throughout the writing process to help writers who struggle to develop their thinking. Citing the common dilemma of a student who did not know what to write about, O'Bryan tried her own interaction with ChatGPT. She asked ChatGPT what she should write about and then shared a quick back-and-forth that resulted in a ChatGPT-generated list of six possible topics. Next, O'Bryan asked ChatGPT about her working topic, and noticed the value it might have for students who were still developing their depth of knowledge and synthesis skills in the context of an assignment and life circumstances. This experience helped her to understand how the tool might be useful in tandem with teacher or peer writing conferences to move individual writers forward in their work.

While MLA noted these kinds of benefits of AI, the organization was clear in its recommendation that ChatGPT should not be considered a co-author. According to the MLA's web page on citing generative AI, they do not recommend citing AI as an author, noting that this is a recommendation that "follows the policies developed by various publishers, including the MLA's journal *PMLA*" (MLA, 2023c). For example, when submitting to *PMLA*, originality is addressed as follows: "Authors must fully cite in the manuscript, at submission, their use of all content (whether text, images, data, or other) created by an AI tool" (MLA, 2023c). It makes sense to cite content generated by AI. However, how much of one's published content can be generated by AI? Should a particular percentage of a published work be non-AI generated, in order for the work to be published in a particular author's name? If AI is used to generate and refine an initial idea, should a unique method of attribution be created to support this specific type of authorship? The issue of agency becomes key when exploring these questions and considering whether it might actually be more ethical to cite AI as a co-author than as a source.

While some might question whether using AI to help develop topic ideas or inquiry questions is less meaningful than composing writing that is the result of a writer's natural curiosity, perhaps this approach is no different than Plato and Socrates privileging orality over writing. After all, this chapter was co-authored by two individuals through a series of conversations, responses to drafted texts, and shared notes from texts written by others. The inquiry evolved through a process that might not look much different - at its core - than one where one of the two authors was an AI tool. Perhaps this image of AI is where it can become an equalizer, as the printing press was, making "widely available" (Bruce, 2003, p. 101) opportunities that were once "held behind locked doors" (p. 101). If all writers have the benefit of a collaborator to support them at every stage of the writing process, then the ethics of using and embracing AI in the writing classroom are even more important. Ultimately, adopting a stance that technologies have, and will continue to influence changes in writing practices for as long as humans have communicated meaning with each other will allow ethical practices to develop over time. With attention to authorship and the dilemmas in attribution that the Internet and AI tools present, writers - and their teachers - can support the development of these practices within a globalized community of writers.

REFERENCES

Barron, B., & Darling-Hammond, L. (2008). *Teaching for meaningful learning: a review of research on inquiry-based and cooperative learning*. George Lucas Education Foundation. https://files.eric.ed.gov/fulltext/ED539399.pdf

Bertrand, M., & Lozenski, B. D. (2023). YPAR dreams deferred? Examining power bases for YPAR to impact policy and practice. *Educational Policy*, *37*(2), 437–462. https://doi.org/10.1177/08959048211019975

Bishop, A. P. (2003). Using the web to support inquiry-based literacy development. In Bruce, B. C. (Ed.). Literacy in the information age: Inquiries into meaning making with *n*ew technologies. International Reading Association.

Bruce, B. C. (Ed.). (2003). Literacy in the information age: Inquiries into meaning making with *n*ew technologies. International Reading Association.

Cammarota, J. (2017). Youth participatory action research: A pedagogy of transformational resistance for critical youth studies. *The Journal for Critical Education Policy Studies*, *15*(2), 188–213.

Cammarota, J., & Fine, M. (2008). *Revolutionizing education: Youth participatory action research in motion*. Routledge.

Connected Learning Alliance. (n.d.) What is connected learning? *About Connected Learning*. Connected Learning Alliance. Accessed July, 21, 2023. https://clalliance.org/about-connected-learning/

Elbow, P. (1973). *Writing without teachers*. Oxford University Press.

Emig, J. (1968). *The composing process of twelfth graders*. NCTE.

Flower, L., & Hayes, J. R. (1981). A cognitive process theory of writing. *College Composition and Communication, 32*, 365–387.

Garcia, A. (2016, May 26). How remix culture informs student writing & creativity. *School Library Journal.* https://www.slj.com/story/how-remix-culture-informs-student-writing-creativity

Garcia, A., & Mirra, N. (2021). Writing toward justice: Youth speculative civic literacies in online policy discourse. *Urban Education, 56*(4), 640–669. doi:10.1177/0042085920953881

Gee, J. (1996). *Social linguistics and literacies: Ideology in discourses* (2nd ed.). Taylor & Francis.

Gibaldi, J. (1999). *MLA handbook for writers of research papers* (5th ed.). Modern Language Association of America.

Heath, S. B. (1983). *Ways with words: Language, life, and work in communities and classrooms.* Cambridge University Press.

Hicks, T., & Turner, K. H. (2011). "That's not writing": Exploring the intersection of digital writing, community literacy, and social justice. [Doi: 10.1353/clj.2012.0000]. *Community Literacy Journal, 6*(1), 55–78.

Hillocks, G. Jr. (1982). Inquiry and the composing process: Theory and research. *College English, 44*(7), 659–673. doi:10.2307/376805

Hillocks, G. Jr. (1995). *Teaching writing as reflective practice.* TC Press.

Hobbs, R. (2010). *Copyright clarity: How fair use supports digital learning.* Corwin.

Hobbs, R. (2019). Meditation: Ethical practices. In Turner, K. H. (Ed.). The ethics of digital literacy: Developing knowledge and skills across grade levels. 155-158. Rowman & Littlefield.

Ito, M., Antin, J., Finn, M., Law, A., Manion, A., Mitnick, S., Schlossberg, D., Yardi, S., & Horst, H. A. (2013). *Hanging out, messing around, and geeking out: Kids living and learning with new media.* MIT Press.

Johnson-Eilola, J. (2004). The database and the essay: Understanding composition as articulation. In *Wysocki, A. F., et al. (2004). Writing new media: Theory and applications for expanding the teaching of composition.* Utah State University Press.

Jungwirth, B. (2003). Information overload: Threat of opportunity? In B. C. Bruce (Ed.), *Literacy in the information age: Inquiries into meaning making with new technologies.* International Reading Association.

Lankshear, C., & Knobel, M. (2003). *New literacies: Changing knowledge and classroom culture.* Open University Press.

Lenhart, A., Arafeh, S., & Smith, A. (2008). *Writing, technology and teens.* Pew Internet and American Life. https://www.pewresearch.org/internet/2008/04/24/writing-technology-and-teens/

Leu, D. J. Jr, Kinzer, C. K., Coiro, J., & Cammack, D. W. (2004). Toward a theory of new literacies emerging from the Internet and other information and communication technologies. In R. B. Ruddell & N. Unrau (Eds.), *Theoretical models and processes of reading* (5th ed., pp. 1570–1613). International Reading Association.

Logie, J. (1998). Champing at the bits: Computers, copyright, and the composition classroom. *Computers and Composition*, *15*(2), 201–214. doi:10.1016/S8755-4615(98)90054-8

Marchetti, A., & O'Dell, R. (2018). *Beyond literary analysis: Teaching students to write with passion and authority about any text.* Heinemann.

McGrail, E., Turner, K. H., Piotrowski, A., Caprino, K., Zucker, L., & Greenwood, M. E. (2021). An interconnected framework for assessment of digital multimodal composition. *English Education*, *53*(4), 277–302.

Miller, A. (2000). *The crucible.* Penguin Classics.

Mills, K. A. (2010). A Review of the digital turn in the new literacy studies. *Review of Educational Research*, *80*(2), 246–271. doi:10.3102/0034654310364401

MLA. (2023a). How do I cite generative AI in MLA style? MLA Style Center. Accessed July 19, 2023. https://style.mla.org/citing-generative-ai/

MLA. (2023b). MLA Style [timeline]. MLA Style Center. Accessed July 19, 2023. https://www.mla.org/About-Us/About-the-MLA/MLA-Archives/Time-Lines/MLA-Style

MLA. (2023c) Submitting manuscripts to PMLA. Modern Language Association. Retrieved July 19, 2023, from https://www.mla.org/Publications/Journals/PMLA/Submitting-Manuscripts-to-PMLA

MLA-CCCC Joint Task Force. (2023). MLA-CCCC joint task force on writing and AI: Overview of the issues, statement of principles, and recommendations. Modern Language Association and Conference on College Communication and Communication. Accessed July 20, 2023. https://aiandwriting.hcommons.org/working-paper-1/

Murray, D. M. (1968). *A writer teaches writing: A practical method of teaching composition.* Houghton Mifflin.

National Association for Media Literacy Education. (2023). Core principles of media literacy education [Fact sheet]. Accessed July 14, 2023. https://namle.net/resources/core-principles/

National Association for Media LIteracy Education. Student Television Network, Media Commission of the National Council of Teachers of English, Action Coalition for Media Education, & Visual Communication Division of the International Communication Association. (2008/2018). *Code of best practices in fair use for media literacy education.* Accessed July 20, 2023. https://ncte.org/statement/fairusemedialiteracy/

National Council of Teachers of English. (2005). *Multimodal literacies.* NCTE Position Statements. Retrieved July 19, 2023. https://ncte.org/statement/multimodalliteracies/

National Council of Teachers of English. (2019). *Definition of Literacy in a Digital Age.* NCTE Position Statements. Accessed July 19, 2023. https://ncte.org/statement/nctes-definition-literacy-digital-age/

National Writing Project. Devoss, D. N., Eidman-Aadahl, E., & Hicks, T. (2010). Because digital writing matters: Improving student writing in online and multimedia environments. Jossey-Bass.

National Writing Project & Nagin. C. (2003). Because writing matters: Improving student writing in our schools. Jossey-Bass.

O'Brien, M. (2023, July 12). Sarah Silverman and novelists sue ChatGPT-maker OpenAI for ingesting their books. APNews. Accessed July 20, 2023. https://apnews.com/article/sarah-silverman-suing-chatgpt-openai-ai-8927025139a8151e26053249d1aeec20

O'Bryan, V. (2023, May 7). Using ChatGPT to get started with the writing process. NCTE. Accessed July 20, 2023. https://ncte.org/blog/2023/05/chatgpt-writing-process/

Ong, W. J. (2002). *Orality and literacy: The technologizing of the word.* Routledge.

Palmeri, J. (2012). *Remixing composition: A history of multimodal writing pedagogy.* Southern Illinois University Press.

Rhodes, J., & Alexander, J. (2014). Reimagining the social turn: New work from the field. *College English, 76*(6), 481–487.

Selfe, C. (2004). Students who teach us. In *Wysocki, A. F., et al. (2004). Writing new media: Theory and applications for expanding the teaching of composition.* Utah State University Press.

Switaj, E. (2015). Web Writing and Citation: The Authority of Communities. In J. Dougherty & T. O'Donnell (Eds.), *Web Writing: Why and How for Liberal Arts Teaching and Learning* (pp. 223–232). University of Michigan Press., doi:10.2307/j.ctv65sxgk.24

The Neglected "R": The Need for Writing Revolution. (2003). College Entrance Examination Board. https://archive.nwp.org/cs/public/download/nwp_file/21478/the-neglected-r-college-board-nwp-report.pdf?x-r=pcfile_d

Turner, K. H. (2013). Error or strength? Competencies developed in adolescent digitalk. In Pytash, K. E. & Ferdig, R. E., Eds. *Exploring Technology for Writing and Writing Instruction*, 114-134. IGI Global.

Turner, K. H. (2019a). Where did I find that?: Helping students develop ethical practices in digital writing. In Turner, K. H. (Ed.). The ethics of digital literacy: Developing knowledge and skills across grade levels. 159-164. Rowman & Littlefield.

Turner, K. H. (Ed.). (2019b). *The ethics of digital literacy: Developing knowledge and skills across grade levels.* Rowman & Littlefield.

Turner, K. H., Abrams, S., Katic, E., & Donovan, M. J. (2014). Digitalk: The what and the why of adolescent digital language. *Journal of Literacy Research*, *46*(2), 157–193. Doi.org/10.1177/1086296X14534061

Turner, K. H., & Hicks, T. (2017). *Argument in the real world: teaching adolescents to read and write digital texts*. Heinemann.

Vicinus, M., & Eisner, C. (2008). *Originality, imitation, and plagiarism: Teaching writing in the digital age*. University of Michigan Press.

Whitney, A. E. (2011). In search of the authentic English classroom: Facing the schoolishness of school. *English Education*, *44*(1), 51–62.

Wysocki, A. F., Johnson-Eilola, J., Selfe, C. L., & Sirc, G. (2004). *Writing new media: Theory and applications for expanding the teaching of composition*. Utah State University Press.

Chapter 8
Metaverse Libraries:
Leveraging Technology for Empowering Digital Citizenship

Mohammad Daradkeh
https://orcid.org/0000-0003-2693-7363
Yarmouk University, Jordan & University of Dubai, UAE

ABSTRACT

As the field of intelligent libraries continues to evolve, Metaverse technology presents a promising opportunity for driving innovation and improving library services. This chapter delves into the potential of Metaverse to transform libraries into fully-intelligent spaces. The authors will explore the underlying technologies and theoretical foundations of Metaverse, and examine how it can be applied to enhance library collection management, reading spaces, cultural promotion, reading experiences, and user services. By harnessing the power of Metaverse, intelligent libraries have the opportunity to create immersive experiences and achieve significant progress in their digital transformations.

INTRODUCTION

The emergence of virtual reality (VR) technology represents a new wave of technological revolution and industrial change, and it is considered a key technology in promoting the development of the digital economy and industrial transformation. As the VR industry continues to grow and expand, there is a vast space for innovation and a great potential for growth. This is highlighted by the World VR Industry Conference on October 20, 2021, which emphasized the need for further "VR+" actions to enrich terminal products, content services, and promote the industrialization and scaling up of VR technology (Zhang et al., 2022). By focusing on "VR+" actions to enrich terminal products, content services, and promote the industrialization and scaling up of VR technology, it is believed that VR can empower digital citizenship by providing users with immersive and interactive experiences that enhance access to information and resources, and promote greater participation and engagement in the digital world.

DOI: 10.4018/978-1-6684-8934-5.ch008

The integration of virtual reality (VR) and metaverse technology in traditional cultural venues such as libraries, museums, and galleries is revolutionizing the way digital citizenship is empowered. VR technology allows for a more immersive and interactive experience for visitors, bringing exhibits and collections to life in a way that was not previously possible. Moreover, the use of VR in education and training is gaining momentum as it enables students to explore and learn in a virtual environment that simulates real-world scenarios. Metaverse technology, in particular, offers a virtual shared space where users can interact with each other and with virtual objects in a way that resembles the real world. The use of metaverse technology in libraries can enable the creation of virtual reading rooms and libraries, providing users with access to a vast collection of books and resources in a virtual environment. Furthermore, it can also enable the creation of virtual events such as book clubs and author talks, providing a new way for libraries to engage with their communities, and thereby empowering digital citizenship.

There are 3 stages in the development of intelligent libraries: pseudo-intelligent, partially intelligent and fully intelligent (Njoku, Nwakanma, Amaizu, & Kim, 2022). From the current stage of development of intelligent libraries, libraries are still in pseudo-intelligent and partially intelligent stages (Alpala et al., 2022). Technologies such as Web 3.0, Internet of Things, big data, and cloud computing (Dai, Wang, & Gao, 2022) are widely used in the intelligent services of libraries, but their intelligent services are limited by time and space. Each technological revolution has brought about disruptive changes to the service models of libraries. From information to knowledge to big data, the service models of libraries have evolved from traditional service models to digital library service models to intelligent library service models.

The advent of metaverse technology has the potential to revolutionize the field of intelligent libraries, creating immersive and fully-intelligent spaces that empower digital citizenship. As libraries embrace this new technology, they must consider how it can be applied to enhance collection management, reading spaces, cultural promotion, reading experiences, and user services. By leveraging the power of metaverse, intelligent libraries have the opportunity to take their digital transformations to the next level, providing users with truly interactive and personalized experiences that drive innovation and improve services. Thus, understanding the capabilities and potential applications of metaverse technology in libraries is crucial for the development and advancement of intelligent library services.

This chapter aims to explore the potential of Metaverse technology in driving the transformation of intelligent library services. The chapter addresses a significant gap in the field of intelligent libraries, particularly in the context of digital transformation and technological innovation. While technologies such as Web 3.0, Internet of Things, big data, and cloud computing have been integrated into library services, the current state of intelligent libraries remains limited by temporal and spatial constraints. This chapter recognizes the emergence of Metaverse technology as a potential game-changer that can revolutionize intelligent library services by creating immersive and fully-intelligent spaces. The gap lies in the lack of comprehensive understanding and exploration of how Metaverse technology can be effectively harnessed to drive library transformation. The chapter aims to bridge this gap by providing in-depth insights into the applications, advantages, and potential implications of Metaverse technology in libraries, ultimately facilitating their transition to fully intelligent spaces that cater to evolving user needs and preferences.

This chapter makes a significant contribution to the field by exploring the potential of Metaverse technology in transforming intelligent library services. It identifies a crucial gap in the current understanding of how Metaverse technology can be integrated into libraries to enhance their offerings and empower digital citizenship. By analyzing the development opportunities and application prospects of Metaverse technology in intelligent libraries, the study aims to drive innovation and optimization of library services. Furthermore, the study delves into the advantages of Metaverse and its underlying technologies, establishing a theoretical foundation for the application of Metaverse in library settings. Through comprehensive insights, the study paves the way for libraries to evolve into fully intelligent spaces, utilizing Metaverse technology to enhance collection resource management, reading spaces, cultural promotion, reading experiences, and specialized user services. By addressing this gap, the study provides a forward-looking perspective that aligns with the ongoing digital revolution, ultimately contributing to the advancement of library services in the contemporary digital landscape.

In the subsequent sections, this chapter will undertake an examination of the developmental prospects inherent in Metaverse technology, as well as its potential applications within the context of intelligent libraries. The overarching objective of this analysis is to foster a culture of innovation and enhance the optimization of library services. Moreover, the chapter will engage in an exploration of the inherent advantages offered by Metaverse technology, including its underlying technical components. Furthermore, a thorough examination of the theoretical rationale underpinning the application of Metaverse technology within library environments will be conducted.

Furthermore, this chapter will offer insightful perspectives on the strategic utilization of Metaverse technology and its associated components, with the aim of facilitating the transition of libraries into fully intelligent entities. By leveraging Metaverse and its conceptual framework, libraries can potentially garner enhanced advantages, particularly in the domains of collection resource management, spatial configuration for reading activities, cultural dissemination, reading experiences, and specialized user services. The eventual aspiration of this chapter is to furnish a comprehensive comprehension of the manner in which Metaverse technology can be effectively harnessed to augment library services, thus positioning libraries at the vanguard of progressive advancements within the landscape of the ongoing digital revolution.

CONNOTATIONS OF METAVERSE AND INTELLIGENT LIBRARIES

Metaverse

Metaverse is a general concept describing future iterations of the Internet, consisting of a continuously shared three-dimensional virtual space connected to a perceptible virtual world (Allam et al., 2022). The term originated in the science fiction novel "Snow Crash" by N. Stephenson, in which the plot takes place in a virtual space where real humans live with virtual people through VR devices (Magalhães et al., 2022). Since the emergence of the metaverse to date, there has not been a unified definition of the metaverse, and different vendors and scholars have their own interpretations of the metaverse. The burgeoning of the metaverse has a significant relationship with Facebook founder M. Zuckerberg's interview at TheVerge, in which he expressed his desire to turn Facebook into a metaverse company (Nguyen, 2022). He also believed that the metaverse was the future of the Internet. Since then, many scholars and entrepreneurs have started to explore the concept of metaverse. Yang et al. (2022) argues that the metaverse needs to achieve a high degree of realism, where people's lives in reality can be mapped to the metaverse space. Cappannari and Vitillo (2022) argued that the metaverse is a virtual space parallel to and independent from the real world, a digital virtual world that is increasingly realistic. Duan et al. (2021), the founder of r. Beamable, divided the metaverse according to different levels. Following the advent of the metaverse, many companies have sought to implement the concept, taking Roblox (Allam et al., 2022), a gaming company listed in the U.S., as an example, which was the first company to build a metaverse ecology. Later, major manufacturers have been investing in building the metaverse.

The metaverse aims to build a sustainable virtual shared space while maintaining the perception and experience of the real world, which requires adhering to the values of co-creation, co-building, sharing, and co-governance, and fully integrating new technologies such as big data, artificial intelligence, virtual reality, 5G, blockchain, and 3D engines (Buhalis, Lin, & Leung, 2022; Chen & Lee, 2021). The technical core of metaverse depends on integration and application, and its underlying technologies are shown in Table 1. Different disciplines have different views on the exploration of metaverse. However, in the field of graphical information, considering only virtualization and digitization is not a prospect for the development of intelligent libraries. The concept of intelligent libraries needs to be gradually realized. Only by continuously introducing the underlying technologies of metaverse and upgrading the intelligent services of libraries can we eventually realize the profound integration of metaverse and libraries.

Table 1. Major underlying technologies of metaverse

Underlying Technologies	Implemented functions
Artificial intelligence, digital twin	Meta-universe ecology
Blockchain	Access Verification
Artificial Intelligence, Cloud Computing	Underlying algorithms
Expanded reality, robotics, brain-computer interface	Virtual Reality Simulation
5G	Unhindered Networks

Intelligent Library

In 2003, M. Aittola of the University Library of Oulu, Finland, introduced the term intelligent library (Hassani, Huang, & MacFeely, 2022), arguing that intelligent libraries are not limited by time and space and can perceive mobile library services that help users to find books and other types of materials in the library by connecting to wireless Internet. Intelligent libraries are a three-dimensional combination of physical space, digital space, and human society, as well as virtual technologies. Core technology clusters such as intelligent technologies underlie their creation and development (Baghalzadeh Shishehgarkhaneh et al., 2022). In the development of intelligent libraries, it undergoes pseudo-intelligence and partial intelligence stages; the fully intelligent library is the final form of intelligent library development (Eom, 2022). Although the fully intelligent library is based on intelligent technology, the core of the fully intelligent library is the integrated use of metaverse underlying technology. It aims at fully intelligent library services and the realization of dual digital libraries. These two goals are also the two stages of metaverse application in intelligent libraries.

The first stage of the fully intelligent library is to use the underlying technologies of the metaverse, such as extended reality and blockchain, to achieve fully intelligent library services, i.e., an unattended library (Eom, 2022). Since the key to the metaverse lies in technological breakthroughs and technology integration, the development of fully intelligent libraries at this stage also lies in the adoption and integration of the underlying technologies of the metaverse. For this stage of development, many libraries have made experimental practices, such as the WEB collection book location system built by Harbin Engineering University Library (Duan et al., 2021) and the University of Miami Library (Yang et al., 2022) using augmented reality technology to identify books. However, these attempts can only partially solve the consultation and service requirements of intelligent libraries without librarians, but cannot truly achieve unattended fully intelligent services.

The second stage of the fully intelligent library is the construction of digital twin libraries to realize the full intelligence of libraries in the real sense. In this stage, the service of fully intelligent library should break through the limitation of time and space. In the metaverse, the fully intelligent library provides users with digital identity verification, and users can freely enter the virtual space of the library with their digital identity without physical restrictions. Meanwhile, the real library's intelligent services reach the stage of full intelligence. With the realization of the morphological changes of the fully intelligent library, its service model will change in the following directions:

(1) Most of the reading services provided by traditional libraries are for individual readers, and there is a lack of interactive communication between readers and patrons. For example, individual readers can enjoy the collection navigation and display services of intelligent libraries using virtual reality technology (Ruiz Mejia & Rawat, 2022; Wang, Yu, Bell, & Chu, 2022), but different readers cannot interact with knowledge through such devices, and the devices are fragmented from each other. In contrast, intelligent libraries emphasize the interconnection of readers using Metaverse devices and technologies to provide a shared reading space for readers. This shared space serves as a dual modeling of the reader and the library, with multiple readers accessing twin digital libraries as digital twins. This is not a conjecture; VR socialization has been proposed for many years (Y. Wang et al., 2022).

(2) The emergence of intelligent libraries means a closer relationship between digital libraries and real libraries. When users enter a digital library, they do not browse as a web page, but as a virtual library in the form of a second life (Salem & Dragomir, 2022). The difference is the addition of physical interaction and real-time rendering, which is lacking in the second life. The digital library is a three-dimensional reconstruction based on the real library. Also, the digital transformation of the collection resources in real libraries will be faster. With the support of the underlying technology of metaverse, the digital library and the real library continuously interact with each other, forming the prototype of the intelligent library. Meanwhile, intelligent libraries can integrate more services, such as health information center, medical libraries, audio-visual entertainment, art galleries and museums, engineering libraries, etc. (Jamil, Rahman, & Fawad, 2022), all of which can be realized in a twin digital library.

(3) The realization of intelligent libraries in the metaverse perspective is not a departure from reality, but a dual sublimation of digital and real libraries using metaverse technology. For unattended services, although libraries have attempted (Bolger, 2021), the real intelligent services have not been realized yet. For the intelligent development of intelligent libraries, the underlying technology of metaverse needs to be continuously adopted and improved. This includes the use of 3D scanning technology, spatial distance and proximity sensing for library navigation, image and text retrieval, real-time voice broadcasting for users (Xiong & Wang, 2022), and intelligent monitoring using face recognition technology and RFID technology (Y. Wang et al., 2022).

METAVERSE-BASED THEORETICAL LOGIC FOR INTELLIGENT LIBRARIES

In the traditional era of literacy, libraries were places where knowledge was stored and paper documents were kept for those who needed to read and learn (Dahan et al., 2022). However, since the middle of the 20th century, human society has entered the process of informatization, that is, the ubiquitous information society (Faraboschi, Frachtenberg, Laplante, Milojicic, & Saracco, 2022). People's requirements for collection resources have become more personalized and diversified, and modern libraries have become comprehensive places for people to learn knowledge and to engage in casual recreation. Although traditional libraries continue to innovate in their services, the optimization of services mostly involves physical space. With the continuous development of mobile libraries and digital libraries, traditional libraries have started to develop into intelligent libraries, but for a long time, the development of digital libraries has lagged far behind physical libraries (Maddahi & Chen, 2022). The arrival of metaverse, on the other hand, will enable libraries to develop more quickly into intelligent and diversified libraries.

In terms of the underlying technology application of metaverse, Siyaev and Jo (2021) proposed blockchain technology to help the equity management of digital libraries, Hassani et al. (2022) proposed the integration of 5G and intelligent libraries. Jin, Xu, and Leng (2022) reviewed and presented an outlook on the application of artificial intelligence in libraries. Viewed from this aspect, metaverse is not a castle in the air for the construction of intelligent libraries, and the technical references behind it have been discussed by many scholars.

For the establishment of digital twin libraries, intelligent libraries pursue authenticity, real-time and interaction among readers. Keshmiri Neghab, Jamshidi, and Keshmiri Neghab (2022) proposed the theory of three-dimensional space and virtual space-time tunnel in 2007. The virtual library in Second Life (Ahn, Kim, & Kim, 2022) also tried spatial interconnection. Although, it lacks real-time interaction with readers and somatic simulation, its 3D modeling technology and digital transformation of collection resources undoubtedly provide many experiences and ideas for the application of metaverse in libraries.

PROSPECTS OF METAVERSE FOUNDATION TECHNOLOGY IN INTELLIGENT LIBRARIES

The metaverse comes alive in 2021, and there is endless curiosity about the metaverse in both the research and practice communities. The library community should remain rational in the face of the transformational impulse and temptation that the metaverse brings to libraries. Metaverse may not be realized in libraries immediately, but exploring its underlying technology in intelligent libraries is one of the directions to advance the library services.

Application Prospect of Virtual Reality in Intelligent Libraries

Metaverse and virtual reality are inseparable, and even virtual reality is the technology underlying metaverse. Virtual reality usually refers to technologies such as VR (virtual reality), AR (augmented reality), and MR (mediated reality). There have been many attempts to apply these three technologies in the construction of intelligent libraries. For example, the library administration of the University of Oklahoma has carried out a "VR" project in the library to present text in the form of 3D images (Vishkaei, 2022). The library of the University of North Carolina at Miami, USA, uses AR technology to allow patrons to see a brief description and history of the book while viewing the shelves (Ukko, Saunila, Nasiri, Rantala, & Holopainen, 2022). In terms of technology, VR, AR, MR, and Metaverse each have their own characteristics, but there is a transition across generations.

The application prospect of virtual reality technology in intelligent libraries is divided into two stages. The first is the application of virtual reality technology in unattended libraries, which includes the integration of collection resources with holographic projections using mixed reality technology. Patrons enjoy the combination of visual and auditory sensations through head-mounted displays while reading books and real-time navigation using AR technology. MR technology and AR applications are used to help patrons pinpoint the location of books and obtain the borrowing status and popularity of books (Buhalis et al., 2022; Popescu, Dragomir, Popescu, & Dragomir, 2022). Second, the establishment of a digital twin library. At this stage, the creation of a digital twin library must be based on the real collection resources. By creating digital avatars for readers, the immersion of readers is enhanced by virtual reality physical devices. The digital avatars and the borrowing habits of readers in the real world promote relevant information to the digital twin library to provide readers with real, comprehensive and diversified intelligent services.

Application of Blockchain Technology in Intelligent Libraries

Blockchain technology is a distributed recording method that allows users to track the origin and ownership of digital tokens [36]. Regardless of the stage of development of intelligent libraries, blockchain technology plays a crucial role. Blockchain technology can determine the first upload time of virtual digital resources in digital twin libraries through timestamps, prove the authenticity of digital resource owners through full-node authentication and tamper-proof functions, and effectively maintain the problem of identifying the property rights of digital resources in digital twin libraries. It can also realize the interlibrary communication of digital libraries, realize the direct cooperation between libraries-author, promote the circulation of digital resources, and realize the block-chaining of library resources by building a federated chain of digital twin libraries (De Ketelaere et al., 2022).

While blockchain is the core technology of the metaverse, non-fungible tokens (NFTs), or non-homogeneous tokens, are the core of blockchain. NFTs are protected and irreplaceable tokens on the blockchain. Since NFTs can map the value of real items, their uniqueness and irreplaceability will provide a reliable basis for people to map things in the real world to the metaverse [38]. In this way, users can obtain the corresponding economic rights and interests in the virtual and real worlds. In intelligent libraries, NFTs can provide non-fakeable identifiers for the collection resources, and these identifiers provide anchor values for the collection resources. The borrowing records, values, and literature sources of the collection resources in the intelligent library are encapsulated by the NFTs, and the circulation reading and promotion use of the books are recorded on the blockchain through the NFTs. The management of books will become easier and simpler, with information about the books being synchronized and displayed through the rapid scanning of the NFTs. The NFTs also empower digital twin libraries with a new form of intellectual property. The owner of the collection is recorded in the NFT, and patents are granted through uniquely identified NFTs. In fact, the first fiction book written using NFTs has also been circulated on the blockchain (Park & Kim, 2022). In the future, digital resources in the form of NFTs may also become part of intelligent library collections.

Application Prospects of Digital Twin Technology in Intelligent Libraries

Digital twin technology is the process of constructing an identical entity in a virtual space by digital means using a physical model to reflect its entire life cycle. Digital twin technology itself is the core technology of digital twin libraries. Therefore, for the purpose of introducing digital twin technology into intelligent libraries, the discussion topic presented in this chapter focuses on the first phase of intelligent libraries; i.e., how digital twin technology can achieve service optimization in unattended libraries.

The goal of the unattended library is to provide fully intelligent self-service for readers, and to improve the borrowing efficiency of readers by constructing a map for planar navigation through digital twin technology, and by selecting different areas of the library in the panoramic modeling and browsing the procedures of the resource catalog of that area. During the borrowing process, the digital twin technology is used to construct a dynamic display platform for the reading data of the collection resources, provide real-time resource recommendations for the readers, and create a twin portrait of the readers based on their reading habits and reading interests (De Ketelaere et al., 2022) to make precise recommendations for their next borrowing experience. Digital twin technology allows readers to obtain more information about the books during the borrowing process and presents the physical objects such as cultural relics and monuments in the books in a graphic form to optimize the reading experience. However, unmanned libraries do not eliminate the need for management staff. Twin librarians can provide consulting services and reading navigation for patrons, while managers can use the twin libraries to monitor and manage the library interior in real time, including facility maintenance and security testing. At the same time, they can enhance the situational awareness of service optimization by dynamically predicting the future resource demand of the library based on the resources borrowed from the twin libraries. In general, the application of digital twin technology in intelligent libraries is shown in Figure 1.

Figure 1. Application scenario of digital twin technology in intelligent libraries

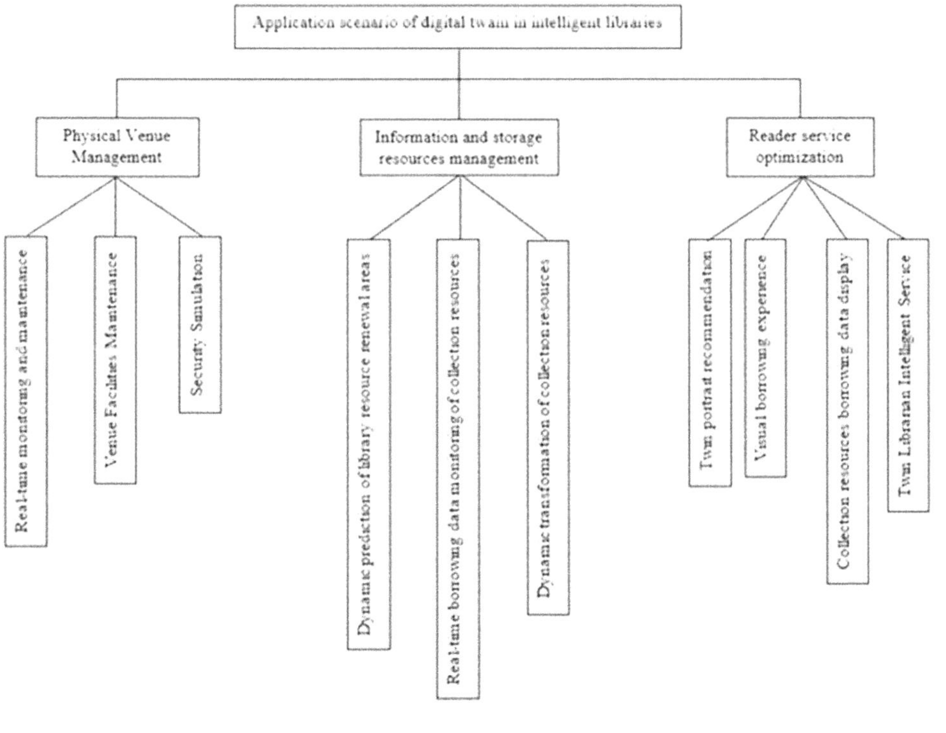

SERVICE TRANSFORMATION OF INTELLIGENT LIBRARIES FROM A METAVERSE PERSPECTIVE

The metaverse is not a monolithic development of blockchain, virtual reality and artificial intelligence, but a synthesis of technological innovations. Therefore, the development of metaverse in intelligent libraries should always be examined with holistic thinking. This chapter examines the application of metaverse in intelligent libraries in terms of collection resource management, shared reading space, special user services, simplified lending methods, optimized reading experience, and cultural promotion, and discusses the service transformation of intelligent libraries built by metaverse. Figure 2 shows the ecology of intelligent library under the perspective of metaverse.

Figure 2. Intelligent library ecology from a metaverse perspective

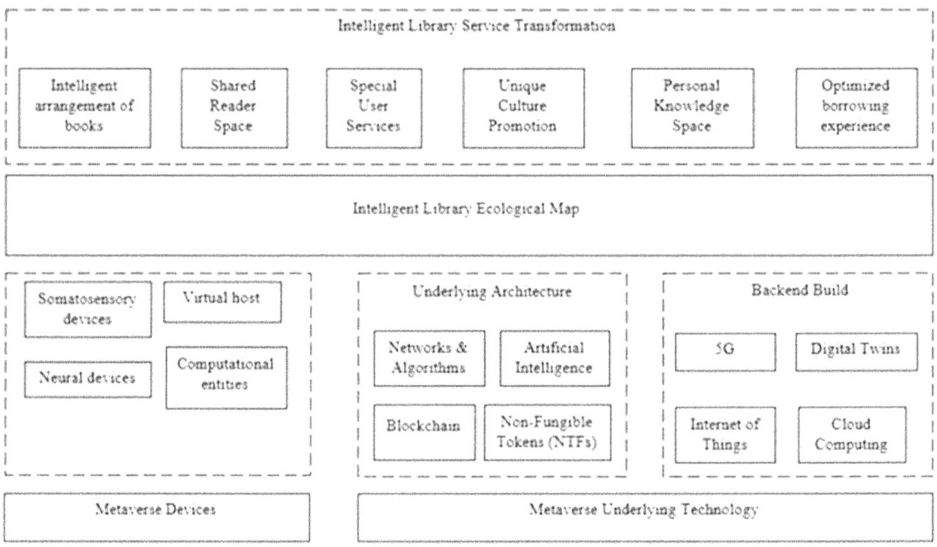

Collection Resource Management: Realizing Intelligent Arrangement of Books

In the metaverse, the clustering, fusion and reorganization of multimodal information resources are realized based on the intelligent library. A three-dimensional model of the library is established using digital twin technology. For the arrangement and classification of multimodal collection resources, the intelligent library is not classified by subject areas, but by readers' needs in the metaverse, and readers' personalized needs become the main way of arranging the collection resources. Based on readers' needs, the intelligent library built in the cloud searches for books in the whole domain and rearranges the books in the whole domain according to their relevance to readers' needs, with books with strong relevance placed at the top of the subject area and books with weak relevance placed at the end. The next time a reader performs a reading, the intelligent library has formed an arrangement of books that meets the reader's requirements. If the reader's needs change, the library performs a second search and rearranges them. The books in the library are arranged differently according to the reader's needs, so that each reader has his or her own intelligent library in the metaverse.

Read Experience Optimization: Enjoy Dynamic Real-World Reading

The evolution of social forms and iterations of literature repositories have led to the transmutation of library mandates and changes in service approaches (Bibri & Allam, 2022). Physical literature will always be fundamental to libraries, but the emergence of metaverse has facilitated the integration of virtual and reality, allowing the information and knowledge implicit in physical literature to be presented to readers in a visual form. The traditional applications of virtual reality technologies are all local virtual and local reality in nature, with problems such as lack of immersion and low frame rates. The emergence of metaverse enables readers to achieve full immersion in the reading process through human-computer interaction technologies such as digital twins, spatial perception, motion capture, gesture recognition, and digital twins, allowing readers to interact with the characters in the book and the knowledge in the book to be presented in an audio-visual manner. The high degree of immersion and interactivity not only satisfies the freshness of readers' reading, but also provides a new path for library culture dissemination.

Cultural Exhibition Promotion: A Novel Operational Model

In the Internet era, the influence of the traditional promotion mode of libraries is gradually decreasing. Only by finding more innovative and unique promotion methods can we attract the public's attention. In metaverse, library managers can spread culture through unique virtual library models, setting library themes regularly, and conducting virtual book exhibitions. At the same time, NFTs on the blockchain can be used to form a value bond with the collection resources for cultural promotion. In fact, this promotion model has appeared in the art world, souvenirs, and fashion world (Tromp, 2022). It is both possible and necessary for libraries to achieve cultural promotion by producing their own unique cultural and creative products and souvenirs, which are tied to NFTs. The introduction of blockchain technology into the art world increases the attention of readers by constantly updating the library's themes in the metaverse and strengthening thematic connections with other fields.

Special User Services: Providing Cultural Services for All

According to the 14th Five-Year Plan for the Protection and Development of Persons with Disabilities released by the State Council in 2021 (Chen & Lee, 2021), there are tens of millions of persons with disabilities in China, most of whom are visually and hearing impaired. Libraries, as public cultural service institutions, are unable to provide services for disabled patrons. In most cases, people with disabilities are limited by many conditions and are mostly in an information disadvantaged position. In order to promote the equalization of public cultural services, it is necessary for public libraries to explore a new model of public cultural intelligent services for people with disabilities (Pamucar, Deveci, Gokasar, Tavana, & Köppen, 2022).

For people with disabilities, Metaverse can provide multiple service models for libraries. For visually impaired patrons, libraries can use intelligent robots to provide voice-assisted services to ensure that visually impaired patrons have a voice to follow throughout the library. At the same time, the intelligent robot can introduce them to the book categories and book contents of each shelf and avoid crowded paths by calculating real-time foot traffic (Erdei, Krakó, & Husi, 2022). For patrons with hearing impairment, visual reinforcement services can be provided using metaverse technology, visual information prompting and visual information retrieval using augmented reality, and text and images can be superimposed on real objects in the library through AR devices. Another type of service is to create an intelligent library in the metaverse that enables people with disabilities to borrow books from their homes. With visual simulation through devices such as holographic projection, odor simulation through devices such as odor sensors, physical simulation through somatosensory devices, and mental simulation through brain-computer interfaces, disabled readers can visit the library in the metaverse, such as creating virtual reader images, lending and returning books, and experiencing virtual reading. It is possible to visit the library in the metaverse, such as creating a virtual reader image, lending and returning books, experiencing a virtual reading community, and recording virtual notes.

Personal Knowledge Space: Readers Become the Subject of Knowledge Creation

The metaverse itself is a continuous open space. Openness and sharing enable metaverse to support user-generated content. This means that libraries that use metaverse as an ecology also generate new ways of knowledge innovation. Patrons can use the collection resources in the library to create artifacts in the metaverse space, forming a personal knowledge space of the metaverse in the library. This space is a virtual space built in the metaverse in the library, and readers create and publish native digital knowledge through this virtual knowledge space, which is created entirely based on virtual things, such as novels and movie works describing virtual worlds. The personal knowledge space of different readers will become an important part of the library in the metaverse, and then the digital creations of readers in the knowledge space are associated with the collection resources of the library in the metaverse with high intensity, which both strengthens the participation of readers in the library in the metaverse and provides a new impetus for library transformation (Shen et al., 2021).

Shared Reading Space: A Connected Online Library for All

Second life social platform brings us the prototype of metaverse, but the core of Second life is the combination of stereoscopic modeling and internet. Readers can only access the virtual library through the button and mouse device, lacking physical interaction, and the reading experience is not much different from the traditional digital library. The intelligent library of metaverse is a high integration of rules and algorithm operation, blockchain and digital currency, and virtual reality, and builds a fully simulated digital twin library based on the concept of digital twin. Readers can enjoy library services at home through somatosensory devices and virtual reality technology. The library in Second life provides ideas for the construction of meta-universe libraries, such as the digital transformation of the collection and the integration of open resources (Bibri, 2022). For the establishment of a metaverse library, the library in Second life can be used as a framework to improve the disadvantages of real-time rendering, physical interaction, and low latency, which are lacking in Second life. Through better underlying algorithmic techniques and physical modeling, the metaverse library can become a universal online library. In this universal online library, a decentralized and multi-level shared space can be established, and the service boundaries of this shared space can be enriched to add more diverse virtual scenarios of learning, office, social, and entertainment to make the service model of the metaverse library more comprehensive.

CONCLUSION

The emergence of the metaverse has brought new ideas for the future change of library landscape. Although metaverse has not yet been brought to fruition, the underlying technology of metaverse is gradually penetrating into our daily lives. However, while the emergence of metaverse may cause human beings, or even the whole society, to face changes in rules, it also undoubtedly brings new opportunities to libraries. In this regard, libraries should always pay attention to new emerging technologies and need to seriously think about the rules, processes, methods, procedures and underlying logic of metaverse applications in libraries, while actively embracing new technologies and continuously introducing their underlying technologies into the construction of intelligent libraries. If metaverse becomes a reality, then intelligent libraries will also be transformed as a result.

REFERENCES

Ahn, S., Kim, J., & Kim, J. (2022). The bifold triadic relationships framework: a theoretical primer for advertising research in the metaverse. *Journal of Advertising*, *51*(5), 592–607. doi:10.1080/00913367.2022.2111729

Allam, Z., Bibri, S., Jones, D., Chabaud, D., & Moreno, C. (2022). Unpacking the ‘15-minute city’ via 6g, IoT, and digital twins: towards a new narrative for increasing urban efficiency, resilience, and sustainability. *Sensors (Basel)*, *22*(4), 1369. doi:10.339022041369 PMID:35214271

Allam, Z., Sharifi, A., Bibri, S., Jones, D., & Krogstie, J. (2022). The metaverse as a virtual form of smart cities: opportunities and challenges for environmental, economic, and social sustainability in urban futures. *Smart Cities*, *5*(3), 771–801. doi:10.3390martcities5030040

Alpala, L., Quiroga-Parra, D., Torres, J., & Peluffo-Ordóñez, D. (2022). Smart factory using virtual reality and online multi user: towards a metaverse for experimental frameworks. *Applied Sciences (Basel, Switzerland)*, *12*(12), 6258. doi:10.3390/app12126258

Baghalzadeh Shishehgarkhaneh, M., Keivani, A., Moehler, R., Jelodari, N., & Roshdi Laleh, S. (2022). Internet of things (IoT), building information modeling (BIM), And Digital Twin (DT) in construction industry: a review, bibliometric, and network analysis. *Buildings*, *12*(10), 1503. doi:10.3390/buildings12101503

Bibri, S. (2022). The social shaping of the metaverse as an alternative to the imaginaries of data-driven smart cities: a study in science, technology, and society. *Smart Cities*, *5*(3), 832–874. doi:10.3390martcities5030043

Bibri, S., & Allam, Z. (2022). The metaverse as a virtual form of data-driven smart urbanism: on post-pandemic governance through the prism of the logic of surveillance capitalism. *Smart Cities*, *5*(2), 715–727. doi:10.3390martcities5020037

Bolger, R. (2021). Finding Wholes in the Metaverse: Posthuman Mystics as Agents of Evolutionary Contextualization. *Religions*, *12*(9), 768. doi:10.3390/rel12090768

Buhalis, D., Lin, M., & Leung, D. (2022). Metaverse as a driver for customer experience and value co-creation: implications for hospitality and tourism management and marketing. *International Journal of Contemporary Hospitality Management*. doi:10.1108/IJCHM-05-2022-0631

Cappannari, L., & Vitillo, A. (2022). XR and metaverse software platforms. In Roadmapping Extended Reality (pp. 135-156). doi:10.1002/9781119865810.ch6

Chen, Q., & Lee, S. (2021). Research status and trend of digital twin: visual knowledge mapping analysis. *International Journal of Advanced Smart Convergence*, *10*(4), 84–97. doi:10.7236/IJASC.2021.10.4.84

Dahan, N., Al-Razgan, M., Al-Laith, A., Alsoufi, M., Al-Asaly, M., & Alfakih, T. (2022). Metaverse framework: a case study on e-learning environment (ELEM). *Electronics (Basel)*, *11*(10), 1616. doi:10.3390/electronics11101616

Dai, Y., Wang, J., & Gao, S. (2022). Advanced electronics and artificial intelligence: must-have technologies toward human body digital twins. *Advanced Intelligent Systems*, *4*(7), 2100263. doi:10.1002/aisy.202100263

De Ketelaere, B., Smeets, B., Verboven, P., Nicolaï, B., & Saeys, W. (2022). Digital twins in quality engineering. *Quality Engineering*, *34*(3), 404–408. doi:10.1080/08982112.2022.2052731

Duan, H., Li, J., Fan, S., Lin, Z., Wu, X., & Cai, W. (2021). *Metaverse for Social Good: A University Campus Prototype*. Paper presented at the Proceedings of the 29th ACM International Conference on Multimedia, Virtual Event, China. 10.1145/3474085.3479238

Eom, S. (2022). The emerging digital twin bureaucracy in the 21st century. *Perspectives on Public Management and Governance*, *5*(2), 174–186. doi:10.1093/ppmgov/gvac005

Erdei, T., Krakó, R., & Husi, G. (2022). Design of a digital twin training centre for an industrial robot arm. *Applied Sciences (Basel, Switzerland)*, *12*(17), 8862. doi:10.3390/app12178862

Faraboschi, P., Frachtenberg, E., Laplante, P., Milojicic, D., & Saracco, R. (2022). Virtual worlds (Metaverse): from skepticism, to fear, to immersive opportunities. *Computer*, *55*(10), 100–106. doi:10.1109/MC.2022.3192702

Hassani, H., Huang, X., & MacFeely, S. (2022). Impactful digital twin in the healthcare revolution. *Big Data and Cognitive Computing*, *6*(3), 83. doi:10.3390/bdcc6030083

Jamil, S., Rahman, M., & Fawad, M. (2022). A comprehensive survey of digital twins and federated learning for industrial internet of things (IIoT), internet of vehicles (IoV) and internet of drones (IoD). *Applied System Innovation*, *5*(3), 56. doi:10.3390/asi5030056

Jin, J., Xu, H., & Leng, B. (2022). Adaptive points sampling for implicit field reconstruction of industrial digital twin. *Sensors (Basel)*, *22*(17), 6630. doi:10.339022176630 PMID:36081088

Keshmiri Neghab, H., Jamshidi, M., & Keshmiri Neghab, H. (2022). Digital twin of a magnetic medical microrobot with stochastic model predictive controller boosted by machine learning in cyber-physical healthcare systems. *Information (Basel)*, *13*(7), 321. doi:10.3390/info13070321

Maddahi, Y., & Chen, S. (2022). Applications of digital twins in the healthcare industry: case review of an IoT-enabled remote technology in dentistry. *Virtual Worlds*, *1*(1), 20–41. doi:10.3390/virtualworlds1010003

Magalhães, L., Magalhães, L., Ramos, J., Moura, L., de Moraes, R., Gonçalves, J., Hisatugu, W. H., Souza, M. T., de Lacalle, L. N. L., & Ferreira, J. (2022). Conceiving a digital twin for a flexible manufacturing system. *Applied Sciences (Basel, Switzerland)*, *12*(19), 9864. doi:10.3390/app12199864

Nguyen, T. (2022). Toward human digital twins for cybersecurity simulations on the metaverse: ontological and network science approach. *JMIRx Med*, *3*(2), e33502. doi:10.2196/33502

Njoku, J., Nwakanma, C., Amaizu, G., & Kim, D. (2022). Prospects and challenges of Metaverse application in data-driven intelligent transportation systems. *IET Intelligent Transport Systems*. doi:10.1049/itr2.12252

Pamucar, D., Deveci, M., Gokasar, I., Tavana, M., & Köppen, M. (2022). A metaverse assessment model for sustainable transportation using ordinal priority approach and Aczel-Alsina norms. *Technological Forecasting and Social Change*, *182*, 121778. doi:10.1016/j.techfore.2022.121778

Park, S., & Kim, Y. (2022). A metaverse: taxonomy, components, applications, and open challenges. *IEEE Access: Practical Innovations, Open Solutions*, *10*, 4209–4251. doi:10.1109/ACCESS.2021.3140175

Popescu, D., Dragomir, M., Popescu, S., & Dragomir, D. (2022). Building better digital twins for production systems by incorporating environmental related functions—literature analysis and determining alternatives. *Applied Sciences (Basel, Switzerland)*, *12*(17), 8657. doi:10.3390/app12178657

Ruiz Mejia, J., & Rawat, D. (2022). Recent advances in a medical domain metaverse: status, challenges, and perspective. Paper presented at the *2022 Thirteenth International Conference on Ubiquitous and Future Networks (ICUFN)*. IEEE. 10.1109/ICUFN55119.2022.9829645

Salem, T., & Dragomir, M. (2022). Options for and challenges of employing digital twins in construction management. *Applied Sciences (Basel, Switzerland)*, *12*(6), 2928. doi:10.3390/app12062928

Shen, B., Tan, W., Guo, J., Zhao, L., & Qin, P. (2021). How to promote user purchase in metaverse? a systematic literature review on consumer behavior research and virtual commerce application design. *Applied Sciences (Basel, Switzerland)*, *11*(23), 11087. doi:10.3390/app112311087

Siyaev, A., & Jo, G. (2021). Towards Aircraft Maintenance Metaverse Using Speech Interactions with Virtual Objects in Mixed Reality. *Sensors (Basel)*, *21*(6), 2066. doi:10.339021062066 PMID:33804253

Tromp, J. (2022). Extended Reality & The Backbone: Towards a 3D Mirrorworld. In Roadmapping Extended Reality (pp. 193-227).

Ukko, J., Saunila, M., Nasiri, M., Rantala, T., & Holopainen, M. (2022). Digital twins' impact on organizational control: Perspectives on formal vs social control. *Information Technology & People*, *35*(8), 253–272. doi:10.1108/ITP-09-2020-0608

Vishkaei, B. (2022). Metaverse: A New Platform for Circular Smart Cities. In P. De Giovanni (Ed.), *Cases on Circular Economy in Practice* (pp. 51–69). IGI Global. doi:10.4018/978-1-6684-5001-7.ch003

Wang, M., Yu, H., Bell, Z., & Chu, X. (2022). Constructing an Edu-Metaverse Ecosystem: A New and Innovative Framework. *IEEE Transactions on Learning Technologies*, 1–13. doi:10.1109/TLT.2022.3226345

Wang, Y., Su, Z., Zhang, N., Xing, R., Liu, D., Luan, T., & Shen, X. (2022). A survey on metaverse: fundamentals, security, and privacy. *IEEE Communications Surveys and Tutorials*, 1–1. doi:10.1109/COMST.2022.3202047

Xiong, M., & Wang, H. (2022). Digital twin applications in aviation industry: A review. *International Journal of Advanced Manufacturing Technology*, *121*(9), 5677–5692. doi:10.100700170-022-09717-9

Yang, C., Tu, X., Autiosalo, J., Ala-Laurinaho, R., Mattila, J., Salminen, P., & Tammi, K. (2022). Extended reality application framework for a digital-twin-based smart crane. *Applied Sciences (Basel, Switzerland)*, *12*(12), 6030. doi:10.3390/app12126030

Zhang, Z., Wen, F., Sun, Z., Guo, X., He, T., & Lee, C. (2022). Artificial intelligence-enabled sensing technologies in the 5g/internet of things era: from virtual reality/augmented reality to the digital twin. *Advanced Intelligent Systems*, *4*(7), 2100228. doi:10.1002/aisy.202100228

Chapter 9
Open Innovation and Crowdsourcing in the Public Sector in Portugal: The Participatory Budget

Susana Bernardino
https://orcid.org/0000-0002-1639-3553
CEOS, ISCAP, Polytechnic of Porto, Portugal

Vera Rocha
ISCAP, Polytechnic of Porto, Portugal

ABSTRACT

Participatory budgeting is a recent mechanism that is related to open innovation and crowdsourcing in public entities and allows the participation of society in governmental interests. This chapter aims to analyse the use of open innovation and crowdsourcing tools in the public sector in Portugal, in the form of participatory budgets, and also to understand the characteristics of the participatory budget processes used, as well as the motivations for their implementation and the difficulties faced. For these purposes, the investigation uses a quantitative methodology, by means of a questionnaire sent to Portuguese municipalities. The research carried out reveals influences between the size of municipalities and the use of participatory budget and its characteristics. The investigation also highlights the main motivations for its implementation, the greatest difficulties experienced, and the perceived barriers to non-implementation.

INTRODUCTION

Innovation is seen as the predisposition of individuals to experiment, be engaged in creative processes and explore new ideas (Moss et al., 2011). Overall, innovation can be seen a process that aims, both technological and tasks development improvement. These improvements can occur in all fields of organizations, such as products, processes, markets or structure (Pinheiro, 2009). In recent years, due to

DOI: 10.4018/978-1-6684-8934-5.ch009

the strong changes experienced in the market, there have been significant changes in the importance attached to the field of innovation and, specifically, within the scope of the organizations management. In this context, new paradigms have emerged in the innovation field, being increasingly common the use of open innovation techniques (Chesbrough, 2003; Pinheiro, 2009).

Open innovation is understood as the use of networks of people with some kind of skills that are available to collaborate with various entities in order to obtain new ideas to innovate. Through open innovation, companies can have access to the know-how embedded in their network, which allows them to innovate and therefore differentiate themselves from other companies on the market. In addition, the interaction and collaboration with the other elements involved in the network allows the ideas to stand out in the markets (Pinheiro, 2009). The network may be composed of elements that are internal or external to the organisation.

Crowdsourcing is also a new concept, increasingly used by organisations, in which different stakeholders get involved in mobilising different types of resources, including informational ones and the generation of ideas for problem solving and strategy definition (Barbosa, 2015).

Participatory budgeting, in turn, is a concept embodied in a process in which the population has a voice in society and in the decisions that are taken with public resources in a given location. The participatory budget has been increasingly used by the public sector all over the world and also in Portugal, which is one of the most preeminent countries in the use of participatory budgets (Sampaio, 2016).

This chapter aims to study the extent to which Portuguese municipalities use participatory budget and, in those cases, how participatory budgeting processes are being implemented. The chapter also seeks to understand the main benefits and difficulties experienced by Portuguese municipalities in using participatory budgets.

The choice of this topic arose primarily due to its importance in the public sector. This is a sector that has several various problems to address and for which participatory budgeting, as an instrument for applying open innovation and crowdsourcing, can have a great impact.

To achieve the proposed research objectives, the chapter is structured in four sections. The first section of the chapter explores the literature review on open innovation, crowdsourcing and participatory budgeting. The second section is describes the methodology chosen for the research. Subsequently, section three presents the results obtained through the quantitative study carried out. Finally, the conclusions, contributions of the study and clues for future research are presented.

1 THEORETICAL FRAMEWORK

1.1 *Open Innovation*

1.1.1 Historical Context and Definition of the Concept

Innovation is important for businesses in terms of capacity to generate a competitive advantage, and is also capable of benefiting society in general (Ferreira, 2018).

For Schumpeter (1934), innovation is a new combination of pre-existing knowledge and skills, which supports the economic development of society. As highlighted by the Organisation for Cooperation and Development, there is a substantial body of evidence that innovation is the dominant factor in national economic growth and international trade patterns (Dias et al., 2018, p.79).

Innovation is a process that involves technological improvements and as well as the optimization on the way tasks are carried out. These improvements can occur in all fields of the company, whether products, processes, markets or organizational structure (Pinheiro, 2009).

In recent years there have been substantial changes in the importance attributed by organisations to innovation, as well as in the way in which it is undertaken. In this context, and as argued by Chesbrough (2003), new paradigms arose in terms of innovation, including the concept of open innovation.

Open innovation is a relatively recent concept. Not being seen as just one thing, it encompasses several elements and can be used in various ways. As such, according to Huizingh (2011) open innovation has an eminently multidimensional nature . The concept of open innovation is seen as a new paradigm in which companies use internal and external knowledge to implement innovative ideas with better technology (Baggio, 2019).

In 1900, the concept of open innovation did not exist in companies, mostly industrial. As science evolved and time went by, companies began to invest in development, through the creation of laboratories. In these laboratories, new ideas emerged, although innovation occurred only inside companies, which could be called as closed innovation. Accordingly, until 1980, companies could create ideas mainly through internal sources (Freire et al., 2022).

In 2000, according to Freire et al. (2022), several forms of collaboration with companies were established through cooperation with other companies, for example, through the establishment of partnerships between companies and universities. In 2003, through Henry Chesbrough, the concept of open innovation arises, which consists of the use of ideas and innovations developed through partnerships by companies, in order to obtain new ideas through external stakeholders, to achieve competitive advantage in the market (Freire et al., 2022).

As with the vast majority of concepts, there is no single and agreed definition of open innovation. The various definitions share similar aspects, although they express different viewpoints. Junior et al. (2007), consider that open innovation is the search for solutions/innovative ideas through a network of people, where various ideas are shared.

According to Chesbrough (2011) the concept of open innovation can be understood as the use of knowledge contributions as a tool to accelerate internal innovation and outputs to expand in different markets. For the author, open innovation assumes that firms can have both internal and external ideas, and could use different approaches to reach the market in a technologically innovative way.

For Bogers et al. (2018), open innovation is innovation process that is widespread in accordance with the flows of knowledge that come from beyond organizational boundaries. In this sense, open innovation could also be seen as the practice of creating collaborative networks to generate ideas through collective connection, using external knowledge in a collaborative and open way (Freire et al., 2022). In the open innovation concept is implicit the innovation through external elements, where any product can be subject of innovation, through internal and external elements to the organization.

The use of open innovation focuses on the development of new partnerships in the field of innovation, with elements that are internal and/or external to companies, in order to leverage the creation of new innovative products or services, in a faster way in companies (Dias et al., 2018).

1.1.2 Use of Open Innovation

Open innovation embrace the opportunity for creativity in new products and services and for the modification of aspects in society, with the collaboration of elements external to the organisations (Freire et al., 2022).

According to Freire et al. in the article prepared for the DAT Journal in 2022, all companies will start using open innovation in the future in their processes.

Chesbrough and Crowthe (2006) consider that the competitive advantage of companies today comes from the methodologies they use, such as open innovation. In this sense, companies are not only dependent on their own research and development departments to create and/or improve their products and services. By using new approach and innovative ideas, the development of products and services occurs more spontaneously and more quickly. The knowledge about the customers needs will be more accurate and closer to the companies due to open innovation.

In order to gain competitive advantage over other organisations, organizations have made some efforts to involve various stakeholders in innovation processes. This collaborative process may include different types of stakeholders in the creative process, such as customers, suppliers or other elements, aiming to generate the best ideas.

The concept of open innovation can be interpreted as the opposite of closed innovation, where new ideas arise only from processes developed within the companies themselves. With the evolution of times and the current social and economic changes, it was crucial to make these innovation processes more open, with greater disclosure to the outside, as this methodology makes possible to obtain new ideas and collaboration through more (geographically) distant people and people with different ways of thinking (Huizingh, 2011). According to Gassman, (2004), the "do-it-yourself" mentality in innovation management has become outdated.

To use open innovation in companies it is necessary to build and define a process with all the necessary procedures to be carried out, that could be subdivided into two main phases. The first phase focus on the necessity to implement the tasks / policies that lead to the introduction of new ideas in the company. The second phase, in turn, is related to the need to conceive the development of the projects of open innovation and how to motivate stakeholders (Huizingh, 2011).

In the development of an open innovation process, is necessary to make decisions about several aspects, such as the responsible, the processes to innovate or the goals to achieve (Huizingh, 2011).

Fetterhoff and Voelkel (2006) present five phases for the introduction of open innovation in companies. These phases are divided between opportunity search, evaluation of market potential, recruitment of elements for the development of ideas, value creation and extension of the innovative offer. In the first phase, companies must search within their own organisation for products or processes that could be improved. Then, in the second phase, they have to check if existing ideas in the market can be used or improved, even in a different way, and how the company can innovate to meet market expectations. In the third phase, after the previous verifications, it is necessary to disseminate the need for innovation within the company and also in platforms that support the search of elements for the development of the desired ideas. With the recruitment of the necessary elements and the clarification of the defined idea, the next step is the creation of value for the company, that occurs through the use of innovative strategies to increase the value of a product or service for those who use it. Finally, the authors mention the stage of extension related of the innovative offer, where innovation must be thought of as a whole and not just as an isolated case, This aims to achieve complementarity of innovative products and services to others.

Wallin and Von Krogh, (2010) also suggest five different stages for open innovation. The first stage is related to the definition of the innovation process, where is necessary to define how the innovation will happen, subsequently, the second stage is related to the identification of relevant contents, i.e., in the company are identified the elements with greater relevance to be used in an innovation process. In a third step, the selection of integration mechanisms, are chosen the way and means through the contributing elements of innovation will be integrated in the company. After that, in the next step are created the mechanisms of innovation management and established the mechanisms to be used for innovation. In the fifth stage, is necessary to establish the balance between incentives and management / control systems.

When companies decide to work with open innovation they have to be aware of the objectives to be achieved, the capabilities they intend to develop and the elements and materials needed for open innovation to happen (Andrade, 2015).

The open innovation model of Fetterhoff and Voelkel, (2006) focuses on the contribution given to companies by external elements, through six dimensions: (i) company, (ii) customer utility, (iii) competition, (iv) market, (v) capital cost and (vi) copyright. With regard to the dimensions "company", it is necessary to take into consideration the strategy defined by the company in innovation matters, which can be internal innovation, external innovation or a combination of both ; the usefulness of the new products or services for the customers is a very relevant aspect, since they are the purchasers of the products, so they must be respected; the attention directed towards the market follows the customer's thinking, so it is necessary to take into account the type of market where the company operates to adapt the innovation strategies; the cost of capital is one of the dimensions with great impact on companies, since they seek the best innovation with the lowest possible cost; for this reason, using open innovation produces advantages; the last dimension is related to copyright, although the idea appears on behalf of a company, it is essential that is clarified how the idea arose to safeguard the network of partners and possible copies (Huizingh, 2011).

Globalization and the diffusion of information and communication technologies in society have enabled innovation to occur in an open manner in various sectors of activity (Pinheiro, 2009). For the use of open innovation it is essential to establish a network of partners and people with interest in collaborating with new ideas for the company, although this may be a slow and lengthy process (Huizingh, 2011). There are authors who argue that this process should be carried out using an intermediary/mediator, where a bridge is established between the interested elements and the company (Lee et al., 2010). Spithoven et al. (2010) also argue that intermediaries are an asset for all types of companies.

Some examples of open innovation methodologies relate to the study of the external context, the search for expert people to solve existing problems, the formation of joint ventures, and participation in networking and networks in order to learn about new ideas (Hagel and Brown, 2008).

Working open innovation in companies, by using networks of people, is something useful, however, not all individuals have innovation capabilities. For this reason, companies must carefully select the elements that will be part of their network. As argued by Andrade (2015), companies that decide to pursue collaborative innovation as a strategy need to be able to develop the capabilities, structures and support processes in a collaborative approach. In fact, in open innovation the relationships between people in the network and the companies, could be good and fruitful just if the objectives and the rules are clearly defined.

1.1.3 Advantages and Limitations in the Use of Open Innovation

Using open innovation is using a new business model as a source of value creation and capture (Dias et al., 2018, p. 83). The use of open innovation encompass some advantages, such as the decrease in the development and launching time of new products, Indeed, by using open innovation, ideas emerge faster and without the need for a slow and lengthy procedure. Open innovation also helps to mitigate the internal risks of innovation, since with the use of innovation partners, the risk of failure is lower and therefore the risk is reduced. Other advantaged could also be related to the improvement in the protection of intellectual property, as there is a record of "ownership" of innovations that prevent possible frauds. Open innovation increase in the company's ability to innovate in the long term and focus on topics such as sustainability (Andrade, 2015).

According to Grassmann and Enkel (2004), the benefits of open innovation is related to the access to knowledge in a complementary way and to unique resources that the company would otherwise not be able to access.

There are several companies that have used open innovation to develop their products and have obtained good results. With the use of open innovation, companies are able to take advantage of already existing technologies abroad, without this implying a great effort at the economic level for the research and development department of the company itself, achieving greater profitability of its resources (Ferreira, 2018). For this reason companies are increasingly looking for cooperative relationships as an alternative in the process of improving their competencies and, consequently, performance (Andrade, 2015, p.40).

Furthermore, open innovation is also able to leverage the innovation process of companies and improving their competitiveness in the market.

Similar to other management tools, open innovation also has some limitations. The success of the use of open innovation will depend on the context/location where it is inserted. This success will depend on internal and external issues. The factors that influence the success of open innovation in companies are not strictly related to internal elements, such as management forms, or greater or lesser openness towards the exterior. Instead the success is also related to external factors, such as the knowledge of the individuals who contribute the ideas, their willingness to participate and the market needs.

Other limitation of open innovation is related to the possible loss of know-how, high costs of coordination, the reduction of control over the activities developed, the difficulty in getting good innovation partners, the large expenditure of time and the lack of financial resources (Andrade, 2015).

According to Dahlander and Gann (2010), firms may start in a situation of cognitive constraint, in which it will not be possible to absorb all the ideas that the partner network has for innovating in the firm.

Another less favourable aspect is related to the possibility of neglect internal capabilities (Spithoven et al., 2013). Moreover, this fact is especially dangerous since research carried out enhances the ability of companies to use external knowledge inside of the company, being of central importance that they do not completely neglect their internal competencies (Ferreira, 2018).

When companies become dependent on innovation from the external context and are constantly transmitting information to their network, they can harm themselves by passing too much irrelevant information, hat could lead to the loss of control of professional secrecy also to lead to the loss of control of their core competencies" (Andrade, 2015).

1.2 Crowdsourcing

1.2.1 Historical Context and Definition of the Concept

Crowdsourcing emerges as an activity that trigger the productive capabilities of consumers in a community for a specific purpose. The first person to refer to the term "crowdsourcing" was Jeff Howe, through an article published in 2006. The author defined crowdsourcing as a new online business model which foster creative solutions of a network of individuals through a call for proposals (Brabham, 2008).

The definition presented by Jeff Howe in 2006 meets the main idea of the concept. Accordingly, crowdsourcing is seen as a range of approaches that have in common to ask some contribution from the population. These contributions may consist only in the collection of feedback, in the creation of a product or service, in helping to solve one or several problems, among other possibilities. These contributions are essential, as they bring together different perspectives that would not be possible to obtain if they were limited to the organisation by itself only (Brabham, 2008). The word results from the junction of two other words, "crowd" and "outsourcing", that literally make crowdsourcing mean to resort to the crowd (Fernandes, 2012).

With the 2008 financial crisis and the instability experienced in recent years, obtaining funding has involved major difficulties for many organisations, whether private or public. In parallel with the economic crisis, the strong globalisation results in a scenario of greater competitiveness for companies, so organisations need new strategies. These new strategies could involve, for example, new production methods, differentiating and sometimes even low-cost processes. To ensure companies obtain what they need, the use of crowdsourcing is a priority for the them, as it could allow them to better achieve the established objectives. As mentioned by Fernandes (2012), many organizations are already using crowdsourcing as a business model and/or as an occasional differentiation strategy.

Jeff Howe, in 2006, considered that crowdsourcing could be defined as the act of a company or institution taking a function once performed by employees and subcontracting it to an undefined (and usually large) network of people in the form of an open call. This can take the form of peer production (when work is carried out collaboratively), but is also often performed by isolated individuals. The crucial prerequisite is the use of the open call format and the large network of potential workers.

A few years later, Estellés (2012), defines crowdsourcing as a type of online participatory activity in which an individual, institution, non-profit organization or company proposes to a group of individuals of varied knowledge, heterogeneity, and number, the voluntary performance of a task through a flexible open call. The crowd could answer to the call, bringing its labour, money, knowledge and/or experience, for the task development.

Pederson et al. (2013) have a different view, according to which crowdsourcing is seen as a model of collaboration enabled by people-centred web technologies to solve individual, organisational and social problems, using a dynamically formed crowd of interested people who respond to an open call for participation.

There are several definitions of crowdsourcing, and following the evolution of society, the concept is changing overtime, both at technological level and in the ways companies use it. Even so, there is a common understanding that crowdsourcing consists of a strategy to attract individuals interested in collaborating with companies. This strategy can be applied in different ways and to solve various problems in multiple business areas (Barbosa, 2015).

Focusing the analysis on the origin of the word, we observe some similarities to the concept that is part of its formation – outsourcing; the difference lies in who is used. On the one hand, in the case of outsourcing, the organisations resort to the help of other organisations, as it requests part of the tasks/activities of the company to be developed externally by another entity, such as, for example, carrying out accounting or manufacturing of some component; on the other hand, crowdsourcing resorts to society or to the communities to obtain collaborations at lower costs (Hammon et al., 2012).

With the development of information and communication technologies, the use of crowdsourcing has been increasing. Indeed, due to the greater availability and quality of infrastructures and collaboration/cooperation spaces, it is easier for any interested person to access the companies' information and engage in the crowdsourcing process, which will be favourable to the emergence of a greater number of crowdsourcing cases (Hammon & Hippner, 2012).

Online is tending to be increasingly used in society, both for personal and professional purposes. In fact, more and more companies use and take advantage of online communities to create new products, to solve problems or to develop new ideas. Online users feel they play an important role, being more and more active in communities and organisations. Companies that resort to this form of resource gathering are making use of crowdsourcing (Fernandes, 2012).

Crowdsourcing tries to bring consumers and organisations closer together and therefore allowing to add more value to the company. However, to enhance the proximity, it is necessary to establish rules or create certain guidelines, seen as a competitive advantage for the company that puts them into practice.

The aforementioned proximity does not only occur between consumers and private companies, it also occurs when crowdsourcing is applied by public entities. As an example, when municipalities decide to use crowdsourcing, they encourage society's interest in participating in these activities and, consequently, increase the connection between both parties. In this case, citizens get closer to their local governments, as they become more aware of the needs and act in areas identified as priorities by the citizens themselves.

1.2.2 Crowdsourcing as a Process

A crowdsourcing process occurs, firstly, through a previous participation invitation to those interested in collaborating, the network members. Three stakeholders are implicit in this process, the crowd (or network), the organizations and the (crowdsourcing) platforms (Barbosa, 2015).

Herein, the network can be seen as a composition of several elements responsible for the execution of the tasks required to solve the problem found (Barbosa, 2015). Therefore, organizations are all entities that intend to use crowdsourcing as a means of solving their problems. The crowdsourcing platforms, in turn, are the virtual places, that build the bridge between the needs of the entities and the proposals presented by the network. Each entity should decide the best platform for itself (Barbosa, 2015). The information technologies represent an important element in the dynamization of this process. As mentioned by Barbosa (2015), the internet is an essential prerequisite of this process because of its characteristics, such as speed, reach or anonymity, that allows increasing the quality, quantity and space of cooperation, coordination and generation of ideas.

Fernandes (2012) identified four steps for the implementation of crowdsourcing. The first step is the identification of the area where external improvements need to occur. This stage is followed by the disclosure of the need on online platforms and the definition of the assumptions / rules of collaboration; subsequently, the third step is the definition of the collaboration methodology (by parts or as a whole) and, finally, the evaluation of the proposals submitted is carried out.

For crowdsourcing to work properly within companies, people (the crowd) must be motivated to participate. According to Osamuyimen (2009), there should be two types of motivation, the motivation by itself and the motivation caused by the remuneration. Extrinsic motivation (involving a given reward) consists of a call to the crowd that can involve tangible prizes, not necessarily money. Intrinsic motivation, as do not involve reward, have to seek the crowd through "social goals" as people crave to participate because they are willing to work towards altruism (Melo et al., 2015). Crowdsourcing is able to facilitate the implementation of good practices in communities through civic participation (Brabham, 2013).

1.2.3 Advantages and Disadvantages Associated With the Use of Crowdsourcing

Barbosa (2015) highlighted some main benefits that may be derived for organizations by using crowdsourcing methods. These benefits could be related to quality, cost reduction, risk reduction, and access to a specific group of people. Regarding quality, the benefits relate to products and services being of higher quality, leveraged though the feedback received from those who consume the products or use the services of a given company/organisation. Because of this consumer participation, the improvement process becomes faster, better in terms of quality as they can meet more effectively the needs of its users and can also lead to lower costs. The cost reduction is one of the attractive factors of using crowdsourcing, although the costs vary depending on the type of crowdsourcing used. When the participants collaborate as a hobby, extra to their main work, hey charge lower values, as typically occur. In this sense, the use of crowdsourcing could become cheaper than hiring professionals or entities to perform the tasks (Barbosa, 2015).

Companies, to obtain solutions, often transfer this possibility to several players, not being dependent only on a unique "supplier". When this occurs, the risk of not getting a good idea is thus much lower. In addition, through crowdsourcing, companies can externalise the risk of failure and remain specialised in certain areas (Barbosa, 2015). Crowdsourcing allows access to a specific group of people, who could provide from different areas and have varied capabilities, For these reasons, crowdsourcing can contribute to greater innovation and a better search for solutions (Barbosa, 2015).

According to Hammon (2012), the advantages crowdsourcing offers to companies are still remarkable, such as access to vast range of skills, improved relationships between company players, increased customer loyalty, anticipation of consumer needs, greater innovation with greater customer focus, shorter time for process execution and cost reduction. Also Melo et al. (2015) suggested that the use of crowdsourcing is an asset for companies, since its benefits are quite significant. For the authors, crowdsourcing will have implications in the implementation prices of the ideas, as the remuneration to the intervening parties is smaller. This method will also reduce the time and money spent on research and, finally, it will provide contact with people with a different mindset to the company, thus adding greater value.

On the other hand, there are some drawbacks in the use of crowdsourcing by organizations. Barbosa (2015) mentioned some of these challenges, namely the management of the contact network, the company's openness to participation through crowdsourcing, the exact definition of the existing problem(s), the manipulation, the difficulty in obtaining diversified opinions and the monetary exploitation of the crowd participants. Managing a network of contacts, implies the creation of a network of people available to actively participate in the search for new ideas and solutions (the crowd). This will be fundamental for the good functioning of crowdsourcing, in the company. This step may be difficult to achieve, since it entails an increase of work to those responsible for crowdsourcing in the company, as they have to be active in groups and define strategies to motivate the network for their participation (Barbosa, 2015).

When a company is predisposed to use crowdsourcing, it has to be aware of the required exposure to the group of people who may help it through their involvement in participative and collaborative processes. AS an example, companies will have to share confidential information, as well as make known its strengths and weaknesses, so that the crowd can know the whole structure and, consequently, design a relevant contribution to the organization (Barbosa, 2015).

When a company needs to fill a gap, it must define the problem in a structured way. The problem raised by the company and must be clearly and objectively defined to provide the expected results. As adverted by Barbosa (2015), a poor problem definition can lead to a high number of solutions or no solution at all.

Crowdsourcing may be implemented using online tools, and some organizations use contests to elect the best solution presented. In this context, it will be necessary to pay attention to the way the voting process is developed, so that the results obtained cannot be subject to any manipulation (Barbosa, 2015).

When using crowdsourcing, it is expected to obtain information provided by individuals with a great cultural diversity and very distinct characteristics. Even so, as mentioned by Barbosa (2015), it is necessary to take care that this diversity within the group may not be achieved due to external factors. The members of the group may not have the same working skills and, in turn, the same response to the problem. Thus, if crowds are too homogeneous and elite in their composition, they do not allow the diversity of opinion to be guaranteed (Barbosa et al., 2015).

Another advantage of crowdsourcing is its cost, which is lower than other options for obtaining similar results. However, this advantage may incur in the (professional) exploitation of the participants, service providers and demonstrators of their talent. Participants in crowdsourcing processes receive a low remuneration for the service provided, especially in comparison to what would be paid to professionals. As such, the intensified use of crowdsourcing tools could have negative consequences in terms of unemployment and work conditions. Furthermore, the low remuneration involved may have consequences on the motivation and involvement of individuals in collaborative process, that could be reflected in a reduction in the quality of the resources obtained (Barbosa, 2015).

Hammon (2012) also alerted to the risk associated with the loss of control of all stages inherent to the project, the constant need for communication with all stakeholders, the difficulty in knowing the capabilities of all those involved, the exact definition of what needs to be improved, and the need to constantly motivate those involved in the process.

In the same sense, for Melo et al. (2015), the ability to participate in crowdsourcing processes may be influenced by the good access to the internet by the interested population, as well as by the skills and knowledge of the individuals involved. Furthermore, the author reinforces the difficulties that may occur in terms of the ability to maintain the motivation of the network.

1.3 Participatory Budget

1.3.1 Genesis and Definition of the Concept

The participatory budget is a very current issue in our society. According to Gonçalves (2009), it is difficult to define the concept of participatory budget due to the different experiences of each location where it is carried out. The concept of participatory budgeting can still be understood as a process in which the population of a locality contributes to decisions about the destination of public resources of the locality/municipality.

Nevertheless, the literature review indicates a slightly different understanding of the concept by some authors. Participatory budgeting can also be seen as a direct, voluntary and universal democratic process, where it is possible to discuss ideas and make decisions about budgets and public policies (Moreira et al., 2016). Citizens, as active members of a community, are not only limited to exercising their right to vote, but get involved in issues related to public spending and the control of government administration. Herein, citizen become to be seen as the main figure in public administration.

Silva, in 2017, refers to participatory budgeting as a structure and process of citizen participation in decision-making on municipal public investments, with three fundamental aspects: (i) Participation of all citizens; (ii) citizens have a fundamental role in defining the rules of democracy; and (iii) Definition of priorities according to criteria established by citizens, taking into account the needs of their community.

The participatory budget is seen by Bernardino and Santos (2020) as a process of permanent intervention by citizens in the management of their municipalities. Souza (2003), in turn, mentions the participatory budget as a process of direct, voluntary and universal democracy, where people can discuss and decide about the budget and public policies.

Sintomer et al. (2012) defined the participatory budget taking into account 5 (five) criteria. According to the authors, firstly the financial issues must be explained clearly, since participatory budget involves scarce resources; secondly, the participation of citizens must take place at local/municipal level with the constitution of a representation; The third criterion is that to be considered a participatory budget, the actions/activities must be repeated over time and not only be a single/isolated action; it is also necessary to paid attention to create condition for citizens participating in participatory budgets, to meet and discuss the needs of the community before presenting them to the municipalities; and, as a final criterion, Sintomer et al. (20) refer that the existence of accountability after the results of the participatory budget is critical.

Dias (2008) referred to the budget from a different perspective. For the author, participatory budgeting implies assigning greater responsibility on citizens, thus creating a greater proximity connection, insofar as they are co-responsible for the decisions taken by the rulers. For Silva (2017) the participatory budget represents for a local authority a new legitimacy and understanding on the part of citizens, a greater co-responsibility in the acceptance of financial cuts imposed on local authorities, Participatory budget also implies an increase in civic awareness and reorganisation of the social fabric on a more constructive and less protesting basis, a better knowledge of the territory, an increase in interest in electoral events and a better relationship between local authorities (Silva, 2017).

For Cabannes (2009), participatory budgeting is a mechanism or process through which the population contributes to decision-making on the destination of part the public resources available to a locality. This is an important mechanism people involvement, so that they can participate actively in the political decisions that will be taken in the municipality.

The participatory budget can be understood as a new way of governance, in which citizens play an active role in both decisions and the definition of priorities for the problems presented by the community. As such, participatory budgeting presupposes two-way communication and strong articulation between the different institutions.

According to Franklin et al. (2009), having active citizens participating in the budget is fundamental to align strategies and plan according to the priorities and values assigned to each region. For De Renzio and Wehner (2017), the participatory budget reflects the existence of greater transparency of the expenses of each municipality, contributing to better budget results and lower costs related to financing. It also reduces corruption, avoiding parallel and hypothetical accounting of the values of each municipality. Costa

(2014) corroborated the same opinion as for him, having citizens more involved in the decision making related to the community resources, helps to get satisfied the most basic needs raised by the population, and therefore strengthening the relationship between the government elected representatives and the population in general. Cabannes (2004) defends the use of the participatory budget as a possibility for a good government strategy. For Dias (2008) participatory budgeting is a mechanism of public governance, in which the citizen participates in local democracy, empowering the representatives he or she elects.

1.3.2 The Development and Implementation of a Participatory Budget

The use of the participatory budget, as a mechanism of civic participation, aims to be a process of knowledge for the inhabitants of the municipality. Herein, they have to know where and how to invest public financial resources. (Neto & Silva, 2013)

According to Cabannes (2009), it is fundamental to have a participatory democracy, referring to the existence of basic principles of participatory budgeting. Santos (2003) presents three fundamental principles for guaranteeing the conditions for success in a participatory budget for a municipality. These are citizen participation, where there is no distinction between the different businesses and citizens of a local community, articulation between representative democracy and participatory democracy, and the definition of investment priorities for the public budget, based on the needs of the population.

To implement a participatory budget process it is necessary to establish some criteria, that according to Silva (2017) involves: (i) public deliberation; (ii) continuous and periodic process; (iii) local organisation (by municipality or parish); (iv) detailed public debate, with a financial and budgetary dimension; and (v) public accountability.

According to Cabannes (2009), it is essential for a participatory budget to have participation, transparency, equality, tolerance, efficiency and effectiveness, impartiality, competitiveness and respect for established agreements. These eight principles, mentioned by the author can be achieved, but they vary in form depending on the locality, because each municipality can achieve the principles through different means.

The first test for the implementation of participatory budgeting took place in Brazil in 1989, and was later applied in South and Central America, Europe, Africa and Asia. All these budgets, despite using the same name, were implemented differently (Moreira et al., 2016).

The implementation of crowdsourcing took place in four phases (Moreira et al., 2016). The first phase, an experimentation phase, took place between 1989 and 1997 in the regions of Porto Alegre (Brazil), Santo André (Brazil) and Montevideo (Uruguay). Between 1997 and 2000, the second phase of the implementation of participatory budgeting took place, where it was intensified in Brazil. A third phase took place from 2000 to 2016, in which the implementation of participatory budgeting was extended to South and Central America and then to Europe, specifically Lisbon. The fourth and final phase is the one we are currently in, in which we seek to build networks of national and international collaboration of participatory budgeting (Moreira et al., 2016, p.13).

For Cabannes and Baierle (2004), the participatory budget has four dimensions. The first dimension is a participatory dimension, in which there are control mechanisms concerning the implementation of the budget and the execution of investments. In Europe, these are not put into practice; citizen participation is used in a consulting perspective, collecting opinions and ideas for defining the priorities of collective investments. Sintomer et al. (2008) presents five objectives for participatory budgeting in local governments:

1. Building a more participatory democracy;
2. Establish a democracy of proximity;
3. Promote a participative modernisation of the Administration itself;
4. Create public-private partnerships with citizen involvement;
5. To promote community development;

The second dimension presented is the dimension related to numbers, i.e., it is a budgetary or financial dimension, as it is essential to have a good budget available to invest in community projects. Local leaders have to make efforts in order to correspond to what they proposed when they were elected, thus achieving good results with citizen participation. Another dimension of a participatory budget concerns legal issues or norms, the normative/legal dimension. For most implementations to take place, it is necessary to draw up internal regulations with all the rules of the participatory budget, but these may vary depending on the place where it is implemented.

At the moment, participatory budgeting is only used at local level, however, it is intended to extend this form/measure of governance to national level, with all citizens being able to intervene and contribute to a better country (Silva, 2017).

Although existing studies indicate the possibility of implementing participatory budgeting in all places, it is plausible to admit that not all places and people may react in the same way to the same inputs. For this reason, for the implementation of participatory budgeting it will be necessary for places to meet some essential basic conditions, such as ideological and political, institutional, financial, technical and even societal.

For Tavares and Carr (2012), the size of the existing population in a given location can influence the participation of these citizens. For the authors this condition is directly proportional to the increasing size of the locality population, because the larger the population, the more difficult their interest in participating in local economic activities, being necessary to implement actions/campaigns and mechanisms of proximity.

There are several mechanisms for citizen participation in local political issues, but there is not just one that is valid and considered "the best" among the others. In other words, the different mechanisms will be useful in different situations, and it is sometimes necessary to use several mechanisms to obtain the desired final solution.

Sintomer et al. (2012) refers to the non-existence of a uniform model of civic participation. This author argues for the existence of six conceptual models, participatory democracy, proximity democracy, participatory modernisation, multi-stakeholder participation, community development and neo-corporativism.

According to the authors, a participatory democracy occurs when participating citizens have deliberative power and mobilise society. Sometimes, this model has repercussions in terms of social justice and the relationship between society and political representatives. From another perspective, a proximity democracy may occur, where informal rules are established for society, but it has low autonomy in decision-making. In this case there is no differentiated form of democracy, only an improvement in communication between society and its political representatives (Costa, 2014). Another model of participation, where the aim is to make the state a more efficient and legitimate element, is participatory modernisation, with the constitution of a policy aimed at politicians and not at society itself. A model, with ideologies contrary to the above, brings together the participation of multiple actors from society, under this form of participation. The various elements of society, such as citizens, private companies or political representatives, play an active role in taking decisions affecting the municipality (Costa, 2014).

There are two other mechanisms of civic participation: community development and neo-corporativism. In the former, citizens participate actively in society through the presentation of projects. In this model, little use is made of municipal politics and there is a great deal of citizen participation. The advantages of community development are seen in places where local government has weak power and society is more independent, giving the community the possibility to manage its local projects. In the second model, neo-corporativism, various organised groups are formed, and there is a need for consultation with the population. Thus, the government consults the group with the best framework, through the mediation of interests, values and demands of the various segments of society (Costa, 2014).

Each of these models is different. However, they complement each other and facilitate the implementation of society's participation in common interests. For this reason, according to Costa (2014) several mechanisms are used for the same issue, as a form of complement.

According to Franklin, Ho and Ebdon (2009), local governments can and should use various methods in order to achieve citizen participation. Each mechanism has strengths and weaknesses when more than one is used for the same purpose. When one has a flaw, the other can fill that limitation and therefore deliver the expected result.

With the use of the participatory budget, two distinct objectives are expected to be achieved: the transmission of an education about participatory budgets and the dissemination of information about decision-making. As regards the education to be transmitted, budgets are complex documents/formalities, involving the participation of several people from the community. Consequently, it is necessary the communication to be transmitted in clearer language, making it more perceptible to the majority. With this participation, citizens can also understand why certain actions/activities were not carried out in the desired time. With the limitation of existing resources, it is difficult to satisfy all the needs. Added to this fact, sometimes, in a hierarchy of established priorities, it is urgent to establish a new one with greater restrictions, given the scarcity of resources in the municipality (Silva, 2017). The other objective, related to the dissemination of information to the population on decision-making, lies in the fact that it is essential to have a fluid communication between the various parties, whether they are members of the political power or citizens of the community. Citizens communicate with their representatives about the ideas and decisions taken and express their opinion about the work being done. In this way, it is possible to create a sense of community (Silva, 2017). The greater the volume of citizens active in these dynamics, the greater the trust felt in their political representatives and public entities.

According to Franklin, Ho and Ebdon (2009), the objectives of using participatory budgeting boil down to putting imagined ideas into practice to build the desired society and using these ideas strategically (Silva, 2017).

In order for the participatory budget to be implemented in a municipality, some conditions for its success must be met. According to Cabannes (2009) there are six necessary conditions. These conditions relate to the existence of willingness on the part of the mayor and his/her members for implementation, i.e. there must first be political will to implement the participatory budget; the need for interest on the part of the population, regarding their involvement in civic participation activities, without the interest of the population in presenting proposals/projects, a participatory budget cannot be carried out The definition of the "rules of the game", i.e. for a good implementation it is necessary for all the assumptions to be defined in order to run as expected; it is also necessary to establish rules regarding issues such as participation, voting, elections and results. It is also necessary to elaborate a process and establish stages; the empowerment of all elements involved, both the population and the municipal authorities need to be equipped with skills/knowledge about participatory budgeting, so that they can participate and evaluate in a constructive and reasoned way. Therefore, it is the will to promote the training of the population and municipal authorities on Participatory Budgeting, in order to promote the empowerment of the population and municipal authorities on the participatory budget (Cabannes, 2009, p. 38); the dissemination of the existence of a participatory budget edition, that is, it is necessary to inform the population of the opening of a civic participation edition, as well as to share the schedule regarding deadlines and timing of phases, so that participation is informed and free of doubts for citizens; and the existence of technical criteria about the infrastructures (existing and "needed") and public services available in each locality. If the above conditions are not met, the implementation of participatory budgeting, according to Cabannes (2009), should not proceed.

1.3.3 The Advantages and Limitations of Using a Participatory Budget

According to Bernardino and Santos (2020), with the use of the participatory budget, citizens are able to make demands and set priorities according to the distribution of existing budgets in the municipality. De selection criteria is based on the understanding of all citizens involved, the order in which the investment(s) should focus is chosen.

For Souza (2001), through participatory budgeting, citizens are protagonists in public administration, not only intervening in elections with the exercise of the duty to vote.

Franklin, Ho and Ebdon, (2009), consider it is possible to achieve local citizen participation through different methods, all of which involve different advantages and disadvantages. Thus, for the authors, citizen participation can be ensured through public meetings. This modality will have as an advantage the better facility in sharing information, such as particular needs, development opportunities or business ideas. However, as a limitation, it may involve a low knowledge of the issues by the citizens. In a different way, citizen participation could be ensured through consultative committees, allowing stimulating learning on issues related to participatory budgets. However, this option will require great efforts in terms of time and cost, and may not be representative. In addition, participation can be promoted through citizen surveys, allowing for the mapping of citizens' needs and desires However, this modality may be costly and offer a limited space for collecting opinions, not being able to transmit them all in depth.

Another author, Cabannes, developed a study in 2015 in Brazilian cities, where it was found that the use of participatory budgeting gave voice to projects related to basic/primary services. In this sense, the participatory budget is very powerful, for putting basic services and needs into practice in the localities where they are implemented.

The participatory budget is designed to be an innovative urban management tool, with the possibility of numerous benefits for municipalities, where the transfer between the two parties is promoted and the involvement/interest of citizens in public decisions and the application of public resources is stimulated.

There are also other benefits, such as an increase in the municipality's revenue and the orientation of investment to satisfy the needs of basic infrastructures in places where they are not provided for. In this way, a better relationship between citizens and those elected to represent them in the municipality may be possible.

Participatory budgeting can be seen positively for the creation of inclusive cities, as it can reach as many people as possible and representing an important tool for the democratisation of cities (Costa, 2014).

Given the aforementioned benefits, an increasing number of cities have implemented participatory budgeting (Costa, 2014). Through the implementation of participatory budgeting, it is possible to notice local citizens who are more committed to the good of their society.

According to an article published in Brazil, on the municipality of Sobral, by Neto et al. (2013) about the implementation of participatory budgeting as a mechanism for civic participation, there are great advantages to using participatory budgeting. These authors highlight four advantages, one of which is greater transparency regarding the municipality's income and expenses; another is the creation of a new/different political culture, based on the participation of the municipal inhabitants and on decisions taken in community; yet another advantage lies in the fact that it allows visible changes at the behavioural level of society and at the political level, since proposals are made to solve collective rather than individual problems. These proposals aim to improve and extend the existing network of services so that everyone has equal access. Another advantage is related to the fact that a greater number of people can adhere to political issues related to the municipality; a final advantage is related to the capacity to increase the amount of money available for the projects/proposals presented by the population over the years (Neto et al., 2013).

It is possible to verify the existence of distinct benefits, both for the locality where participatory budgets are carried out and for its citizens or for the local administration - the organisation responsible for implementing the participatory budget. In the case of benefits for the city and its citizens, according to Cabannes (2009), one of the main benefits of participatory budgeting is to contribute to deepening the exercise of democracy through the dialogue that public authorities establish with citizens. There is also transparency in the amounts of funds used, a promotion of social justice and a prioritisation of the really important social causes, since they are presented by the population. According to the author, "citizens, actively participating in the Participatory Budgeting process, define their priorities and, in doing so, have the opportunity to significantly improve their quality of life within a relatively short period of time" (Cabannes, 2009, p. 29).

In spite of the advantages presented by the implementation of the participatory budget, this mechanism also has some limitations along the way. Some of these limitations are related to the lack of connection between the citizens, that could hinder their participation/collaboration in the municipality's new projects/ideas (Silva, 2017).

Not all municipalities have a significant number of young people or people with higher education. The low schooling of the resident population could constraint the process, as the language used is not easy to deal with for all elements, the population and the lack of understanding about the decision-making or the methods used by the local government elements could arise (Silva, 2017).

Participatory budgeting allows projects to be put to the vote, so that they can be implemented with the help of funds made available by public entities. However, the projects must comply with a series of requirements presented by the municipalities, which implies structuring, critical thinking and several hours of work by the elements involved in the project. This whole process is seen as time-consuming and may penalise participation. Some authors mention the exaggerated expenditure of time as a limitation (Silva, 2017).

The culture rooted in the local population may lead to the creation of resistance on their part, not accepting new ideas transmitted by the rulers. This culture can be considered a barrier to the participation in the participatory budget (Neto et al., 2013).

Other limitation on the use of participatory budget could also be related to the existing administrative structure in the municipality. This structure may sometimes not be capable of meeting the expectations of civic participation methodologies, making it necessary to make some modifications. Therefore, for some municipalities, it is crucial to readapt their internal structure. The structure itself is also pointed out as a difficulty, taking into account the technical means, namely the existing human resources. The human resources may not be endowed with all the specific knowledge required to use the mechanisms, and it is therefore necessary to invest in the training so that they can respond to any doubts that the population may have. As the participatory budget is a relatively recent topic in our society, the appearance of various doubts and questions is a natural fact (Neto et al., 2013).

2. RESEARCH METHODOLOGY AND CONTEXT

This research aims to analyse the use of open innovation and crowdsourcing by the public sector in Portugal, namely through the use of participatory budgets. The aim is to understand how participatory budgets are used by the different municipalities in Portugal, and to find out about the advantages and difficulties perceived by municipalities when they adopt them. The chapter also aims to understand the characteristics of the participatory budgets already implemented in the various Portuguese municipalities, and to what extent they are influenced by the characteristics of the territory.

The study consider as unit of analysis the municipalities in Portugal. Currently, Portugal has 308 municipalities, divided between the mainland with 278, Madeira with 11 and the Azores with 19. The participatory budget was first implemented in Portugal in 2002 in the municipality of Palmela. After two years, in 2004, it was implemented in three other cities: Serpa, Avis and Santiago do Cacém. In 2012, the data recorded refer that the participatory budget was used in 71 Portuguese municipalities (Costa, 2014).

For the study it was opted to use quantitative method. Primary data collection was done through a questionnaire survey in online format.

The questionnaire survey was designed based on the existing literature review and the research objectives, and is composed of four groups. The first group contains questions on the characterisation of the municipalities, namely the district, the municipality and the respective number of inhabitants. The second group of questions concerns the development and implementation of participatory budgeting. In the third group, an assessment is made of the advantages perceived through the use of participatory budgeting; in contrast, the fourth group seeks to find the greatest difficulties experienced.

The questionnaire was made available on the Google Forms platform. The questionnaires were distributed by email to the country's Municipal Councils. Firstly, through a desk research in the internet the emails address of all Portuguese municipalities was performed. After that, an email was sent, explaining the research purposes and the link the municipalities should use to answer the questionnaire. The period for data collection took place between 12 August 2022 and 12 September 2022. In a first approach it was not possible to obtain a considerable number of answers, so a survey of the municipalities already participating was carried out and a reinforcement email was sent to the non-participating municipalities. This online questionnaire resulted in 101 responses. The data obtained were analyzed using the statistical software SPSS, version 26.

3. PRESENTATION OF RESULTS

3.1 Characterization of the Sample

According to Table 1, we can observe the different districts that had participated in the questionnaires and corresponding percentage.

Table 1. Characterization of the sample

	Frequency	Percentage
	District	
Azores	8	7,9
Aveiro	5	5,0
Beja	7	6,9
Braga	5	5,0
Bragança	4	4,0
Coimbra	4	4,0
Evora	4	4,0
Faro	5	5,0
Guard	5	5,0
Leiria	8	7,9
Lisbon	9	8,9
Wood	4	4,0
Portalegre	4	4,0
Porto	7	6,9
Santarém	8	7,9
Setúbal	5	5,0
Vila Real	3	3,0
Viseu	6	5,9
Total	101	100,0
	Number of Inhabitants	
< 10,000	39	38,6
Between 10,000 and 30,000	31	30,7
Between 30,000 and 60,000	16	15,8
Between 60,000 and 90,000	4	4,0
Between 90,000 and 120,000	2	2,0
Between 150,000 and 200,000	7	6,9
More than 200,000	2	2,0
Total	101	100

Source: Own Elaboration

The analysis of Table 1 also reveals the existence of differences in the size of the municipalities. The sample is mainly composed of municipalities with less than 10 thousand inhabitants, corresponding to 38.6% of the sample, while the municipalities with between 10 thousand and 30 thousand inhabitants have a percentage of 30.7%, corresponding, together with the smaller municipalities, to the majority of the sample. As the number of inhabitants of a municipality increases, so does its participation in the study. With between 30 to 60 thousand inhabitants we have 16 municipalities, followed by 4 municipalities with a population between 60 to 90 thousand inhabitants, 2 municipalities with 90 to 120 thousand inhabitants, 7 with between 150 to 200 thousand inhabitants and only 2 municipalities with more than 200 thousand inhabitants, belonging to the largest regions of the country.

3.2 Development and Practice of Participatory Budgeting

The study seeks to understand the Portuguese reality, assessing the municipalities that do or do not develop participatory budgets and, if so, to know what characteristics are presented by the participatory budgets already developed. Herein some characteristics are analysed, as shown in Table 2.

We observe that 51.5% of the sample having municipalities in the sample have already developed some type of participatory budget and 48.50% not having implemented participatory budgets.

Table 2. Characteristics of participatory budgets

	Frequency	Percentage
	Development of participatory budgeting	
Yes	52	51,5
No	49	48,5
1st edition of the participatory budget		
1996	1	1,0
Between 2008 and 2014	19	37,3
Between 2015 and 2022	31	60,8
Frequency of the participatory budget		
Annual	41	78,8
Biannual	5	9,6
No frequency defined	6	11,5
Members of the team responsible for the participatory budget		
Up to 3 elements	31	59,6
Between 3 and 5 elements	13	25,0
More than 5 elements	8	15,3
Budgetary allocation		
Up to 250 thousand euros	39	78,0
More than 250 thousand euros	11	22,0
Provision of financial resources		
City council	52	51,5
Parish councils	51	50,5
Municipal enterprises	52	51,5
Other public sources of funding	52	51,5
Other patrons	52	51,5
Private companies	52	51,5
Proposals received by the municipality		
Up to 10 proposal	18	34,6
Between 10 and 5s0	28	53,8
More than 50	6	7,7
Area of the proposals submitted		
Support to vulnerable groups	1	1,9
Well-being of the population	16	30,8
Culture and Sport	3	5,7
Circular economy	1	1,9
Public space	1	1,9
Social inclusion	2	3,8
Road, traffic and mobility infrastructures	1	1,9
Technological innovation	1	1,9

Continued on following page

Table 2. Continued

	Frequency	Percentage
	Development of participatory budgeting	
Youth	1	1,9
Civil protection	1	1,9
Infrastructure rehabilitation	22	42,3
Sustainability	2	3,8
Proposals approved by the municipality		
Up to 10 proposals	42	80,8
More than 10 proposals	10	19,2
Projects approved by the municipality		
Between 1 and 3	24	47,1
Between 3 and 10	23	45,1
More than 10	4	7,8
Means of dissemination of the participatory budget		
Municipality website	36	35,6
Site created for participatory budget	30	29,7
Social networks for participatory budgeting	33	32,7
Local newspaper	8	7,9
Local radio	10	9,9
Flyers	6	5,9
Outdoors	12	11,9
Social networks of the municipality	12	11,9
Request disclosure	6	5,9
Other	4	4,0
Investment providers for the participatory budget		
Municipal council	52	93,83
Parish councils	3	2,78
Municipal companies	2	1,54
Other public funding sources	1	0,62

Source: own preparation

The implementation of the participatory budget did not occur on the same date in all municipalities. In order to obtain a better reading and analysis, the results were regrouped in three, establishing one group in 1996, another between 2008 and 2014, and a third group between 2015 and 2022. The first occurrence, according to the answers obtained and the data in Table 2, was in 1996, with only 1%. Between the years 2008 and 2014, 37.3% of the municipalities started participatory budgeting. Most municipalities, 60.8%, have implemented participatory budgets recently, that is, between 2015 and 2022.

The first participatory budget occurred in 1996, and the most recent municipality to implement it did so in 2022. Most implementations took place between 2014 and 2015, i.e. for most municipalities the implementation of the participatory budget is quite recent.

For the municipalities that have already implemented the participatory budget (N=51), we tried to find out how often this had occurred. The analysis of Table 2 indicates that most municipalities replied that they have annual editions of the participatory budget, which corresponds to 78.8% of the respondents; 9.6% of the municipalities indicate that they have editions of the participatory budget every two years, i.e. a biannual frequency, and 11.5% of the municipalities do not present a frequency defined for the occurrence of participatory budgets.

The members of the municipal councils responsible for the implementation of the participatory budget have different functions; however, it was found that most of them have functions related to management. However, the composition of these teams varies. In most cases (59.6%), they consist of around 3 elements. Some municipalities have teams with more than 3 elements and less than 5 elements (25% of the sample), and only 15.3% of the municipalities have teams with more than 5 elements.

With regard to the budget allocation involved in participatory budgets, the answers obtained were grouped into two categories, one comprising allocations of up to 250 thousand euros, observed in 78% of the municipalities, and the other allocations of more than 250 thousand euros (observed only in 22% of the municipalities).

With regard to the ways the financial resources necessary for participatory budgets were obtained, we observe that the resource most frequently used is the Municipal Councils, with 93.8%. The remaining entities, in turn, have little or no financial support for the implementation of participative budgets.

According to the answers obtained in the questionnaires, and as shown in Table 2, the largest number of proposals submitted by citizens for evaluation is between 10 and 50 proposals, comprising 53.8% of the sample. Some municipalities, 34.6%, only manage to receive 10 proposals within the scope of the participatory budget created. Rare are the places where more than 50 proposals are submitted per participatory budget, but these still represent around 7.7% of the sample. Analysing the proposals submitted by the population within the scope of the participatory budgets created by the municipalities, it can be observed that the most relevant areas of proposals submitted to be considered in participatory budgets are the rehabilitation of infrastructure, with 42.3%, and the well-being of the population, with 30.8%. The areas where there are fewer proposals are related to the circular economy, public space, support for vulnerable groups, road infrastructure, traffic and mobility, technological innovation, youth, and civil protection, each one of these areas representing 1.9%.

After the submission of the proposals, the municipality must proceed, or not, to their approval, depending on the observation of the eligibility criteria defined by the municipality. The analysis carried out indicates that in most municipalities, 80.8%, the number of proposals approved is lower than 10.

Considering the number of projects effectively approved during each participatory budget, approval is between 1 and 3 projects in 47.1% of the municipalities, with similar percentages for 3 to 10 projects in 45.1%, and more than 10 projects approved in 7.8% of the municipalities.

The means of dissemination most used to inform about the participatory budget are the municipality's institutional website, with 69%, social networks created specifically for the participatory budget, with 63%, and websites designed for the same purpose, with 58%. The municipality's social networks are also used in some cases, with a weight of 23%, as well as dissemination by means of posters. The remaining means of dissemination, such as local radio, local newspaper, dissemination by other actors, flyers and other options, are used only by a small number of municipalities, their representation being less than 20%.

After presenting the characteristics associated with the development of participatory budgeting in the different municipalities, we aimed to analyse the characteristics observed in participatory budgets according to the characteristics of the municipalities, particularly in terms of the size of the municipalities.

The smaller the municipality, the less participation in participatory budgets it has (Table 3). In other words, in smaller municipalities, it is more common for no participatory budget to be implemented. On the other hand, in large municipalities, with more than 10,000 inhabitants, 61.3% have a participatory budget, while in small municipalities, with less than 10,000 inhabitants, only 35.9% have a participatory budget.

Table 3. Analysis of the participatory budget according to size and location

	Size of the Municipality	
	<10 Thousand	>10 Thousand
	%	%
Develops participatory budgeting		
Yes	35,9	61,3
No	64,1	38,7
1st edition of the participatory budget		
1996	0	2,6
Between 2008 and 2014	15,4	44,8
Between 2015 and 2022	84,6	52,6
Frequency		
Annual	78,6	78,9
Biannual	7,1	10,5
No frequency defined	14,3	10,5
Team members		
Up to 3 elements	64,3	57,9
Between 3 and 5	21,4	26,3
More than 5	14,3	15,8
Budgetary allocation		
Up to 250 thousand euros	100	69,4
More than 250 thousand euros	0	30,6
Received proposals		
Up to 10	57,1	26,3
Between 10 and 50	42,9	57,9
More than 50	0	15,8
Approved projects		
Up to 5	76,9	60,5
More than 5	23,1	39,5
Proposals adopted		
Up to 10	92,9	76,3
Between 10 and 50	7,1	18,4
More than 50	0	15,4

Source: own preparation

For the municipalities that already use participatory budgets, we intended to understand when the first edition of this budget took place. Although municipalities with more than 10,000 inhabitants have a greater implementation weight between 2015 and 2022, smaller municipalities have increased implementation from 10.5% (between 2008 and 2014) to 35.5%. The tendency is for more and more municipalities to use participatory budgets, independently of their size.

The frequency of participatory budgets is not affected by the size of the municipalities, as most of them take place annually, regardless of the number of inhabitants. In small municipalities, 78.6% have an annual frequency, 7.1% are held biannually, and 14.3% have no defined frequency. For municipalities with more than 10,000 inhabitants, the figures are similar: 78.9% have an annual frequency, 10.5% have a biannual frequency and the same percentage do not have a defined frequency of participatory budgets.

The teams responsible for the implementation of participatory budgets are mostly made up of around three people, as shown in Table 3, corresponding to 64.3% in small municipalities and 57.9% in larger municipalities. It is not possible to find a pattern that is influenced by the municipal dimension.

Also related to the size of the municipalities, or their number of inhabitants, is the budget allocation involved in participatory budget projects. In this variable, there is an influence between the number of inhabitants and the funds available. In large municipalities, with more than 10,000 inhabitants, 30.6% of these have more than 250,000 euros, while in small municipalities, those with less than 10,000 inhabitants, none has an allocation of more than 250,000 euros. Thus, it correlates, the greater the number of inhabitants, the greater the available budgetary allocation.

In relation to proposals presented to municipalities, Table 3 shows a relationship with the size of the district where the participatory budgets took place. The relationship found is similar to what was perceived before. -Smaller districts have a smaller number of proposals presented to municipalities, and consequently, fewer projects and proposals approved. As the size of a municipality increases, the number of proposals increases, as well as its projects and proposals approved. However, even in municipalities with more than 10,000 inhabitants, the proposals submitted rarely exceed 50.

3.3 Perceived Advantages of Using Participatory Budget

By analysing Table 4, it can be seen that practically all the options presented to the respondents, with the exception of the last one, were considered advantages arising from the implementation of the participatory budget. It should be noted that these questions were only applied to municipalities that had indicated that they used participatory budgets (N=51). It is worth highlighting "Increased proximity between public entities and citizens" and "Promotion of more active citizenship on the part of residents", with 33.7% of respondents considering that they totally agree with its relevance. It should also be noted that the fact that the participative budget allows for "Increased levels of satisfaction of employers and other local agents" and "Increased tourist visibility of the municipality" is not highly valued by municipalities, with 19.8% of answers in point 3 of the Likert scale and 11.95 in point 4 of the scale; in the second, 18.8% is in option 3, 10.9% in option 2, 7.9% in option 1, considering a greater disagreement than agreement, in the positive part of the scale, options 4 and 5, there are only 9.9% and 3%, respectively.

Table 4. Perceived advantages of participatory budgeting

	%1	%2	%3	%4	%5	Average	Standard Deviation
Increased proximity between public entities and citizens	0	0	1,0	16,8	33,7	4,63	0,525
Improved communication with the population	0	1,0	8,9	23,8	17,8	4,13	0,768
Improving the transparency of the municipality	2,0	0	8,9	16,8	22,8	4,16	0,987
Improving the level of citizen satisfaction	0	0	8,9	21,8	18,8	4,20	0,728
Increased legitimacy of the municipality among the population	2,0	1,0	15,8	17,8	13,9	3,80	1,000
Increase citizens' commitment to the municipality	0	1,0	12,9	20,8	15,8	4,02	0,812
Development of more altruistic behaviour by citizens	1,0	4,0	17,8	13,9	13,9	3,71	1,026
Promoting more active citizenship on the part of residents	0	0	12,9	12,9	25,7	4,25	0,837
9.Source of creativity and generation of new ideas for the municipality	0	2,0	7,9	20,8	19,8	4,16	0,834
Tapping into the ideas and knowledge of citizens	0	0	7,9	27,7	14,9	4,14	0,664
11. it allows to take advantage of the ideas and knowledge of other local actors	0	2,0	9,9	24,8	13,9	4,00	0,800
12.Allows a more efficient use of the resources that the municipality has at its disposal	1,0	9,9	16,8	13,9	8,9	3,39	1,060
It allows a better analysis of the most promising ideas/ projects	0	1,0	14,9	20,8	13,9	3,94	0,810
It allows better knowledge of the problems that affect the population	1,0	2,0	12,9	14,9	14,9	4,00	1,000
Increased tourist visibility of the municipality	7,9	10,9	18,8	9,9	3,0	2,78	1,119
Increase the level of satisfaction of employers and other local agents	3,0	9,9	19,8	11,9	5,9	3,16	1,065
Improving the quality of life of the population	1,0	3,0	15,8	14,9	15,8	3,82	1,014

Note: 1 Strongly disagree, 2 Disagree, 3 Neither disagree nor agree, 4 Agree, 5 Strongly agree
Source: own preparation

In order to better understand the results obtained, the data were reduced by means of factor analysis.

The factor analysis by principal components, using the *varimax* rotation method, allowed the reduction of the information on the perceived benefits of the implementation of the participatory budget into 4 principal components, as shown in Table 5.

Table 5. Factor analysis of the advantages

	Loading	Eigen Values	% of Variance	% Cumulative Variance	Cronbach's Alpha
Component 1: Communication and transparency		8,719	51,286	51,286	0,800
Increased proximity between public entities and citizens	0,674				
Improved communication with the population	0,703				
Improving the transparency of the municipality	0,821				
It allows better knowledge of the problems that affect the population	0,751				
Component 2: Citizen involvement		2,145	12,615	63,901	0,866
Improving the level of citizen satisfaction	0,572				
Increase citizens' commitment to the municipality	0,738				
Development of more altruistic behaviour by citizens	0,743				
Promoting more active citizenship on the part of residents	0,857				
9.Source of creativity and generation of new ideas for the municipality	0,543				
Component 3: Visibility and resource optimisation		1,242	7,309	71,210	0,890
Increased legitimacy of the municipality among the population	0,539				
12.Allows a more efficient use of the resources that the municipality has at its disposal	0,736				
Increased visibility for tourism in the municipality	0,896				
Increase the level of satisfaction of employers and other local agents	0,877				
Improving the quality of life of the population	0,586				
Component 4: Idea Generation		1,035	6,087	77,297	0,854
Tapping into the ideas and knowledge of citizens	0,828				
11. it allows to take advantage of the ideas and knowledge of other local actors	0,932				
It allows a better analysis of the most promising ideas/projects	0,587				

Source: own preparation

The main component 1, called "Communication and transparency," comprises the variables "Increased proximity between public entities and citizens," "Improved communication with the population," "Improved transparency of the municipality" and "Allows better knowledge of the problems that affect the population".

Component 2 is called "Citizen involvement" and comprises the variables "Improved level of citizen satisfaction," "Increased commitment of citizens to the municipality," "Development of more altruistic behaviour by citizens," "Promotion of more active citizenship on the part of residents" and, "Source of creativity and generation of new ideas for the municipality".

Component 3, called "Visibility and optimization of resources", comprises the variables "Increased legitimacy of the municipality among the population", "Allows a more efficient use of the resources that the municipality has at its disposal", "Increased tourist visibility of the municipality", "Increased level of satisfaction of employers and other local agents" and "Improved quality of life of the population".

The fourth main component is designated as "Generating ideas" and comprises the variables "Enables to take advantage of the ideas and knowledge of citizens", "Enables to take advantage of the ideas and knowledge of other local actors" and "Enables a better analysis of the most promising ideas/projects".

After that we proceeded to calculate an index for each of the variables obtained in the principal component analysis, based on the weighted average factor of each variable in the principal component. This procedure allowed to understand the average value attributed by the municipalities to each of the dimensions. The results indicate that the most valued benefit is "Communication and transparency", followed by the "Citizen involvement" component, as indicated in Table 6. The least valued component is the third, "Visibility and resource optimisation".

Table 6. Average of the components of the advantages

	Overall Average	Size of the Municipality	
		<10 Thousand (Average Value)	> 10,000 (Average Value)
Component 1: Communication and transparency	4,225	4,14	4,25
Component 2: Citizen involvement	4,05	4,08	4,04
Component 3: Visibility and resource optimisation	3,316	3,39	3,29
Component 4: Idea Generation	4,033	4,07	4,02

Source: own preparation

Comparing the means observed in each of the principal components in the different municipalities, according to their size, it can be seen that the advantage with the greatest expression is "Communication and transparency", regardless of the size of the municipality,

3.4 Difficulties Perceived by Municipalities With Implemented Participatory Budget

An analysis of Table 7 shows that the greatest difficulties experienced by municipalities are those related to "Difficulties in mobilising the population for participation in the participatory budget", with 21.8% of the answers located at point 4 of the Likert scale and 9.9% at point 5 of the scale. It is also worth mentioning the "Difficulties in mobilising other local actors for participation in the participatory budget", with 21.8% of the answers located at point 4 of the scale and 7.9% at point 5. With regard to the "Difficulties in attracting participants for the preparation of proposals", 18.8% and 6.9% of the respondents adhered to the highest points of the scale, respectively 4 and 5 points. Most of the options presented to respondents were not considered a difficulty associated with the implementation of the participatory budget by municipalities, since most of the answers are located in points 1 and 3 of the 5-point Likert scale. For the reply "Insufficient financial resources for the implementation and development of the programme" an average value of 3.04 was obtained. The same occurs, in a similar way, for the criterion "Absence of teams with the *know-how* required for the implementation and development of the participatory budget in the municipality", with a mean value of 2.49. The criterion "Activities related to transparency" has a low average value of 2.10, i.e. it is not considered to be a difficulty.

Table 7. Difficulties perceived by the participatory budget

	%1	%2	%3	%4	%5	Average	Standard Deviation
Insufficient financial resources for the implementation and development of the programme	6,9	11,9	11,9	11,9	7,9	3,04	1,296
Lack of human resources for the implementation and development of the programme	6,9	7,9	12,9	15,8	6,9	3,16	1,255
Lack of material resources for the implementation and development of the programme	10,9	12,9	11,9	11,9	3,0	2,67	1,227
Insufficient technological resources for the implementation and development of the programme	14,9	13,9	12,9	7,9	1,0	2,33	1,125
Lack of teams with the necessary *know-how* for the implementation and development of participatory budgeting in the municipality	13,9	10,9	14,9	8,9	2,0	2,49	1,189
Bureaucracy associated with the preparation and development of the participatory budget	6,9	13,9	14,9	10,9	4,0	2,82	1,161
Time needed for the preparation and development of the participatory budget	8,9	5,9	10,9	19,8	5,0	3,12	1,275
Preparation of communication materials	8,9	18,8	16,8	5,0	1,0	2,41	0,963
Activities related to transparency	17,8	13,9	14,9	4,0	0	2,10	0,985
Difficulties in mobilising the population to participate in the participatory budget	3,0	6,9	8,9	21,8	9,9	3,57	1,136
Difficulties in mobilising other local actors for participation in the participatory budget	3,0	7,9	9,9	21,8	7,9	3,47	1,120
Difficulties in attracting participants to prepare proposals	4,0	5,9	14,9	18,8	6,9	3,37	1,113
Difficulties in attracting people willing to participate in the voting process	5,0	6,9	12,9	19,8	5,9	3,29	1,154
Difficulties in the analysis and selection of projects	10,9	9,9	9,9	15,8	4,0	2,84	1,302
Difficulties associated with monitoring the implementation of the selected proposals	9,9	8,9	18,8	9,9	3,0	2,75	1,163
Unawareness on the part of the population of the concept of participatory budgeting	5,0	7,9	16,8	15,8	5,0	3,16	1,120
Unawareness or difficulty on the part of the population in using digital tools	3,0	12,9	15,8	15,8	2,0	3,02	1,000
Lack of time for the population to get involved in the participatory budget	5,9	12,9	15,8	11,9	4,0	2,90	1,136
Low number of inhabitants in the municipality	26,7	3,0	8,9	7,9	4,0	2,20	1,429
Low level of education of the resident population	22,8	5,0	14,9	6,9	1,0	2,18	1,212
Average age of the municipality's inhabitants	17,8	8,9	15,8	5,9	2,0	2,31	1,191

Note: 1 Strongly disagree, 2 Disagree, 3 Neither disagree nor agree, 4 Agree, 5 Strongly agree
Source: own preparation

In order to better understand the results obtained, the data were reduced by means of factor analysis. The factor analysis by means of principal components, using the *varimax* rotation method, allowed the reduction of the information on the perceived benefits of the implementation of the participatory budget into 5 principal components, as shown in Table 8. This analysis was only carried out on municipalities with a positive response to the implementation of the participatory budget.

Table 8. Factor analysis of difficulties

	Loading	Eigen Values	% of Variance	% Cumulative Variance	Cronbach's Alpha
Component 1: Difficulties in mobilising stakeholders		6,791	33,955	33,955	0,911
Difficulties in mobilising the population to participate in the participatory budget	0,884				
11.Difficulties in mobilising other local actors for participation in the participatory budget	0,844				
12.Difficulties in attracting participants to prepare proposals	0,859				
Difficulties in attracting people willing to participate in the voting process	0,856				
Component 2: Population Characteristics		3,449	17,244	51,199	0,863
Lack of time for the population to get involved in the participatory budget	0,541				
19.Low number of inhabitants in the municipality	0,870				
Low level of education of the resident population	0,908				
21.Average age of the municipality's inhabitants	0,888				

Continued on following page

Table 8. Continued

	Loading	Eigen Values	% of Variance	% Cumulative Variance	Cronbach's Alpha
Component 3: Implementing the participatory budget process		2,252	11,262	62,461	0,800
Lack of human resources for the implementation and development of the programme	0,551				
9.Activities related to transparency	0,500				
Difficulties in the analysis and selection of projects	0,801				
15.Difficulties associated with monitoring the implementation of the selected proposals	0,714				
Component 4: Initiating the participatory budget process		1,395	6,973	69,434	0,855
Bureaucracy associated with the preparation and development of the participatory budget	0,696				
Time needed for the preparation and development of the participatory budget	0,849				
Preparation of communication materials	0,787				
Unawareness on the part of the population of the concept of participatory budgeting	0,651				
Component 5: Insufficient resources		1,232	6,158	75,592	0,831
Insufficient financial resources for the implementation and development of the programme	0,809				

Table 8. Continued

	Loading	Eigen Values	% of Variance	% Cumulative Variance	Cronbach's Alpha
Lack of material resources for the implementation and development of the programme	0,815				
Insufficient technological resources for the implementation and development of the programme	0,690				
5.Lack of teams with the necessary *know-how* for the implementation and development of participatory budgeting in the municipality	0,708				

Source: own preparation

The main component 1, called "Difficulties in mobilising participants", comprises the variables "Difficulties in mobilising the population for participation in the participatory budget", "Difficulties in mobilising other local actors for participation in the participatory budget", "Difficulties in attracting participants for the preparation of proposals" and "Difficulties in attracting people available to participate in the voting process".

The main component 2, designated as "Characteristics of the population", comprises the variables "Lack of time by the population for involvement in the participatory budget", "Low number of inhabitants of the municipality", "Low schooling of the resident population" and "Average age of the municipality's inhabitants".

Component 3, called "Implementation of the Participatory Budgeting Process", considers the variables "Scarcity of human resources for the implementation and development of the programme", "Activities related to transparency", "Difficulties in the analysis and selection of projects" and "Difficulties associated with monitoring the implementation of selected proposals".

Component 4, designated as "Initiation of the participatory budget process", contains the variables "Bureaucracy associated with the preparation and development of the participatory budget", "Time required for the preparation and development of the participatory budget", "Preparation of communication materials" and "Unawareness on the part of the population of the concept of participatory budget".

The last main component, component 5, is called "Insufficient resources" and includes the variables "Insufficient financial resources for the implementation and development of the programme", "Lack of material resources for the implementation and development of the programme", "Insufficient technological resources for the implementation and development of the programme" and "Lack of teams with the necessary know-how for the implementation and development of the participatory budget in the municipality".

After the analysis, an index was calculated for each of the variables obtained in the principal component analysis, based on the weighted average of the factor of each variable in the principal component, which allowed an understanding of the average value attributed by municipalities to each of the dimensions. The results indicate that the most valued difficulty is "Difficulties in mobilising participants", as shown in Table 9; in second place is component 4 - "Initiation of the participatory budget process". Despite being a difficulty felt, the least valued component is component 2 - "Characteristics of the population".

Table 9. Mean of the components of difficulties

	Overall Average	Size of the Municipality	
		<10 Thousand Inhabitants (Average Value)	> 10 Thousand Inhabitants (Average Value)
Component 1: Difficulties in mobilising stakeholders	3,427	3,61	3,36
Component 2: Population Characteristics	2,342	3,34	2
Component 3: Implementing the participatory budget process	2,76	2,78	2,75
Component 4: Initiating the participatory budget process	2,871	2,54	2,98
Component 5: Insufficient resources	2,649	2,77	2,61

Source: own preparation

If we evaluate the mean value of the components extracted from the difficulties according to the size of the municipalities, we can see that, regardless of the size of the municipality, the greatest difficulty is reflected in component 1 - "Difficulties in mobilising participants". The component that is least valued by municipalities with less than 10,000 inhabitants is component 2, which refers to the characteristics of the municipal population.

3.5 Difficulties Perceived by Municipalities That Have Never Implemented Participatory Budgeting

In relation to the municipalities that have not implemented participatory budgeting, we intented to analyse the main the barriers to that fact.

According to Table 10, only the variable "Time needed for the preparation and development of the participatory budget" is considered a possible impediment, but its average rating is 3, with 15.8%. The remaining response options are more disagreed with by the respondents with regard to the fact that they may have constituted an impediment to the implementation of participatory budgets. This is the case, for example, with the variable "Insufficient financial resources for the implementation and development of the programme", with an average of 2.43 on the Likert scale. The same occurs with "Difficulties in attracting participants for the preparation of proposals", with an average of 2.52, and for the variable "Lack of time on the part of the population to get involved in the participative budget", with an average value of 2.25, contributing to the logic that it is not considered a barrier.

The mean value of these criteria does not exceed 3, corroborating the fact that none of the criteria is considered an impediment to the implementation of participatory budgets.

Table 10. Barriers to the implementation of participatory budgeting

	%1	%2	%3	%4	%5	Average	Standard Deviation
Insufficient financial resources for the implementation and development of the programme	15,8	5,0	13,9	5,9	3,0	2,43	1,301
Lack of human resources for the implementation and development of the programme	11,9	3,9	14,9	8,9	5,9	2,87	1,375
Lack of material resources for the implementation and development of the programme	16,8	3,0	15,8	7,9	1,0	2,40	1,232
Insufficient technological resources for the implementation and development of the programme	14,9	5,9	15,8	5,9	2,0	2,42	1,215
Lack of teams with the necessary *know-how* for the implementation and development of participatory budgeting in the municipality	12,9	5,0	14,9	8,9	4,0	2,70	1,314
Bureaucracy associated with the preparation and development of the participatory budget	12,9	5,0	14,9	9,9	1,0	2,57	1,208
Time needed for the preparation and development of the participatory budget	9,9	3,0	15,8	10,9	5,9	3,00	1,317
Preparation of communication materials	14,9	6,9	13,9	7,9	0	2,34	1,140
Activities related to transparency	16,8	9,9	12,9	4,0	0	2,09	1,030
Difficulties in mobilising the population to participate in the participatory budget	8,9	9,9	12,9	7,9	4,0	2,73	1,246
Difficulties in mobilising other local actors for participation in the participatory budget	9,9	10,9	9,9	10,9	3,0	2,69	1,258
Difficulties in attracting participants to prepare proposals	10,9	10,9	12,9	5,9	3,0	2,52	1,210
Difficulties in attracting people willing to participate in the voting process	13,9	7,9	13,9	5,9	3,9	2,47	1,254
Difficulties in the analysis and selection of projects	16,8	6,9	12,9	5,9	1,0	2,25	1,184
Difficulties associated with monitoring the implementation of the selected proposals	14,9	9,9	9,9	6,9	1,0	2,28	1,182
Unawareness on the part of the population of the concept of participatory budgeting	11,9	8,9	10,9	7,9	4,0	2,61	1,316
Unawareness or difficulty on the part of the population in using digital tools	12,9	11,9	8,9	6,9	3,0	2,43	1,265
Lack of time for the population to get involved in the participatory budget	14,9	10,9	10,9	5,9	1,0	2,25	1,144
Low number of inhabitants in the municipality	20,8	7,9	8,9	4,0	2,0	2,05	1,219
Low level of education of the resident population	19,8	9,9	7,9	5,0	1,0	2,02	1,151
Average age of the municipality's inhabitants	18,8	7,9	9,9	5,0	2,0	2,16	1,238

Note: 1 Strongly disagree, 2 Disagree, 3 Neither disagree nor agree, 4 Agree, 5 Strongly agree
Source: own preparation

The data were then reduced once again by means of factor analysis. This option allowed the reduction of the information concerning the barriers to the implementation of the participatory budget into two main components, as shown in Table 11. This analysis only focused on municipalities with a negative response to the implementation of the participatory budget.

The main component 1, called "Difficulties internal to the municipality," comprises the variables "Insufficient financial resources for the implementation and development of the programme," "Shortage of human resources for the implementation and development of the programme, ""Lack of material resources for the implementation and development of the programme," "Insufficient technological resources for the implementation and development of the programme," "Lack of teams with the know-how required for the implementation and development of the participatory budget in the municipality," "Bureaucracy associated with the preparation and development of the participatory budget," "Time required for the preparation and development of the participatory budget," "Preparation of communication materials," "Activities related to transparency" and "Difficulties associated with monitoring the implementation of the selected proposals".

Component 2, called "Difficulties related to the exterior," comprises the variables "Difficulties in mobilising the population for participation in the participatory budget," "Difficulties in mobilising other local actors for participation in the participatory budget," "Difficulties in attracting participants for the preparation of proposals," "Difficulties in attracting people available to participate in the voting process," "Difficulties in the analysis and selection of projects," "Unawareness on the part of the population of the concept of participatory budget," "Unawareness or difficulty on the part of the population in using digital tools," "Lack of time on the part of the population to get involved in the participatory budget," "Low number of inhabitants of the municipality," "Low level of education of the resident population" and also, "Average age of the municipality's inhabitants".

Table 11. Factor analysis of the barriers

	Factorial Loading	Eigen Values	% of Variance	% Cumulative Variance	Cronbach's Alpha
Component 1: Difficulties internal to the municipality		13,860	65,998	65,998	0,950
Insufficient financial resources for the implementation and development of the programme	0,714				
Lack of human resources for the implementation and development of the programme	0,846				
Lack of material resources for the implementation and development of the programme	0,804				
Insufficient technological resources for the implementation and development of the programme	0,823				
Lack of teams with the necessary *know-how* for the implementation and development of participatory budgeting in the municipality	0,752				
Bureaucracy associated with the preparation and development of the participatory budget	0,837				
Time needed for the preparation and development of the participatory budget	0,703				
Preparation of communication materials	0,793				
Activities related to transparency	0,776				
Difficulties associated with monitoring the implementation of the selected proposals	0,682				
Component 2: External related difficulties		2,199	10,473	76,471	0,962
Difficulties in mobilising the population to participate in the participatory budget	0,895				
Difficulties in mobilising other local actors for participation in the participatory budget	0,833				
Difficulties in attracting participants to prepare proposals	0,906				
Difficulties in attracting people willing to participate in the voting process	0,940				
Difficulties in the analysis and selection of projects	0,630				
Unawareness on the part of the population of the concept of participatory budgeting	0,756				
Unawareness or difficulty on the part of the population in using digital tools	0,714				
Lack of time for the population to get involved in the participatory budget	0,649				
Low number of inhabitants in the municipality	0,636				
Low level of education of the resident population	0,693				
Average age of the municipality's inhabitants	0,778				

Source: own preparation

After that, we have performed the calculation of an index for each of the variables obtained in the principal components analysis, based on the weighted average factor of each variable in the principal component. These allowed to understand the mean value assigned by municipalities to each of the dimensions. The results indicate that the most valued difficulty is "Difficulties internal to the municipality," and by contrast, the least valued is "Difficulties related to the exterior," according to table 12.

According to Table 12, taking into account the size of the municipalities, in municipalities with less than 10,000 inhabitants, the biggest perceived barrier to the non-implementation of participatory budgeting is related to issues/ difficulties internal to the municipality, and the same occurs in municipalities with more than 10,000 inhabitants.

Table 12. Mean of the components of the barriers

	Overall Average	No. of Inhabitants	
		<10 Thousand Inhabitants (Average Value)	> 10 Thousand Inhabitants (Average Value)
Component 1: Difficulties internal to the municipality	2,484	2,73	2,25
Component 2: External related difficulties	2,396	2,56	2,24

Source: own preparation

3.6 Discussion of Results

The main objective of this chapter was to understand how the participatory budget is used in the different municipalities, as well as to what extent its implementation is conditioned by its characteristics, namely its size. The next step was to understand the advantages of the implementation of the participatory budget and the major difficulties experienced, in order to understand whether these practices of *open innovation* and *crowdsourcing* are positive for public entities and for society.

A literature review was carried out on the main concepts addressed to understand them and help interpret the data obtained. In the analysis of variables, a quantitative approach was followed with the use of a questionnaire survey to all Portuguese municipalities. 101 responses were obtained.

With the analysis above we highlight the cause-effect relationship between the size of municipalities, through their number of inhabitants, with the implementation of participatory budgets, with the budget allocation available for participatory budgets, with the proposals presented by the population, with projects approved by municipalities and with proposals approved. The conclusion is similar for all the variables mentioned here: the larger the size of the municipality, the greater the use of participatory budgets, the greater the budget allocation available for these projects, and the greater the number of proposals presented and projects and proposals approved for implementation.

Before the study was carried out, it was expected that there would be an increase in the values of available budgets, team members, proposals presented for voting, projects and proposals approved in municipalities with a higher average number of inhabitants, as these have a greater weight in terms of size, and it was expected that there would be greater values available in the different variables.

It was also possible to find explanatory factors for the advantages of the use of participatory budgets, the difficulties experienced and the obstacles to the implementation of mechanisms of civic participation, in this case the participatory budget. With regard to the advantages, four components were found: "Communication and transparency", the component related to "Citizen involvement", "Visibility and optimisation of resources" and the last component "Generation of ideas." The "Communication and transparency" component has a greater impact on the advantages perceived by municipalities.

In relation to difficulties, five explanatory components were identified: "Difficulties in mobilising participants", "Characteristics of the population", difficulties related to "Implementation of the participatory budget process", "Initiation of the participatory budget process" and the "Insufficient resources" component. The greatest difficulty lies in the component "Difficulties in mobilising participants". Sometimes the dissemination and the entire process is prepared by the entities responsible, but the message is not picked up in the best way by the population, and there are still difficulties in obtaining participants in these social initiatives.

Similarly, for the municipalities that have not yet implemented the participatory budget, there are two components that are perceived as barriers to implementation, which are related to factors internal and external to the municipality: the component "Difficulties internal to the municipality" and the component "Difficulties related to the exterior." The biggest barrier felt is related to internal issues of the municipalities, i.e. the population is not seen as a barrier, but bureaucratic and internal issues.

Andrade (2015), in the literature review, mentioned the increase of projects in the area of sustainability through the use of *open innovation*; however, the study carried out showed that the proposals presented focus mostly on the rehabilitation of infrastructures and the well-being of the population. In line with Cabannes (2015), the participatory budget gives voice to projects related to primary services.

According to Andrade (2015), the success of *open innovation* ideas will depend on internal and external factors. The author also mentions difficulties in obtaining participants, and this is one of the components of the difficulties perceived by municipalities. The fact that the participatory budget is a long and time-consuming process to implement is in line with what is mentioned by Moura et al. in their 2008 study.

Crowdsourcing enables the use of specific groups of people to obtain ideas, as mentioned by Barbosa (2015). The analysis carried out allows to understand that citizens participate with ideas, and two of the major advantages pointed out by the use of participatory budgeting are the generation of ideas and the involvement of citizens. This is also mentioned by Franklin, Ho and Ebdon (2009) as an advantage in the use of participatory budgets.

Hammon (2012), mentions the existence of an improvement in the relationships between those involved in the companies. This is also one of the advantages found in the study, where through the participatory budget it is verified that there is an increase in proximity, and one of the components perceived as most beneficial is communication and transparency. Another advantage perceived is the visibility and optimisation of resources, corroborated in the literature by Melo et al. (2015), referring to less time and money spent.

The characteristics of the population are pointed out as a difficulty in the use of participatory budgeting. As mentioned by Barbosa (2015), the characteristics of the network (the *crowd)* will influence the achievement of results in a positive or negative way. The fact that *online crowdsourcing* platforms are used, as mentioned by the same author, is considered a barrier to the implementation of participatory budgets, being seen as an external factor for implementation, through the access difficulties experienced by some members of society to access these platforms.

4. CONCLUSION

The use of participatory budgeting is seen quite positively and with low-relief financial implications. Its use is quite important for the improvement and solidification of political relations with the population, allowing to foster a larger process of transformation of government practices.

The use of participatory budgeting in Portugal has evolved considerably, particularly in municipalities, where it is most widely used. For this reason, the aim of this research was to carry out a survey on the factors that lead to its implementation and to find out the reasons for the non-implementation of participatory budgeting or the difficulties experienced in its implementation.

The contributions made in this chapter relate to the survey on the use of participatory budgeting by municipalities in Portugal, an understanding of the characteristics of participatory budgeting and its implementation. It also seeks to understand the advantages and limitations perceived by the use of participatory budgets. This chapter aims to achieve an analysis of the use of *open innovation* and *crowdsourcing* in the public sector in Portugal, through the use of participatory budgets. This chapter also seeks to understand the influence of the characteristics of participatory budgets already implemented, their perceived advantages and the difficulties experienced. Moreover, the chapter also allowed to understand the influence of municipal characteristics on the implemented participatory budgets.

The main advantages found by the use of participative budgets are summarised in terms of communication and transparency between the municipality and its citizens, the involvement of citizens in the political issues of society, the visibility and optimisation of existing resources and the way of generating ideas, where the population is used as a mechanism for generating ideas for the municipality. With regard to the difficulties experienced with the implementation of the participative budget, these are summarised in five categories, namely with the mobilisation of citizens for participation in these initiatives, with characteristics of the population of each municipality, such as age and education, with bureaucratic difficulties for the initiation and implementation of the process and with the insufficiency of resources existing in some localities. In general terms, the perceived benefits tend to be greater than the limitations pointed out by the municipalities, which means that this is a mechanism with potential for the development of municipalities.

The analysis carried out shows that 49 municipalities have not adhered to the participatory budget, and we tried to understand the reasons felt to be barriers to this. The analysis sought to understand the reasons that are perceived as barriers to this. Two explanatory components resulted, related to external and internal factors, with internal factors being the greatest barrier to the non-implementation of participatory budgets.

The literature studied on the main concepts of *open innovation*, *crowdsourcing* and participatory budgeting is still quite limited, and there are still very few studies that seek to systematise the use of participatory budgets by municipalities in Portugal. This research makes it possible to broaden existing knowledge on participatory budgets in Portugal. Participatory budgeting is a mechanism of recent implementation, being implemented over the last eight years in most municipalities, and therefore reveals the existence of little literature in this field. For those in charge of municipalities and public entities responsible for initiatives such as participatory budgeting, this study conveys more information regarding points for improvement, such as the difficulties experienced by those who have already implemented participatory budgeting. It is also important for municipalities and public authorities to understand the main barriers that inhibit the implementation of participatory budgets. This study fosters a solidified implementation with fewer barriers, since the essential characteristics are already known and where it is necessary to work harder so that the same difficulties are not perceived. The barriers perceived can be overcome because they are already known and can be worked on, allowing them to be surpassed. For citizens, the study is also important, as participatory budget is one of the instruments that can be used to improve their life in the surrounding society, supported on the usefulness of their participation and fostering their activism in the municipality. Participatory budgeting in municipalities seeks to respond to the needs of the population and increase proximity between the population and their leaders. This path towards a future with more participatory budgets implemented is a strategy for the development of municipalities and, consequently, of society, breaking the barriers of the traditional approaches and making all those involved closer and more active. In the current context, it is crucial to use different means of financing and to obtain the population's needs in an agile and non-timely manner, so that the human being is more communicative, enlightened and empathetic. Dubious situations and information that is not easily accessible, lead to misinformation and generate contrary feelings in the population.

The methodology used in this study, through a questionnaire survey, was the most adequate for the production of a portrait of the current situation of Portuguese municipalities regarding the implementation and development of participatory budgets. In a future investigation, other variables could also be added, such as more in-depth reasons for non-implementation and the bureaucracies involved. It is also important to understand the obstacles that come from the government and the ways in which civic participation initiatives can be increased. The theme could also be studied through a different research method. By using more in-depth questionnaires or interviews with the heads of the municipalities, more detailed information can be expected. This study requires a large amount of time, as these are public entities, that entail lengthy and bureaucratic procedures. In the future, it would also be positive to carry out a study that focuses on citizens and their perceptions regarding the development of participatory budgets.

REFERENCES

Andrade, M. C. F. (2015). Evidências teóricas para compreensão da inovação aberta (open innovation) nas organizações. *Prespectivas em Gestão & Conhecimento*, *5*(1), 31-42. https://periodicos.ufpb.br/ojs2/index.php/pgc/article/view/19222

Baggio, D., Gavronski, I., & Lima, V. Z. D. (2019). Inovação aberta: Uma vantagem competitiva para pequenas e médias empresas. *Revista Ciências Administrativas*, *25*(3), 3. Advance online publication. doi:10.5020/2318-0722.2019.7507

Barbosa, A. M. A. (2015). *Recomendação de Boas Práticas no Crowdsourcing – um Caso de Estudo* [Dissertação de mestrado, Faculdade de Engenharia da Universidade do Porto]. Repositório Aberto da Universidade do Porto.

Bernardino, S. J. Q., & Santos, J. F. (2020). Crowdsourcing ideas for public investment: the experience of youth participatory budgeting in Portugal. In Crowdasset – Crowdfunding for policy makers (pp. 353–379); Worcropld Scientific Publishing, https://doi.org/ doi:10,1142/11485

Bogers, M., Chesbrough, H., & Moedas, C. (2018). Open Innovation: Research, Practices, and Policies. *California Management Review*, *60*(2), 5–16. doi:10.1177/0008125617745086

Brabham, D. C. (2008). Crowdsourcing as a Model for Problem Solving: An Introduction and Cases. *Convergence (London), 14*(1), 75–90. doi:10.1177/1354856507084420

Brabham, D. C. (2013). Using crowdsourcing. *Government.* IBM Center for the Business of Government.

Cabannes, Y. (2009). *72 Perguntas frequentes sobre o orçamento participativo.* UN-Habitat.

Cabannes, Y. (2015). The impact of participatory budgeting on basic services: Municipal practices and evidence from the field. *Environment and Urbanization, 27*(1), 257–284. doi:10.1177/0956247815572297

Cabannes, Y., & Baierle, S. (2004). Financiamento local e orçamento participativo. Seminário de Lançamento da Rede URB-AL n.o 9. Porto Alegre, Brasil.

Chesbrough, H. (2003). The era of open innovation. *MIT Sloan Management Review, 44*(3), 35–41.

Chesbrough, H. (2011). Everything you need to know about open innovation. *Forbes.* https://www.forbes.com/sites/henrychesbrough/2011/03/21/everything-you-need-to-know-about-open-innovation/

Chesbrough, H. W., & Crowther, A. K. (2006). Beyond high tech: Early adopters o open innovation in other industries. *R & D Management, 36*(3), 229–236. doi:10.1111/j.1467-9310.2006.00428.x

Costa, C. A. S. (2014). *Mecanismos de participação cívica: orçamento participativo.* [Dissertação de Mestrado, Escola de Economia e Gestão da Universidade do Minho]. Repositorium.

Dahlander, L., & Gann, D. M. (2010). How open is innovation? *Research Policy, 39*(6), 699–709. doi:10.1016/j.respol.2010.01.013

De Renzio, P., & Wehner, J. (2017). The impacts of fiscal openness. *The World Bank Research Observer, 32*(2), 185–210. doi:10.1093/wbro/lkx004

Dias, J. C. V., Sant'Anna, L. S., & Portilho, R. M. R. (2018). Open Innovation e o Papel Instrumentalizador dos Contratos. *RFD-Revista da Faculdade de Direito da UERJ., 0*(34), 77–100. doi:10.12957/rfd.2018.18126

Estellés-Arolas, E., & González-Ladrón-de-Guevara, F. (2012). Towards an integrated crowdsourcing definition. *Journal of Information Science, 38*(2), 189–200. doi:10.1177/0165551512437638

Fernandes, D. B. G. (2012). *O Crowdsourcing: Quando? Como? E porquê?* [Projeto de Mestrado em Gestão. ISCTE Business School. Instituto Universitário de Lisboa]. Repositório ISCTE.

Ferreira, M. I. R. (2018). *A Inovação Aberta em Portugal – O nível de recetividade das PME* [Dissertação de mestrado]. Católica Porto Business School.

Fetterhoff, T. J., & Voelkel, D. (2006). Managing open innovation in biotechnology. *Research Technology Management, 49*(3), 14–18. doi:10.1080/08956308.2006.11657373

Franklin, A., Ho, A., & Ebdon, C. (2009). Participatory budgeting in midwestern states: democratic connection or citizen disconnection? *Public Budgeting & Finance, 29*(3), 52–73. doi:10.1111/j.1540-5850.2009.00936.x

Freire, R. S., Falcão, F. S., & Rocha, A. C. B. (2022). Conceitos históricos e técnicas da inovação aberta – uma revisão sistemática. *DAT Journal, 7*(1), 322–340. doi:10.29147/datjournal.v7i1.546

Fritz, S., McCallum, I., Schill, C., Perger, C., Grillmayer, R., Achard, F., Kraxner, F., & Obersteiner, M. (2009). GeoWiki.Org: The use ofcrowdsourcing to improve global land cover. *Remote Sensing (Basel), 1*(3), 345–354. doi:10.3390/rs1030345

Gassmann, O., & Enkel, E. (2004). Towards a theory of open innovation: three core process archetypes. *R&D Management Conference*, 1–18.

Gonçalves, A. T., 2009. *Orçamento Participativo, Um Contributo para a Construção da Cidadania e da Democracia nos Níveis Locais.* Instituto da Segurança Social, IP.

Hagel, J., & Brown, J. S. (2008). Creation Nets: Harnessing the Potential of Open Innovation. *Journal of Service Science, 1*(2), 27–40. doi:10.19030/jss.v1i2.4293

Hammon, L., & Hippner, H. (2012). Crowdsourcing. *Business & Information Systems Engineering, 4*(3), 163–166. doi:10.100712599-012-0215-7

Howe, J. (2006). *The Rise of Crowdsourcing.* Wired. https://www.wired.com/2006/06/crowds/

Huizingh, E. K. R. E. (2011). Open innovation: State of the art and future perspectives. *Technovation, 31*(1), 2–9. doi:10.1016/j.technovation.2010.10.002

Junior, D. J. K., Ireland, R. D., & Snow, C. C. (2007). Strategic Entrepreneurship, Collaborative Innovation, and Wealth Creation. *Strategic Entrepreneurship Journal, 01*(3-4), 371–385. doi:10.1002ej.20

Kogut, B., & Metiu, A. (2001). Open-source software development and distributed innovation. *Oxford Review of Economic Policy, 17*(2), 248–264. doi:10.1093/oxrep/17.2.248

Lee, S., Park, G., Yoon, B., & Park, J. (2010). Open innovation in SMEs—An intermediated network model. *Research Policy, 39*(2), 290–300. doi:10.1016/j.respol.2009.12.009

Melo, C. B. B., Nascimento, J. C. H. B., Melo, M. A., Bernardes, J. R., & Sousa, W. D. (2015). Crowdsourcing como uma ferramenta à inovação estratégica empresarial: Uma revisão de literatura. *Revista de Empreendedorismo. Inovação e Tecnologia, 1*(1), 13–24. doi:10.18256/2359-3539/reit-imed.v1n1p13-24

Moreira, P., Dantas, F., & Martin, C. (2016). Algumas contribuições de Milton Santos para a compreensão do plano diretor participativo das cidades. *Caminhos de Geografia, 16*(55), 174-185

Moss, T., Short, J., Payne, T., & Lumpkin, G. (2011). Dual identities in social ventures: An exploratory study. *Entrepreneurship Theory and Practice, 35*(4), 805–830. doi:10.1111/j.1540-6520.2010.00372.x

Neto, J. C. A. C., & Silva, M. E. R. (2013). Orçamento participativo: decisão de muitos, benefícios de todos. *SANARE – Revista De Políticas Públicas, 2*(3). https://sanare.emnuvens.com.br/sanare/article/view/70

Osamuyimen, S., Juan, M. H., & Melissa, S. (2009). Designing crowdsourcing community for the enterprise. *Proceedings of the ACM SIGKDD,* 50-53. 10.1145/1600150.1600168

Pedersen, J., Kocsis, D., Tripathi, S., Tarrell, A., Weerakoon, A., Tahmasbi, N., Xiong, J., & Deng, W. (2013). Conceptual foundations of crowdsourcing: a review of is research. *HICCS-46 Hawaii International Conference on System Sciences.* 10.1109/HICSS.2013.143

Pinheiro, L. (2009). *Open Innovation e a redescoberta da importância das relações Universidade-Empresa: o caso paradigmático da Brisa* [Dissertação de mestrado. Faculdade de Economia da Universidade do Porto]. Repositório Aberto da Universidade do Porto.

Sampaio, R. (2016). e-Participatory Budgeting as an initiative of e-requests: Prospecting for leading cases and reflections on e-Participation. *Revista de Administração Pública, 50*(6), 937–958. doi:10.1590/0034-7612152210

Schumpeter, J. (1934). *The theory of economic development: An inquiry into profits, capital, credit, interest, and the business cycle.* Transaction Publishers.

Silva, M. G. V. S. (2017). *Implementação do orçamento participativo nos municípios portugueses* [Tese de Mestrado. Universidade do Minho]. Repositorium.

Sintomer, Y., Herzberg, C., & Röcke, A. (2008). Participatory budgeting in europe: potentials and challenges. *International Journal of Urban and Regional Research, 32*(1), 164–178. doi:10.1111/j.1468-2427.2008.00777.x

Sintomer, Y., Herzberg, C., Röcke, A., & Allegretti, G. (2012). Transnational models of citizen participation: the case of participatory budgeting. *Journal of Public Deliberation, 8*(2), 9. doi:10.16997/jdd.141

Souza, C. (2001). Construção e consolidação de instituições democráticas: Papel do orçamento participativo. *São Paulo em Perspectiva, 15*(4), 84–97. doi:10.1590/S0102-88392001000400010

Spithoven, A., Clarysse, B., & Knockaert, M. (2010). Building absorptive capacity to organise inbound open innovation in traditional industries. *Technovation, 30*(2), 130–141. doi:10.1016/j.technovation.2009.08.004

West, J., & Gallagher, S. (2006). Challenges of open innovation: the paradox of firm investment in open-source software. *R & D Management, 36*(3), 319–331. doi:10.1111/j.1467-9310.2006.00436.x

ADDITIONAL READING

Balestrin, A., & Verschoore, J. (2010). Aprendizagem e inovação no contexto das redes de cooperação entre pequenas e médias empresas. *Organizações & Sociedade, 17*(53), 311–330. doi:10.1590/S1984-92302010000200005

Bernardino, S. J. Q., & Santos, J. F. (2018). Unleashing the intelligence of cities by social innovation and civic crowdfunding: an exploratory study. *International Journal of Technology and Human Interaction.* https://EconPapers.repec.org/RePEc:igg:jthi00:v:14:y:2018:i:2:p:54-68

Bernardino, S. J. Q., & Santos, J. F. (2020). Gender differences and generation of ideas on civic crowdsourcing. In *Approaches to Alternative Entrepreneurship Opportunities* (pp. 353–378). IGI Global. doi:10.4018/978-1-7998-1981-3.ch017

De Vries, M., Nemec, J., & Špaček, D. (2022). *International trends in participatory budgeting between trivial pursuits and best practices.* Palgrave Macmillan. doi:10.1007/978-3-030-79930-4

Gadja, O., Marom, D., & Wright, T. (2020). *CrowdAsset: Crowdfunding for policymakers*. World Scientific Publishing Co.

Wampler, B., McNulty, S., & Touchton, M. (2021). Participatory budgeting. In *Global Perspective*. Oxford University Press.

KEY TERMS AND DEFINITIONS

Crowdsourcing: The process of collecting different types of resources through the sum of contributions obtained from a significant group of people, whereby each person's individual contribution is relatively small to the whole

Idea Generation: Step in the innovation process that includes the recognition and discovery of new possibilities.

Innovation: Process of actively seeking and developing new ways of doing things.

Municipalities: Territory with legal personality and a certain autonomy, consisting of administrative and political bodies.

Open Innovation: Innovation method in which the process of ideas generation is not restricted to the borders of the organisation, involving and calling the outside world in the process of generation and co-contruction of the entity.

Participatory Budgeting: Instrument for involving individuals in the decision-making process regarding resource allocation, who are called upon to contribute to the generation of new project ideas and in the selection of the most promising projects that are worthy of receiving funds for their implementation.

Public Sector: Territory with legal personality and a certain autonomy, consisting of administrative and political bodies.

Chapter 10
The Role of New Media Literacy and Critical Thinking on Predicting Digital Citizenship:
A Cross-Sectional Study With Turkish University Students

Ali Orhan
https://orcid.org/0000-0003-1234-3919
Zonguldak Bülent Ecevit University, Turkey

ABSTRACT

This chapter aimed to investigate the predictive role of new media literacy levels and critical thinking dispositions of university students on their digital citizenship levels. The study was carried out with 124 university students studying in different departments of a state university in Türkiye, and the data were gathered using New Media Literacy Scale, Sosu Critical Thinking Dispositions Scale, and Digital Citizenship Scale. It was concluded that university students had high new media literacy levels and critical thinking dispositions while they presented moderate digital citizenship levels. Also, university students' new media literacy levels and critical thinking dispositions had a moderate and positive relationship with their digital citizenship levels. It was also concluded that university students' new media literacy levels and critical thinking dispositions significantly predicted their digital citizenship levels, and they together explained 17% of the total variance on their digital citizenship levels.

INTRODUCTION

The concept of citizenship has changed a lot in the last decades. While traditional citizenship is seen as a concept related to only public participation in the political processes (Westheimer & Kahne, 2004), today conceptualization of citizenship has become to include a wide range of civic behaviors like taking a role in voluntary activities and community, working to solve social problems and to enhance the quality of

DOI: 10.4018/978-1-6684-8934-5.ch010

community life, and trying to find solutions for social injustices (Levine, 2007; Thorson, 2012; Zukin et al., 2006). Besides, with the rapid development in digital and internet technologies, citizenship practices have been transformed into digital citizenship while they were limited to the physical environment of individuals in the past (Berson & Berson, 2003; Choi & Park, 2023; Hermes, 2006; Kahne & Sporte, 2008; Zaff et al., 2008). New citizenship practices emerged thanks to rapid advancements in new digital technologies such as interaction with political parties, learning news, starting campaigns, organizing voluntary activities on new media platforms, etc. (Dow et al., 2021; Hermes, 2006). As opposed to only voting in the elections in the past, individuals can show their reactions in different ways on various online platforms today (Erdem et al., 2023). They can now have a chance to start online campaigns and organize people to protest something they do not like. It can be said that new digital and internet technologies allow individuals to demonstrate citizenship practices on digital media platforms in this century (Ribble & Park, 2019). Therefore, digital citizenship has started to be a vital competency in this century to enhance the quality of the social life and workforce (Erdem et al., 2023).

Digital citizenship can be defined as the appropriate, ethical, responsible, safe, and empowered internet and digital technologies use (Mattson, 2017; Ribble et al., 2004). According to Oxley (2010), digital citizenship refers to behavioral practices that ensure participating in the digital world in a legal, responsible, and ethical way. In a similar way, digital citizenship is defined by Altınay Gazi (2016) as a set of behaviors and practices which may fertilize the development of individuals and protection of social values in the digital world. While Ohler (2011) defines digital citizenship as a kind of character education in this digital world, Jones and Mitchell (2016) see it as tolerant and respectful behaviors towards other people which can make a positive contribution to civic engagement and they define it as using digital and internet technologies for civic engagement activities. Ribble (2017) proposes six digital citizenship elements which are digital commerce, digital communication and collaboration, digital etiquette, digital fluency, digital health and welfare, digital law, digital rights and responsibility, and digital security and privacy. In other words, it can be said that digital citizenship includes the concepts of safety, responsibility, security, and rights in the digital world (Ribble & Bailey, 2011) and it is a mixture of technical, ethical, political, and social factors (Ribble, 2012). Also, Choi (2016) states that digital citizenship comprises five elements, namely, internet political activism, technical skills, critical perspectives, networking agency, and local/global awareness.

Being a qualified digital citizen is not only about using the internet and digital technologies sufficiently but also about using it responsibly and safely because digital citizens are individuals who can show the norms of responsible and appropriate behaviors while using digital technologies (Ribble & Miller, 2013). ISTE (2023) defines digital citizens as individuals who know their responsibilities, rights, and possible opportunities of living in the digital world, possess the necessary abilities to work and learn in this digital world, and can participate, produce and consume in the digital world in a responsible, legal, safe, and ethical way. Digital citizens can use the internet and digital technologies well and can contribute to the digital world effectively by using these technologies, are aware of possible dangers in the digital world, respect ethical principles and human rights while participating in the digital world, and behave responsibly and appropriately in this world (Greenhow et al., 2009; Orhan, 2023a; Richards, 2010). Also, they are regular and effective internet users (Choi & Cristol, 2021; Mossberger et al., 2012) and can manage their time in the digital world effectively and protect themselves against the possibly harmful information which can spread really fast across online media (Ribble, 2008). Adequate digital citizens are new media and technology literate, can perceive the reality of the digital world and culture, can communicate and participate in this digital world in a secure way, recognize digital privacy, use social media actively, possess the necessary abilities to access, create, use, and evaluate information in the digital world, can make reasonable decisions regarding their behaviors on new media platforms, question the accuracy of each information on these platforms, and look for sound evidence before trusting the veracity of this information (Choi, 2016; Işıklı, 2015; Mason, 2008). They, who are full members of the digital world, also combine knowledge and attitudes with digital skills to participate in social life, use digital technologies to perform their citizenship practices, contribute and demand participatory democracy, and for economic purposes (Işıklı, 2015; Mossberger et al., 2007). Therefore, it can be said that digital citizenship requires responsible and empowered technology use (Ribble & Miller, 2013), ethical, responsible, and safe usage of digital technologies (Frau-Meigs et al., 2017), being intentional consumers and producers on new media platforms (Lauricella et al., 2020), and being adequate critical thinkers.

New Media Literacy and Digital Citizenship

As a consequence of the rapid development of digital technologies, accessing new information and delivery of this information to hundreds of thousands of people is easy. However, having the technological devices and being able to use them is not enough to reach reliable and valid information on the internet as well as to participate actively in the digital world. Individuals have to possess some media literacy skills to produce and/or consume content on new media platforms including the internet (Eshet, 2012). This rapid development in internet and mobile technologies has also transformed the term of media into new media. Instead of old printed text and analog broadcasting, online publication and digital media platforms are popular today which means that not only authorized people (e.g. newspaper owners, website editors, etc.) but also regular internet users may produce content to show their ideas, ideologies, and social values on digital media (Koç & Barut, 2016). Therefore, it can be said that the distinction between the producers and consumers of the content on new media, which refers to technology-based socio-cultural platforms on which digitally coding and distribution of messages by any users are possible, has been blurred.

The rise of new media has also triggered the birth of new media literacy (NML). It is an umbrella term which includes various types of literacies like audiovisual (electronic media), classic (e.g., writing), information, and digital literacies and it is related to a set of some vital skills which help individuals access, analyze, evaluate, criticize, produce and/or participate media content in a responsible and effective way (Hobbs & Jensen, 2009; Lee et al., 2015). In other words, NML is about some skills which are employed by individuals during their interactions with media content and individuals have two different roles as consumers and producers during their interaction with media content (Hsiao et al., 2023; Toffler, 1981). While traditional media literacy was seen as having skills related to accessing and understanding media content (i.e., consuming it), NML also includes (re)producing media content and sharing it with other people (i.e., producing media). NML includes both some functional literacies that are related to individuals' ability to know how media content is created, disseminated, or commercialized (Thoman & Jolls, 2004) and some critical literacies which are related to individuals' ability to understand the media content and to evaluate its credibility, accuracy, and usefulness (Buckingham, 2003).

New media-literate individuals are able to consciously use different media platforms, (re)produce new content on these platforms, recognize different kinds of media content, and critically examine its types, effects, and the hidden messages they have (Kellner & Share, 2007). Therefore, it can be said that NML equips individuals with the required abilities for actively accessing, decoding, understanding, evaluating, and analysis of the hidden messages of various types of media content (Buckingham, 2015).

In other words, individuals take a critical standpoint and make independent judgments on the accuracy of media content (Leaning, 2017). Therefore, NML transforms individuals into active consumers of media content instead of passive consumers who accept the veracity of media content without thinking (Hobbs, 2017). In short, individuals with high NML are able to create innovative media content, interact with other users on media platforms, actively participate in content creation, understand potential biases and underlying messages of media content, and investigate different types of media content critically (Jenkins et al., 2006; Kellner & Share, 2005). They are also conscious of the power of mediated messages of the media content and consequences of them and they know that these media contents can be rapidly spread to large groups of people. That is to say, new media-literate individuals can consciously and responsibly produce and consume media content on new media platforms.

As digital citizenship includes using digital technologies ethically, responsibly, appropriately, and safely (Ribble & Miller, 2013), NML is an important skill to be intentional users and consumers of digital technologies and new media content, and to be digital citizens (Gallagher, 2014; Lauricella et al., 2020). Choi et al (2017) include media and information literacy among the four dimensions of digital citizenship along with participation and engagement, critical resistance, and digital ethics. Therefore, we can say that NML is important for being adequate digital citizens. Through their NML skills, individuals can consciously use different tools and media platforms, evaluate and analyze media content critically, (re)produce media content to express their beliefs, opinions, or views, participate in civic and social life in the digital world, claim and protect their rights and show respect to other people's rights, and solve problems on digital media platforms (Leaning, 2017; Jenkins et al., 2006; Kellner & Share, 2005; Hobbs, 2017) which are also important practices of being digital citizens (Ribble & Bailey, 2011; Ohler, 2011; Jones & Mitchell, 2016). Thanks to NML skills, individuals can engage in civic processes and advocate their beliefs which makes them informed digital citizens (Mihailidis & Thevenin, 2013). As high NML skills equip individuals with the ability to critically and consciously examine media content and its messages in the digital and online world, NML may support individuals' civic participation and promote digital citizenship (McGrew, 2020; Lauricella et al., 2020). Individuals use their NML skills to investigate, create, share, and interpret civic and digital messages and this can contribute to the enhancement of their democratic capacities and civic life on the internet (Hobbs et al., 2013), and hence, they become promising digital citizens.

In their study aiming to examine the relationship between practices of active citizenship and different aspects of media literacy, Davidson and Silkane (2019) concluded that active citizenship and citizen engagement are significantly predicted by critical prosumption and functional prosumption aspects of NML. Haskey-Valerius et al. (2022) found that preservice teachers think that digital citizenship and media literacy are deeply connected in their qualitative study investigating the relationship between media literacy and digital citizenship based on preservice teachers' perspectives. Erdem et al. (2023) carried out a study to investigate the relationship between digital citizenship and media literacy and the mediating effect of digital literacy in this relation and they found that there is a direct relation between media literacy and digital citizenship. In their study, Kahne et al., (2012) concluded that media literacy education enhances young people's political engagement on online platforms which can be seen as a critical indicator of digital citizenship. Similarly, Park et al. (2021) found that media literacy education significantly affected individuals' digital citizenship skills in their study. Also, Beatty (2019) concluded that NML is effective to enhance the engagement of students on social media platforms which results in the development of voice that encourages a higher level of online civic engagement as well as a high level of digital citizenship.

Critical Thinking and Digital Citizenship

Critical thinking (CT) can be briefly defined as a logical, reasonable, and reflective thinking process employed by individuals to make a judgment on the accuracy and quality of ideas, arguments, and knowledge (Beyer, 1995) and to decide what to believe or do (Ennis, 1993). Through CT, individuals can make reasonable and accurate decisions on the accuracy of other people's ideas, arguments, or opinions as well as about their behaviors and own ideas. Also, CT refers to evaluating the source of information/ knowledge, analyzing its validity and reliability, and drawing inferences appropriately (Paul, 1990). Therefore, individuals are able to acquire the most accurate and useful information, question and examine all ideas and assumptions, defend their own ideas, and protect themselves against false information through CT (Epstein & Kernberger, 2012; Facione, 1990) which may be defined as a rational and logical way of dealing with opinions, assumptions, arguments, or ideas (Ruggerio, 1990). Critical thinkers are inquisitive, open-minded, and reasonable individuals and they always question, evaluate, and examine all ideas, arguments, and assumptions, and they are willing to reconsider their ideas after this evaluation and examination process (Facione, 1990).

Today, individuals are expected to express their opinions, ideas, or views, make decisions, accept/ refuse other people's arguments, views, or opinions, discuss with them, shape arguments and a personal perspective, evaluate ideas, opinions, or arguments critically, and make his/her views public to be adequate digital citizens (Ten Dam & Volman, 2004) and this makes CT an important skill for being digital citizens. Also, digital citizens are full members of online communities and have to build cooperative and collaborative practices in the digital world (Choi et al., 2017). Besides, they are ethical individuals who manage their actions and consequences and understand the risks and benefits of easy access to information (Ohler, 2010; Mossberger et al., 2007). Therefore, it can be said that CT is required to be an adequate digital citizen because it equips individuals with the necessary skills to behave ethically, responsibly, appropriately, and safely in the digital world. Digital citizenship is also about building relationships in online communities, having acceptance and autonomy, using support and services, understanding social roles and shared values, and fulfilling civil responsibilities and rights in the digital world (MacIntyre et al., 2022). CT allows individuals to participate in the community actively, respect other people's views, opinions, or ideas, develop social and moral values, and maintain a safe environment in the digital world. Also, critical thinkers are willing to question all ideas, assumptions, and arguments and they are open to divergent ideas (Facione, 1990). Therefore, they can defend their arguments, discuss with other people, question other people's arguments, and are good at listening to other people and conflict resolution (Ruggerio, 1988). Besides, CT is effective on individuals' communication skills and the ability to influence other people. Therefore, CT can be seen as a critical aspect of digital citizenship because it helps individuals engage in a democratic and pluralistic society and it empowers individuals to influence other people and society (Puolimatka, 1995). In today's digital world, individuals are no longer passive consumers of information but active communicators and producers of this information and they have a chance for practicing analytical and active information acquisition and having an influence using various online platforms through their digital citizenship skills. Also, they need to effectively reflect on information and their behaviors regarding their civil rights and responsibilities (Halstead & Pike, 2006). Therefore, it can be said that being a digital citizen necessarily requires high CT skills and dispositions.

Assante et al. (2022) concluded that CT dispositions are significantly related to digital citizenship in their study aiming to examine the effect of personal values and CT dispositions on students' digital citizenship using structural equation modeling. Similarly, Xu et al. (2019) concluded that digital citizenship was significantly predicted by CT along with some other 21st century skills like creative thinking, collaboration, problem-solving, etc. in their study with 962 college students. Also, Diaz et al. (2022) found a significant correlation among different variables including CT and digital citizenship in their study with 432 university students.

The Current Study

The rapid development in digital and internet technologies has triggered the moving of citizenship practices into the digital world in today's modern society and digital platforms and tools have become one of the most important and widely used environments for citizenship practices. Therefore, digital citizenship is an important concept today and it requires behaving safely, legally, and ethically in this digital world (Ribble, 2017). Digital citizenship has an effective role on the positive development of individuals and societies (Milenkova & Lendzhova, 2021) and digital citizens have a positive role in building effective communities (Bennett, 2008). Therefore, the enhancement of digital citizenship skills should be included among the most important aims of today's education systems and these skills should be supported as a part of 21st century curriculums (Bennett et al., 2016; Oudeweetering & Voogt, 2018; Ribble, 2017; Smith & Storrs, 2023).

As the importance of digital citizenship is clear for modern societies, investigating factors that may be effective on promoting it is important. Previous literature has indicated that CT and NML may play a critical role on how people participate in the digital world and fulfill their civic responsibilities. Also, previous research has provided evidence for the positive role of NML (Kahne et al., 2012; Davidson & Silkane, 2019; Haskey-Valerius et al., 2022; Erdem et al., 2023; Park et al., 2021; Beatty, 2019) and CT (Assante et al., 2022; Diaz et al., 2022; Xu et al., 2019) on digital citizenship. However, there has not been much research regarding the predictive power of NML and CT dispositions on digital citizenship levels of university students. Therefore, it can be said that NML and CT dispositions are vital requirements to be adequate digital citizens and should be promoted through training (Gallagher, 2014; Hobbs, 2010) and there is a clear need to investigate the predictive role of NML and CT dispositions on digital citizenship because despite existing theoretical support for the relationship among these variables, empirical evidence regarding the predictive role of NML and CT dispositions on digital citizenship is limited. Therefore, the aim of this study was to investigate whether university students' NML and CT dispositions are significant predictors of their digital citizenship levels. To this end, the following questions were sought:

1. What are the levels of university students' NML, CT dispositions, and digital citizenship?
2. Are university students' NML and CT dispositions significant predictors of their digital citizenship levels?

METHOD

A cross-sectional survey design was employed in this study. University students' digital citizenship level was the dependent variable of the study while their NML and CT dispositions were predictor variables.

Study Group

The study group consists of 124 university students (64 female and 60 male) studying in different departments of a state university in Türkiye. The mean age of the students was 20.71 (SD=1.97) ranging from 18 to 26. A-priori power analysis with G*Power 3 software program (Faul et al., 2007) revealed that a minimum of 107 data was required to conduct a linear multiple regression analysis (alpha=0.05; power=0.95; 2 predictor variables) for a medium effect size (f^2=0.15). Therefore, the sample size of 124 in this study was enough.

Data Collection Tools

New Media Literacy Scale (NMLS)

NML levels of university students were determined using the NMLS developed by Koç & Barut (2016). The NMLS, which uses a 5-point Likert scale (from strongly disagree to strongly agree), consists of 35 items and four sub-dimensions which are functional consumption, critical consumption, functional prosumption, and critical prosumption. The reliability value of the sub-dimensions of the NMLS ranged between 0.85 and 0.93 and it was found as 0.95 for the total scale in Koç & Barut's (2016) study while it was 0.94 for the total scale in this study.

Sosu Critical Thinking Dispositions Scale (CTDS)

CT dispositions of university students were determined using the CTDS which uses a 5-point Likert scale (from strongly disagree to strongly agree). Sosu (2013) developed the scale and Orhan (2023b) adapted it into Turkish. It has 11 items and two sub-dimensions (critical openness and reflective skepticism). Turkish adaptation study which was carried out with two independent samples has revealed that the factor structure of the Turkish scale is consistent with the original scale. The reliability value of the CTDS was 0.92 for the first sample and 0.94 for the second sample in the adaptation study while it was 0.88 in this study.

Digital Citizenship Scale (DCS)

Digital citizenship levels of university students were determined using the DCS. Choi et al. (2017) developed the scale and Erdem and Koçyiğit (2019) adapted it into Turkish. The scale, which uses a 7-point Likert scale (from strongly disagree to strongly agree), consists of 18 items and five sub-dimensions which are internet political activism, technical skills, local/global awareness, critical perspective, and networking agency. The reliability value of the sub-dimensions of the DCS ranged between 0.61 and 0.93 and it was found as 0.87 for the total scale in the adaptation study while it was calculated as 0.84 for the total scale in this study.

Data Collection

Ethical committee approval was obtained from ZBEU and the data for the present study were gathered in the fall term of the 2022-2023 academic year. All students were informed of privacy and confidentiality issues and their right to withdraw from the study anytime they want. The instruments were completed in about 20-25 minutes.

Data Analysis

After checking all of the variables to ensure that the dataset did not have any missing data, skewness and kurtosis values were investigated and it was revealed that the dataset showed normal distribution (see Table 1). Next, Z transformation and Mahalanobis Distance values were examined to check possible outliers per variable and multivariate outliers and it was found that there were not any influential outliers in the dataset. Besides, an investigation of Pearson correlation, VIF, CI, and tolerance values revealed that there was no high correlation among the variables. Descriptive statistics, Pearson correlation, and multiple linear regression with enter method were used while analyzing the data.

Results

Descriptive statistics for the NMLS, CTDS, and DCS can be seen in Table 1.

Table 1. Descriptive statistics and skewness and kurtosis values for the NMLS, CTDS, and DCS

	Skewness		Kurtosis			
	Statistic	**Std. Error**	**Statistic**	**Std. Error**	$\bar{X}$	**sd**
New media literacy	0.122	0.217	-0.397	0.431	3.23	0.06
Critical thinking dispositions	-0.741	0.217	-0.064	0.431	3.96	0.06
Digital citizenship	0.161	0.217	-0.063	0.431	3.81	0.08

University students had high NML levels ($\bar{X}$=3.23) and CT dispositions ($\bar{X}$=3.96). However, we can say that their digital citizenship levels ($\bar{X}$=3.81) were relatively low because the students' DCS mean scores can range between one (lowest) to seven (highest).

Table 2. Inter-correlations among the NMLS, CTDS, and DCS scores

	Critical Thinking Dispositions	Digital Citizenship
New media literacy	0.191*	0.286**
Critical thinking dispositions	1	0.357**
Digital citizenship	-	1
Note: * p< 0.05; ** p< 0.01		

Table 2 shows that university students' NML levels ($r = 0.286$) and CT dispositions ($r = 0.357$) had a moderate and positive relationship with their digital citizenship levels. Also, it was revealed that university students' NML levels were weakly and positively correlated to their CT dispositions ($r = 0.191$).

Table 3. Multiple linear regression results

	B	Std. Error	β	t	p
Constant	1.184	0.523	-	2.265	0.02
New media literacy	0.300	0.111	0.226	2.691	0.00
Critical thinking dispositions	0.419	0.112	0.314	3.730	0.00
$R=0.420$, $R^2=0.177$, $F_{(2, 123)}= 12.975$, $p<0.01$					

As shown in Table 3, university students' NML levels ($\beta=0.226$ $t_{(124)=}2.691$, $p<0.05$) and CT dispositions ($\beta=0.314$, $t_{(124)=}3.730$, $p<0.05$) significantly predicted their levels of digital citizenship ($R=0.420$, $R^2=0.177$, $p<0.01$). Also, one-way ANOVA test results indicated that the established regression model was significant ($F_{(2,123)}=12.975$, $p<0.01$). Therefore, it can be said that university students' NML levels and CT dispositions were significant predictors for their digital citizenship levels and 17% of the total variance on their digital citizenship levels were explained by these two variables. Also, CT dispositions of university students ($\beta=0.314$) had a relatively higher effect on digital citizenship levels than NML levels ($\beta=0.226$).

DISCUSSION

The aim of this study was to examine the predictive power of university students' NML levels and CT dispositions on their digital citizenship levels. It was found that university students had high NML levels and CT dispositions while they presented moderate digital citizenship levels. This study also revealed that university students' NML levels and CT dispositions had a moderate and positive relationship with their digital citizenship levels. Also, it was revealed that NML levels of university students were weakly and positively correlated to their CT dispositions.

This study also showed that NML levels and CT dispositions of university students significantly predicted their levels of digital citizenship and 17% of the total variance on their digital citizenship levels were explained by these two variables. Previous literature has provided similar results regarding the positive effect of NML (Kahne et al., 2012; Davidson & Silkane, 2019; Haskey-Valerius et al., 2022; Erdem et al., 2023; Park et al., 2021; Beatty, 2019) and CT (Assante et al., 2022; Diaz et al., 2022; Xu et al., 2019) on digital citizenship. Therefore, it can be said that the results of this study are in line with the results of previous literature.

NML is vital for producing and consuming new media content intentionally and using digital technologies effectively (Frau-Meigs et al., 2017; Gallagher, 2014; Lauricella et al., 2020), and hence, it is also important to be digital citizens because digital citizenship includes using digital technologies ethically, responsibly, appropriately, and safely (Ribble & Miller, 2013). Digital citizens spend a lot of time on different media platforms, use different online tools, (re)produce content to express their beliefs and opinions on these various new media platforms, participate in civic and social life in the digital world, claim and protect their rights, show respect to other people's rights, and they are intentional consumers of the produced content on various new media platforms and always willing to investigate and analyze content on these platforms (Ribble & Bailey, 2011; Ohler, 2011; Jones & Mitchell, 2016). Therefore, it can be said that they need to have high NML skills for these because they are also important practices of being new media-literate individuals (Hobbs, 2017, Leaning, 2017; Jenkins et al., 2006). High NML skills are required for investigating, creating, sharing, and interpreting digital and civic messages (Hobbs et al., 2013), critically and consciously examining media content and the messages they convey (Lauricella et al., 2020; McGrew et al., 2017, McGrew, 2020), and engaging in civic life and advocating beliefs (Mihailidis & Thevenin, 2013) in the digital world. Therefore, it can be said that new media-literate individuals have high democratic capacities and can significantly contribute to civic life on the internet, and are promising and informed digital citizens (Hobbs et al., 2013; Mihailidis & Thevenin, 2013).

Digital citizens, who are ethical individuals and full members of online communities, need to build relationships with other people and collaborative and cooperative practices, have acceptance and autonomy, use support and services, understand social roles and shared values, and fulfill civil responsibilities and rights in the digital world (Choi et al., 2017; MacIntyre et al., 2022). Also, they are expected to manage their actions and consequences, understand the risks and benefits of the digital world (Ohler, 2010; Mossberger et al., 2007), express their opinions and ideas, evaluate content on media platforms critically, and make decisions regarding the veracity of them, investigate arguments, views, or opinions of other people and decide to accept or refuse them, shape their own arguments and have a personal perspective, and make their views public in the digital world (Ten Dam & Volman, 2004). Therefore, we can say that CT is a critical aspect of being digital citizens and individuals are required to possess high CT dispositions and skills to be adequate digital citizens because CT equips individuals with the required abilities to participate in the community actively, respect other people's views, opinions, or ideas, develop social and moral values, and maintain a safe environment in the digital world. In addition, as critical thinkers have better communication skills and the capacity to influence other people, can defend their arguments, discuss with other people, question other people's arguments, and are open to divergent ideas, willing to question all ideas, assumptions, and arguments, and good at listening to other people and conflict resolution (Facione, 1990; Ruggerio, 1988), it can be said that it helps individuals engage in a democratic and pluralistic society, empowers individuals to influence other people and the society, and enables them to behave ethically, responsibly, appropriately, and safely in the digital world and to reflect on behaviors regarding their civil rights and responsibilities (Halstead & Pike, 2006; Puolimatka, 1995), and hence, it is necessarily required to be adequate digital citizens.

In short, digital citizenship includes not only appropriate, responsible, and empowered technology use (Ribble & Park, 2019) but also some abilities to perceive realities of the digital culture and world, communicate in a secure way in this world, use new media platforms effectively, recognize and respect digital privacy, and contribute to and demand participatory democracy (Işıklı, 2015) and it has an effective role on the positive development of individuals and societies (Milenkova & Lendzhova, 2021) and is an important concept for today's modern societies. This study showed that university students' NML levels and CT dispositions were significant predictors of their digital citizenship levels. This result which is confirmed by previous literature and coincides with the theoretical background shows the significant role of NML and CT dispositions in promoting digital citizenship behaviors. We can say that individuals with high NML and CT are expected to be better digital citizens. Therefore, NML and CT dispositions should be implemented during the effort of enhancing the digital citizenship skills of individuals and learning and teaching activities should be established with close relations with the enhancement of NML and CT in supporting the enhancement of digital citizenship skills. Also, it can be suggested that enhancing NML and CT of university students should be one of the most important aims of higher education curriculums because equipping university students with high NML and CT skills will not only make them successful individuals in 21st century but also positively contribute to increasing their digital citizenship skills. This study has made a significant contribution to the literature and extended it by presenting evidence regarding the predictive power of NML and CT dispositions on digital citizenship because there has not been much research regarding the predictive power of NML and CT dispositions on digital citizenship.

LIMITATIONS AND IMPLICATIONS FOR FURTHER RESEARCH

This study has revealed some important results regarding the predictive power of NML and CT dispositions on digital citizenship. However, it has several limitations. As the data were gathered from university students who were recruited from a single university in Türkiye, the study group can be shown as the first limitation. Therefore, it can be suggested that future studies should be carried out with other sample groups including participants from different education levels. Second, self-report quantitative tools were used to collect the data in this study and social desirability can affect the data collected using these tools negatively. Therefore, future studies can employ qualitative or mixed methods to deeply examine the predictive power of NML and CT dispositions on digital citizenship.

REFERENCES

Altınay Gazi, Z. (2016). Internalization of digital citizenship for the future of all levels of education. *Education in Science*, *41*(186), 137–148. doi:10.15390/EB.2016.4533

Assante, G. M., Popa, N. L., & Momanu, M. (2022). How personal values and critical dispositions support digital citizenship development in higher education students. *Frontiers in Psychology*, *13*(990518), 1–11. doi:10.3389/fpsyg.2022.990518 PMID:36267083

Beatty, K. (2019). *Developing student voice through Twitter: Using media literacy education to inspire civic engagement* [Unpublished master's thesis]. University of Rhode Island.

Bennett, L. B., Aguayo, R. C., & Field, S. L. (2016). At home in the world: Supporting children in human rights, global citizenship, and digital citizenship. *Childhood Education*, *92*(3), 189–199. doi:10.1080/00094056.2016.1180892

Bennett, W. L. (2008). Changing citizenship in the digital age. In W. L. Bennett (Ed.), *Civic life online: Learning how digital media can engage youth* (pp. 1–24). MIT Press.

Berson, I. R., & Berson, M. J. (2003). Digital literacy for effective citizenship. *Social Education*, *67*, 164–167.

Beyer, B. (1995). *Critical thinking*. Phi Delta Kappa Educational Foundation.

Buckingham, D. (2003). *Media education: Literacy, learning and contemporary culture*. Polity Press.

Buckingham, D. (2015). Defining digital literacy: What do young people need to know about digital media? *Nordic Journal of Digital Literacy*, *10*, 21–35. doi:10.18261/ISSN1891-943X-2015-Jubileumsnummer-03

Choi, M. (2016). A concept analysis of digital citizenship for democratic citizenship education in the internet age. *Theory and Research in Social Education*, *44*(4), 565–607. doi:10.1080/00933104.2016.1210549

Choi, M., & Cristol, D. (2021). Digital citizenship with intersectionality lens: Towards participatory democracy driven digital citizenship education. *Theory into Practice*, *60*(4), 361–370. doi:10.1080/00405841.2021.1987094

Choi, M., Glassman, M., & Cristol, D. (2017). What it means to be a citizen in the internet age: Development of a reliable and valid digital citizenship scale. *Computers & Education*, *107*, 100–112. doi:10.1016/j.compedu.2017.01.002

Choi, M., & Park, H. J. (2023). Korean adolescents' profiles of digital citizenship and its relations to internet ethics: Implications for critical digital citizenship education. *Cambridge Journal of Education*, *53*(4), 567–586. doi:10.1080/0305764X.2023.2191929

Davidson, A., & Silkane, V. (2019). The relationship between media literacy and civic participation among young adults in Latvia. In S. Kurbanoğlu, S. Spiranec, Y. Ünal, J. Boustany, M. L. Huotari, E. Grassian, D. Mizrachi, & L. Roy (Eds.), *Information literacy in everyday life* (pp. 103–112). Springer. doi:10.1007/978-3-030-13472-3_10

Diaz, J. R. D., Ucharima, I. H., & Egas, M. M. S. (2022). Digital competencies as pedagogical predictors for learning to learn. *Revista De La Universidad Del Zulia*, *13*(38), 643–656.

Dow, B. J., Johnson, A. L., Wang, C. S., Whitson, J., & Menon, T. (2021). The COVID-19 pandemic and the search for structure: Social media and conspiracy theories. *Social and Personality Psychology Compass*, *15*(9), 1–22. doi:10.1111pc3.12636 PMID:34512798

Ennis, R. H. (1993). Critical thinking assessment. *Theory into Practice*, *32*(3), 179–186. doi:10.1080/00405849309543594

Epstein, R. L., & Kernberger, C. (2012). *Critical thinking*. Advanced Reasoning Forum.

Erdem, C., & Koçyiğit, M. (2019). Exploring undergraduates' digital citizenship levels: Adaptation of the digital citizenship scale to Turkish. *Malaysian Online Journal of Educational Technology*, *7*(3), 22–38. doi:10.17220/mojet.2019.03.003

Erdem, C., Oruç, E., Atar, C., & Bağcı, H. (2023). The mediating effect of digital literacy in the relationship between media literacy and digital citizenship. *Education and Information Technologies*, *28*(5), 4875–4891. doi:10.100710639-022-11354-4

Eshet, Y. (2012). Thinking in the digital era: A revised model for digital literacy. *Issues in Informing Science and Information Technology*, *9*, 267–276. doi:10.28945/1621

Facione, P. A. (1990). *Critical thinking: A statement of expert consensus for purposes of educational assessment and instruction - executive summary - the Delphi report*. The California Academic Press.

Faul, F., Erdfelder, E., Lang, A. G., & Buchner, A. (2007). G* power 3: A flexible statistical power analysis program for the social, behavioral, and biomedical sciences. *Behavior Research Methods*, *39*(2), 175–191. doi:10.3758/BF03193146 PMID:17695343

Frau-Meigs, D., Brian, O. N., Soriani, A., & Tome, V. (2017). Digital citizenship education: Overview and new perspectives. Council of Europe Publishing.

Gallagher, F. (2014). Media literacy education: A requirement for today's digital citizens. In D. B. S. Abreu & P. Mihailidis (Eds.), *Media literacy education in action: Theoretical and pedagogical perspectives*. Routledge.

Greenhow, C., Robelia, B., & Hughes, J. E. (2009). Learning, teaching, and scholarship in a digital age Web 2.0 and classroom research: What path should we take now? *Educational Researcher*, *38*(4), 246–259. doi:10.3102/0013189X09336671

Halstead, M., & Pike, M. (2006). *Citizenship and moral education: Values in action*. Routledge. doi:10.4324/9780203088593

Haskey-Valerius, B., Gould, H., & von Gillern, S. (2022). Preservice teachers' conceptualizations of media literacy, digital citizenship, and their relationship. In E. Langran (Ed.), *Proceedings of society for information technology & teacher education international conference* (pp. 1378–1383). Association for the Advancement of Computing in Education.

Hermes, J. (2006). Citizenship in the age of the internet. *European Journal of Communication*, *21*(3), 295–309. doi:10.1177/0267323106066634

Hobbs, R. (2010). *Digital and media literacy: a plan of action*. The Aspen Institute.

Hobbs, R. (2017). Teaching and learning in a post-truth world. *Educational Leadership*, *75*(3), 26–31.

Hobbs, R., Donnelly, K., Friesem, J., & Moen, M. (2013). Learning to engage: How positive attitudes about the news, media literacy, and video production contribute to adolescent civic engagement. *Educational Media International*, *50*(4), 231–246. doi:10.1080/09523987.2013.862364

Hobbs, R., & Jensen, A. (2009). The past, present and future of media literacy education. *The Journal of Media Literacy Education*, *1*(1), 1–11. doi:10.23860/jmle-1-1-1

Hsiao, K. L., Ku, Y. Y., & Lee, Y. T. (2023). Online first). Development and validation of a scale for assessing university students' new online media and reporting literacy. *Education and Information Technologies*. Advance online publication. doi:10.100710639-023-11730-8

Işıklı, S. (2015). Digital citizenship: An actual contribution to theory of participatory democracy. *Online Academic Journal of Information Technology*, *6*(18), 21–38. doi:10.5824/1309-1581.2015.1.002.x

ISTE (International Society for Technology in Education). (2023). *ISTE standards for students*. Retrieved from https://www.iste.org/standards/iste-standards-for-students

Jenkins, H., Clinton, K., Purushotma, R., Robison, A. J., & Weigel, M. (2006). *Confronting the challenges of participatory culture: Media education for the 21st century*. The MacArthur Foundation.

Jones, L. M., & Mitchell, K. J. (2016). Defining and measuring youth digital citizenship. *New Media & Society*, *18*(9), 2063–2079. doi:10.1177/1461444815577797

Kahne, J., Lee, N. J., & Feezel, J. T. (2012). Digital media literacy education and online civic and political participation. *International Journal of Communication*, *6*, 1–14.

Kahne, J. E., & Sporte, S. E. (2008). Developing citizens: The impact of civic learning opportunities on students' commitment to civic participation. *American Educational Research Journal*, *45*(3), 738–766. doi:10.3102/0002831208316951

Kellner, D., & Share, J. (2005). Toward critical media literacy: Core concepts, debates, organizations, and policy. *Discourse (Abingdon)*, *26*(3), 369–386. doi:10.1080/01596300500200169

Kellner, D., & Share, J. (2007). Critical media literacy, democracy, and the reconstruction of education. In D. Macedo & S. R. Steinberg (Eds.), *Media literacy: A reader* (pp. 3–23). Peter Lang Publishing.

Koç, M., & Barut, E. (2016). Development and validation of New Media Literacy Scale (NMLS) for university students. *Computers in Human Behavior*, *63*, 834–843. doi:10.1016/j.chb.2016.06.035

Lauricella, A. R., Herdzina, J., & Robb, M. (2020). Early childhood educators' teaching of digital citizenship competencies. *Computers & Education*, *158*(103989), 1–9. doi:10.1016/j.compedu.2020.103989

Leaning, M. (2017). *Media and information literacy: An integrated approach for the 21st century*. Elsevier Science & Technology.

Lee, L., Chen, D. T., Li, J. Y., & Lin, T. B. (2015). Understanding new media literacy: The development of a measuring instrument. *Computers & Education*, *85*, 84–93. doi:10.1016/j.compedu.2015.02.006

Levine, P. (2007). *The future of democracy: Developing the next generation of American citizens*. University Press of New England.

MacIntyre, G., Cogan, N., Stewart, A., Quinn, N., O'Connell, M., & Rowe, M. (2022). Citizens defining citizenship: A model grounded in lived experience and its implications for research, policy and practice. *Health & Social Care in the Community*, *30*(3), 695–705. doi:10.1111/hsc.13440 PMID:34155710

Mason, M. (2008). *Critical thinking and learning*. Blackwell Publishing. doi:10.1002/9781444306774

Mattson, K. (2017). *Digital citizenship in action empowering students to engage in online communities*. International Society for Technology in Education.

McGrew, S. (2020). Learning to evaluate: An intervention in civic online reasoning. *Computers & Education*, *145*, 1–13. doi:10.1016/j.compedu.2019.103711

McGrew, S., Ortega, T., Breakstone, J., & Wineburg, S. (2017). The challenge that's bigger than fake news: Civic reasoning in a social media environment. *American Educator*, *41*(3), 4–39.

Mihailidis, P., & Thevenin, B. (2013). Media literacy as a core competency for engaged citizenship in participatory democracy. *The American Behavioral Scientist*, *57*(11), 1611–1622. doi:10.1177/0002764213489015

Milenkova, V., & Lendzhova, V. (2021). Digital citizenship and digital literacy in the conditions of social crisis. *Computers*, *10*(40), 1–14. doi:10.3390/computers10040040

Mossberger, K., Tolbert, C. J., & Hamilton, A. (2012). Broadband adoption| measuring digital citizenship: Mobile access and broadband. *International Journal of Communication*, *6*(37), 2492–2528.

Mossberger, K., Tolbert, C. J., & McNeal, R. S. (2007). *Digital citizenship: The Internet, society, and participation*. MIT Press. doi:10.7551/mitpress/7428.001.0001

Ohler, J. (2010). *Digital community, digital citizen*. Corwin Press. doi:10.4135/9781452219448

Ohler, J. (2011). Digital citizenship means character education for the digital age. *Kappa Delta Pi Record, 47*(sup1), 25-27.

Orhan, A. (2023a). Fake news detection on social media: The predictive role of university students' critical thinking dispositions and new media literacy. *Smart Learning Environments*, *10*(29), 1–14. doi:10.118640561-023-00248-8

Orhan, A. (2023b). Investigating psychometric properties of the Turkish version of Sosu Critical Thinking Disposition Scale: Evidence from two independent samples. *International Journal of Psychology and Educational Studies*, *10*(2), 348–359. doi:10.52380/ijpes.2023.10.2.1017

Oudeweetering, K., & Voogt, J. (2018). Teachers' conceptualization and enactment of twenty-first century competences: Exploring dimensions for new curricula. *Curriculum Journal*, *29*(1), 116–133. doi: 10.1080/09585176.2017.1369136

Oxley, C. (2010). *Digital citizenship: Developing an ethical and responsible online culture*. International Association of School Librarianship.

Park, Y. M., Chae, J. H., Kim, S. K., & Kwon, H. S. (2021). The effects of media literacy education and its influence on digital citizenship: Focusing on CMF education programs in Korea. *Journal of Information Systems*, *30*(3), 113–135.

Paul, R. (1990). *Critical thinking: What every person needs to survive in a rapidly changing world*. Center for Critical Thinking and Moral Critique.

Puolimatka, T. (1995). *Democracy and education: The critical citizen as an educational aim*. Suomalaisen Tiedeakatemia.

Ribble, M. (2008). Passport to digital citizenship. *Learning and Leading with Technology*, *36*(4), 14–17.

Ribble, M. (2012). Digital citizenship for educational change. *Kappa Delta Pi Record*, *48*(4), 148–151. doi:10.1080/00228958.2012.734015

Ribble, M. (2017). *Nine elements: Nine themes of digital citizenship*. Retrieved from https://www.digitalcitizenship.net/nine-elements.html

Ribble, M., & Bailey, G. D. (2011). *Digital citizenship in schools*. International Society for Technology in Education.

Ribble, M., Bailey, G. D., & Ross, T. W. (2004). Digital citizenship: Addressing appropriate technology behavior. *Learning and Leading with Technology*, *32*(1), 6–9.

Ribble, M., & Miller, T. (2013). Educational leadership in an online world: Connecting students to technology responsibly, safely, and ethically. *Online Learning : the Official Journal of the Online Learning Consortium*, *17*(1), 137–145. doi:10.24059/olj.v17i1.310

Ribble, M., & Park, M. (2019). *The digital citizenship handbook for school leaders: Fostering positive interactions online*. International Society for Technology in Education.

Richards, R. (2010). Digital citizenship and web 2.0 tools. *Journal of Online Learning and Teaching*, *6*(2), 516–522.

Ruggerio, V. R. (1988). *Teaching thinking across the curriculum*. Harper & Row.

Ruggerio, V. R. (1990). *Beyond feelings: A guide to critical thinking*. Mayfield Publishing Company.

Smith, E. E., & Storrs, H. (2023). Digital literacies, social media, and undergraduate learning: What do students think they need to know? *International Journal of Educational Technology in Higher Education*, *20*(1), 1–33. doi:10.118641239-023-00398-2 PMID:37220550

Sosu, E. M. (2013). The development and psychometric validation of a Critical Thinking Disposition Scale. *Thinking Skills and Creativity*, *9*, 107–119. doi:10.1016/j.tsc.2012.09.002

Ten Dam, G., & Volman, M. (2004). Critical thinking as a citizenship competence: Teaching strategies. *Learning and Instruction*, *14*(4), 359–379. doi:10.1016/j.learninstruc.2004.01.005

Thoman, E., & Jolls, T. (2004). Media literacy: A national priority for a changing world. *The American Behavioral Scientist*, *48*(1), 18–29. doi:10.1177/0002764204267246

Thorson, K. (2012). What does it mean to be a good citizen? Citizenship vocabularies as resources for action. *The Annals of the American Academy of Political and Social Science*, *644*(1), 70–85. doi:10.1177/0002716212453264

Toffler, A. (1981). *The third wave*. Morrow.

Westheimer, J., & Kahne, J. (2004). Educating the "good" citizen: Political choices and pedagogical goals. *PS, Political Science & Politics*, *37*(2), 241–247. doi:10.1017/S1049096504004160

Xu, S., Yang, H. H., & Zu, S. (2019, July). *An investigation of 21st-century digital skills on digital citizenship among college students. International Symposium on Educational Technology*, Czech Republic. 10.1109/ISET.2019.00056

Zaff, J. F., Malanchuk, O., & Eccles, J. S. (2008). Predicting positive citizenship from adolescence to young adulthood: The effects of a civic context. *Applied Developmental Science*, *12*(1), 38–53. doi:10.1080/10888690801910567 PMID:22837638

Zukin, C., Keeter, S., Andolina, M., Jenkins, K., & Delli Carpini, M. X. (2006). *A new engagement?: Political participation, civic life, and the changing American citizen*. Oxford University Press. doi:10.1093/acprof:oso/9780195183177.001.0001

Chapter 11
The Future Is Now:
Considerations of Conducting Qualitative Research in Digital Spaces

Crystal L. Beach
https://orcid.org/0000-0003-0590-1329
University of Georgia, USA

ABSTRACT

In today's digital era, it is important to examine the ways researching in digital spaces differs from the general practice of conducting qualitative research. For this reason, this chapter will consider the work done by researchers to include the ethical considerations, affordances, and constraints of conducting qualitative research in digital spaces. Then, with these points in mind, it will include a specific discussion on the research literature concerning how literacy researchers have studied adolescents' digital work both in and outside of educational spaces. Finally, implications will be shared on the future of research in digital spaces, such as with AI, and what is said and unsaid within the research literature about conducting qualitative research in digital spaces.

A few years back I started on an incredible journey: *Becoming 3lectric* (www.becoming3lectric.com). Becoming 3lectric (B3) is a digital research project that was co-created with my colleague and then advisor, Donna Alvermann, as a chance to venture on "a journey beyond and between the planes of 'reality' — a journey of Becoming." Essentially B3 explores how the researchers and research participants connect through "digitalized flows of energy that refuse to separate ways of knowing from ways of becoming" (Alvermann & Beach, 2014). These digitalized flows stem from our blogs, fragments, and remixes (the latter two where project participants submitted their own explorations of digital media creation and production). Since its inception, B3 quired approximately 85 research participants, which quickly (and continually) resulted in a lot of data to navigate within its digital space.

As B3 continued, I questioned the many ways in which researching in the digital sphere was very different from face-to-face interactions. Yet, much of my reading up to that point had focused more on the tools necessary to conduct digital research (Bazeley & Jackson, 2013; Friese, 2014; Paulus, Lester, & Dempster, 2014; Silver & Lewins, 2014). However, after working on B3, my attention focused more

DOI: 10.4018/978-1-6684-8934-5.ch011

on the hows, whys, and implications of qualitative research in digital spaces and what a research project, like B3 for example, can help add to this emerging field.

For this reason, I will begin by examining the ways in which researching in digital spaces alters from the general practice of conducting qualitative research face-to-face. Then, I will discuss how researchers have included ethical considerations, affordances, and constraints in this realm since qualitative research is participant-based. Next, I will discuss what literacy researchers have studied regarding adolescents' digital work both in and out of classroom spaces to support the need for discussing qualitative research in digital spaces. Finally, implications of what is said and unsaid within the research literature, as well as what the future with AI might entail, will be noted.

IS THE DIGITAL DIFFERENT?

Communication of research findings and interpretations to traditional forms of dissemination have shifted "the discourses that define what it means to be an educator, student, or education researcher" (Voithofer, 2005, p. 10). In regards to qualitative research, it is no surprise, then, that many researchers are navigating new research realms within the digital, including methods, data representations, and what it means to participate versus observe (Kozinets, 2010), p. 5). Yet, while there are many similarities that are regularly noted between conducting research face-to-face versus conducting research in digital spaces, the differences between the two are just as important to note and create unique considerations.

Markham (2013) points out that inquiry today "is not only about simplifying and narrowing, but generating layers upon layers of information units that influence our interpretations" (p. 74). Generating layers of interpretation is especially done easily and quickly in the digital sphere when one considers how some of these outcomes seem new, such as through remixing; however, "elements are being combined [that] are borrowed from other sources" (Markham, 2013, p. 75). This point is needed when we begin to consider how the digital sphere can be what Arora (2012) calls a "cyber-playground" for researchers in which we take a lot from the physical world that we can "learn, adopt, and transfer" into the online sphere (p. 612). Thus, there are still ways in which digital research is like face-to-face research because no matter the location, the researcher is still using methods pulled from traditional notions of what it means to conduct qualitative research.

Yet, there are important differences to consider when researching in digital spaces as Kozinets (2010) notes. His focus centers on ethnography and *netnography*—"a specialized from of ethnography adapted to the unique computer-mediated contingencies of today's social worlds" (p. 1). Specifically, he feels the biggest differences between the contexts of face-to-face versus digital social interactions for researchers comes down to four key areas: "adaptation to various technological media; participation under optional conditions of anonymity; vastly enhanced cultural accessibility, and automatic archiving of exchanges" (p. 72).

In order to understand the differences that Kozinets (2010) offered researchers, one can consider Hine's (2015) "E^3 Internet." Hine (2015) coined the *E^3 Internet* as embedded, embodied, and everyday (p. 32). Essentially Hine (2015) suggests that the E^3 Internet provides a way to develop an ethnographic strategy for beginning to understand the challenges and affordances when one applies ethnographic principles to the digital sphere (p. 32). She says: "an ethnographer, even in the Internet age, continues to develop a distinctive form of knowledge by being, doing, learning, and practicing, and by a close association with those who do so in the course of their everyday lives" (p. 21). This example further supports the notion that while qualitative research in digital spaces is similar to research of the past, again, there are still significant considerations that one must include without losing the focus on the knowledge one hopes to gather. With these considerations in mind, it is necessary to examine some of the affordances, constraints, and ethical considerations involved with conducting digital qualitative research.

Affordances

There are many affordances that researching in the digital sphere includes, such as increased accessibility to participants. The fact that anyone can jump on the Internet and engage with anyone at any time (and any where) through social media networks, for example, shows how quickly connections and communities can form in the digital sphere based on "shared interests, ambient affiliation and/or the extension of offline groups" (Seargeant & Tagg, 2014, p. 13).

Though one might think that potential participants would be leery of research advances on social media, boyd (2015) notes that she found within most of her digital studies that it was much easier to connect and start communicating online more quickly with participants than it was to try to setup times to meet face-to-face (p. 5). First, however, boyd (2015) immersed herself into her field (p. 6) and "lurked" (p. 5), or watched from the sidelines of social networking sites without participating, to get an idea of how to better engage her research participants: teens. Here, gaining access, engaging with, and actually starting one's research more quickly with digital participants may suggest that there is in fact a greater cultural accessibility due to digital communities and their publicly generated data when it comes to conducting digital research versus face-to-face research (Kozinets, 2010, p. 72; Paulus, Lester, & Dempster, 2014, p. 194).

For example, James and Busher's (2006) examined participant views of taking part in email-based research. Their results showed that many of the advantages outweighed the disadvantages in their study according to participant response. For example, out of 11 participants, five reported that they were more thoughtful/reflexive in their answers because it was written communication and two participants said that they were also able to think more thoroughly by providing a delayed response to the question versus the immediacy needed in face-to-face discussions, whereas only one participant noted that they were unable to give full thought within this online setting (p. 407).

Though this study involved a small number of participants in which all but one preferred the online format, the authors' study continues to support the idea of ease in connecting and communicating with participants online and that participants may also prefer digital communication as well. In addition, this method allows researchers to continue to automatically archive exchanges (Kozinets, 2010, p. 72) that may not be immediately available in face-to-face settings.

As noted above, access is a huge affordance that researchers will encounter within the digital sphere. Today, access not only helps researchers gain insight into specific communities, but mobility is key to their success. Here, mobility in the digital sphere ties into these specific concepts: 1) mobile technologies that overcome physical distance, 2) mobile technologies that allow participants' selves to move (and change) from online to offline spaces, and 3) mobile technologies (specific devices) that participants engage with on-the-go, which overcomes time or digital network barriers.

Mobility that overcomes physical obstacles. To begin, technologies help researchers "manage large, spread-out groups with relative ease" (Paulus, Lester, Dempster, 2014, p. 194). In fact, Markham (2005) notes that due to the "distance-collapsing capacity of the Internet," the researcher can now "include people previously unavailable for study" (p. 801). For example, in an interpretive case study

I recently worked on, I was able to reach out to literacy educators from all over the United States and across multiple time zones. In addition, boyd (2015) noted that she "casually engaged with teens in 21 continental U.S. states and the District of Columbia in a wide array of communities—rich and poor, urban, suburban, and rural, north, south, east, and west" (p. 6). This point is especially exciting when considering how some obstacles to participants dissolve when location is no longer an issue thanks to the mobile affordances of the digital sphere.

Hine (2015) also supports this point by stating that using the digital sphere in this way is "akin to hanging out in a public setting, catching the prevailing cultural currents, and listening to the way that people talk about a topic when, unlike in an interview [face-to-face], they are not being asked to generate a formal account of their relationship with the [researcher's] topic" (p. 165). Thus, researchers are able to view these digital communities/participants in many of the same ways they would in public settings. However, they now have greater access to even more participants and data, including potential insider access, than they would have had previously due to the co-construction of shared, digital spaces under investigation (Markham, 2005, p. 802). In other words, when researching in the digital sphere, "boundaries are not so much determined by 'location' as they are by 'interaction'" (Markham, 2005, p. 802), and being mobile further helps researchers with their work.

Mobility that influences on/offline selves. In addition, Page (2013) discussed what happens when participants not only embody mobility within their digital interactions, but how their identities may shift, too, between online and offline spaces. She uses the example of Lonelygirl15 who has been studied by numerous scholars. Lonelygirl15 claimed to be sixteen-year-old Bree, but was really nineteen-year-old, New Zealand born actress Jessica Rose (p. 170). In her digital narratives on YouTube, Lonelygirl15 supports yet another example Kozinets (2010, p. 72) mentions in that participants may engage with researchers anonymously or under pseudonyms in ways that give them a chance to self-expose particular self-representations on-the-go (meaning movement between on and offline selves). And, while there are obviously ethical considerations that come into play here (that will be discussed later in this chapter), I would like to re-emphasize that Page's (2013) example shows that within digital research, participants' mobility, or movements between on and offline spaces, allows one to potentially see layers of identities across a variety of networks (Alvermann et al., 2012) that could not be previously seen in face-to-face only settings.

Mobility through device. Another example, by Leander and Vasudevan (2011), highlights work with multimodal texts that are mobile (through technological device) and must be analyzed "on-the-go" for this reason. Their participant, Joey, was a young man with a strong knowledge in digital media production and distribution through his PSP (portable PlayStation gaming system). Leander and Vasudevan (2011) discuss how Joey's work extended beyond what was on his screen and immediate location and time to others all over when he uploaded to his MySpace page. Thus, "Joey's use of the PSP reveals the movement that mobile technologies afford for the cultural production of identity across multiple space-times" (p. 137). In regards to digital qualitative research, specifically, this study shows that through Joey's work, Leander and Vasudevan (2011) were able to see how culture moved beyond the physical page and became "interpreted in the stream of moving bodies, technologies, landscapes, and events" (p .139) to transcend time and space by being on-the-go.

Through the examples of Lonelygirl15 and Joey, one can see that by adapting to media to overcome distance, being mobile (through devices and on/offline selves) and having an immediate archive of actions and productions (Kozinets, 2010, p. 72), the researchers were able to see more deeply how the participants' choices impacted their digital identities and cultural interactions—choices that would not have been available with a print-only format.

Moving Beyond the Traditional Within the Digital

Thus, through choice and adaption, researchers are able to see how rich layers of interpretation are available to them within the digital sphere. In addition, I believe that Markham's (2013) note that the concepts of "generating, playing, borrowing, moving, and interrogating" usually associated with remixing in the digital sphere can be applied to research as it provides an opportunity for researchers to have "freedom to innovate when exploring contexts that defy easy encapsulation" (p. 78), such as digital contexts. In other words, the freedom and affordances of working in digital spaces has a lot of potential in helping researchers with gaining more depth in their inquiries and even being more creative in what "counts" as data beyond print-centric or face-to-face artifacts.

As the "technical and social barriers to [digital] participation have come down, and as the participatory expectation built around social media has increased, so more and more of life's trivial mundanities have become reflected online" (Hine, 2015, p. 164). In addition, as noted previously, there is a plethora of ways in which researchers can interact in the digital "vast information sphere [which] contributes directly to the construction of the structures that eventually get labeled 'field' or 'data'" (Markham, 2005, p. 802). Again, with this point in mind, researchers gain several affordances through digital qualitative research that they would not have within face-to-face research. However, there are still several constraints that should be examined as well to provide a better picture of what researching in digital spaces includes.

Constraints

While discussions concerning qualitative research in digital spaces have suggested that access to participants and to data have increased in many positive ways (boyd, 2015; Paulus, Lester, & Dempster, 2014; Seargeant & Tagg, 2014), one must also realize that the potential for endless variations and access to data can be daunting as well because they are in fact just "partial traces of interactions" at any one time (Markham, 2013, p. 77). In other words, too much data can often weigh down researchers because studies into digital communities are always "on" and never motionless.

The problem with too much data. For example, even if a researcher is focused on one particular event or day within an online community, that information is still held in the digital sphere—potentially to be found at a later time (through the help of a hashtag or search engine, for example) and used however the new user would like. Thus, the data never stops growing. For this reason, the researcher needs to enter the digital space with a specific goal, or cutoff, to help alleviate this potential issue.

In fact, Paulus, Lester, and Dempster (2014) mention "researchers of all kinds are exploring the possibilities of how to make sense of the massive amounts of data being automatically generated by virtual worlds and social media such as Second Life, Twitter, Facebook, and other computing activities" (p. 193). Merchant (2010) also supports the problem of massive amounts of data in regards to young people's desire to leave "some sort of presence on an SNS [social network systems]" (p. 59). With status updates, and geotracking technologies, for example, it is no surprise to see the numerous ways in which participants consistently engage (and manipulate) their digital selves and work in online spaces around the clock that bypasses face-to-face interactions. In other words, the researcher's data is always on-the-go, changing, and bypassing set time limits that face-to-face interactions would have.

The problem with misrepresentation. In addition, Markham (2005) discusses how difficult it is begin to make sense of participant's computer-mediated constructions of self. Specifically, in the digital sphere, the researcher's construction must be more deliberate and conscious because one may not know who one is engaging with in the online realm (Markham, 2005, p. 794), which again will be discussed later concerning ethical considerations.

However, in regards to constraints, Page's (2013) discussion on Lonelygirl15 and the implied problems of anonymity come into play. If one is focusing on a specific community to acquire a specific knowledge set, then finds out the participant was not truthful in his or her representation, then one quickly sees how information would not be helpful to one's overall inquiry. While communicating one's identity always has potential for misrepresentation—in digital or face-to-face settings—it becomes trickier in digital spaces due to the affordances one has to construct one's self (Deumert, 2015; Markham, 2005; Merchant, 2010).

The problem with the unknowns. Furthermore, Markham (2005) writes: "in the virtual field, as one interacts with anonymous participants, tracks disjointed, non-linear, multiple participant conversations, and analyzes hundreds of screens worth of cultural texts, one can begin to feel like the Internet might cause more headaches than it cures" (p. 799). In other words, one can see once again how research in the digital space becomes an overwhelming field.

Also, "the challenge for the qualitative researcher in the computer-mediated environment is to attend to the details of how one is going about the process of getting to know something about the context and the persons being studied" (Markham, 2005, p. 796). This might mean engaging with tools, spaces, or media that are not always convenient or comfortable to the researcher. While one might argue that this is nothing new to the field of qualitative research, the fact that the digital sphere is constantly "on" and constantly changing at the click of a button suggests that the turnover of what is being studied has the power to change just as the researcher is gaining knowledge within the previous setting.

The problem with resistance. Yet another key constraint with conducting digital research (and the resulting digital artifacts from that research) is alluded to by Paulus, Lester, and Dempster (2014) as they point out that it is not academia that has been most active with this research, but industry and private citizens (Paulus, Lester, & Dempster, 2014, p. 193). Part of the hesitation within academia may stem around some of the issues Hagood and Skinner (2014) note in their commentary on the changing nature of literacies' research, education, and publication.

Specifically, Hagood and Skinner (2014) state that there are still issues with including digital artifacts in research articles. Some of these issues include "anonymity, which may or may not be addressed in participant permission through Institutional Review Boards, to storage of data in cross platforms (print article versus online article), and to preparation time for publication process which can seem deterrent for authors and cause them to move back into traditional article representations" (p. 7). While these issues do not mean that all researchers are not engaging in the digital sphere, they do remind one of the need to have publications for tenured-research positions, which these issues can lead one to see how the field of academia still remains quite traditional in many ways.

However, it is important to note that now "we are living in an exponential time, and it is imperative that we join with our [academic] colleagues and students across boundaries to work collaboratively [on our research]" (Johnson, 2014, p. 1120). The possibilities and innovation within the digital sphere provides a unique landscape from which researchers have much to learn. For example, through B3, our digital remix data supported our belief that "as a form of multimodal writing, [it] offers students opportunities to improve communicating in or across mediums that connect to aspects of the world around them through rich literacy practices" (Beach, Alvermann, Loomis, Wright, & Hutcherson Price, 2023,

p. 104). B3, then, reminded us of the (inter)connectedness of the world around us which was made possible through digital qualitative research. Of course, though, even with the new, endless possibilities with research within the digital sphere, researchers must also be sure to focus on ethical considerations as they navigate what it means to conduct qualitative research today.

Ethical Considerations

As shown above, navigating online communities and spaces can be overwhelming for researchers. On top of that, within the digital sphere it is often even more critical that researchers examine the ethical considerations that they may face.

Considering issues of consent. When it comes to considering participation, consent is obviously a huge factor when it comes to any research, however, it can become even trickier within the online sphere due to anonymity and self-construction(s) (as previously noted). While informed consent can cause participants to lack trust in a researcher (Macfarlane, 2014, p. 21) due to the fear of losing credibility in one's community, for example, many Institutional Review Boards make sure that researchers include a very detailed outline of how one plans to protect participants, especially when search engines can bring up usernames and status updates immediately.

Furthermore, when one deals with participants under the age of 18, boyd (2015) feels that researchers have both a legal and ethical responsibility to obtain parental permission (p. 9) even in online spaces in which participants' information is public to all. Thus, informed consent for research in digital spaces has a variety of layers that make it more difficult to navigate at times, especially when the researcher cannot support this consent with a physical, face-to-face interaction.

Considering issues of privacy. In addition, there are issues of public versus private spaces within online communities that can be viewed with an ethical lens. For example, even though participants' information may be visibly public to everyone, anywhere around the world, boyd (2015) acknowledges that she could not violate what teenagers were hearing with regards to "creepy strangers" online by her solely lurking (watching online without actually directly participating with subjects) in the digital sphere (p. 5).

Furthermore, Baker's (2013) study also brings up the issues of moving from public to private spaces as her study began as face-to-face interactions, but ended up on Facebook for ease of communication. In her study, she looked at students' literacy practices and perceptions of these practices moving from one level of study to another; however, though on Facebook she had access to participants' private thoughts through status updates, she decided to focus only on their academic literacy practices and discussions on their academic transition over their personal information in order to strengthen relationships with the participants (p. 140). Here, it is necessary to note that Baker's (2013) common informational ground (focused on academic literacy practices/transitions) as needed to navigate and negotiate the data that was forthcoming and to also value the participants' use of Facebook as a private space, too.

Considering issues of search engines. Another ethical consideration of conducting digital qualitative research is the issue surrounding data. With various search engines available, the Internet seems like an infinite database of anything and everything one could imagine. For this reason, boyd (2015) "purposefully obscure[s] data" (p. 10). In order to delete all identifying information and provide participant anonymity, names are changed, photos are blurred, and quotes are altered from profiles. This ensures that data cannot be traced back to the individual from where it originated; however, it may never be possible to completely avoid complete anonymity in digital spaces.

Considering issues of editing. Markham (2005) reminds us that even the process of editing is an ethical, conscious choice researchers make. When researchers obscure data, they are choosing how they are representing another's work which may not be how the author intended the representation to be. Again, while this is not necessarily only prevalent within digital research, "computer-mediated environments seem to highlight this dilemma of research reporting because it's so clear that text can be the primary, if not sole means of producing and negotiating self, other, body, and culture" (Markham, 2005, p. 811).

Since there are more ways to represent participants through the digital based on the numerous ways they are engaging within the digital, researchers should make intentional choices with their participant consent forms which can help in this area. However, they must also be sure that while they are trying to help in one area (shown by boyd's [2015] choice to obscure data) that they do not unintentionally misrepresent their data, which can be easier within the digital realm thanks to the variety of digital media available if one is not careful.

In the end, even though there is more to explore and more connections than researchers ever may have imagined within the digital realm, it is increasingly more important for them "to tell stories about how that feels, to explore the consequences and unanticipated connections that arise" (Hine, 2015, p. 182). In other words, researchers have an ethical responsibility to be sure to explore the tensions they are experiencing to better note how the digital may differ from traditional research practices in face-to-face settings.

WHERE'S THE PROOF?

Above, some of the affordances, constraints, and ethical considerations that researchers face in the digital sphere have been outlined. While "every community has its own flavor, biases, and limitations" (boyd, 2015), the digital realm is a vibrant space from which researchers garner many forms of knowledge.

In regards to literacy research, and specifically adolescent literacy, there is incredible potential to gain valuable information as "teen life is increasingly intertwined with technology, but the traces that teens leave through technology are not rich enough to convey their practices" (boyd, 2015, p. 16). In addition, because the Internet is a contested space, researchers need to better understand how online and offline selves may differ (Baker, 2013, p. 132), especially regarding students' in (traditionally offline) and out of school (typically online) literacy practices are impacted.

In-School Examples of Research in Digital Spaces

Vasudevan (2010) said, "too often, the digital proclivities of youth are framed in the media and curricula through lenses of fear (e.g., cyberbullying), protectionism (e.g., Internet predators), frivolity (e.g., social networking), or appropriation (e.g., uncritically assigning rap lyrics as a medium of math problem solving)" (p. 44). Perhaps this is why most schools lack the serious attention to students' informal learning spaces due to the nature of the learning (Merchant, 2010, p. 66). Yet since the questions, of "'where,' 'how,' and 'whom' remain important factors" (Vasudevan, 2010, p. 45), online or offline, literacy researchers need to consider when they are investigating youth's language use, especially through digital media. For this reason it is now more important than ever for researchers to consider how adolescents' digital literacy practices are viewed in schools.

Some research studies, however, do branch into this realm of possibility within the classroom. For example, Dezuanni (2010) discusses youth's digital media literacy practices, specifically through production and learning about video games, and the connection they have to their identities. This study came about at an independent, Catholic boys' school through a "Video Games Immersion Unit" (p. 129), in which each student was required to blog after each learning session (p. 131). Students were given the chance to create their own pseudonyms, thus mimicking the process of identity creation they would have in informal learning spaces (p. 134). In addition, by participating in other forms of online communication, such as chat, forum discussions, or email, students were able to connect to their "everyday media experiences and to develop digital media literacies that go beyond the use of technologies to include experiences that might otherwise be unavailable to the students" (p. 138).

Thus, Dezuanni's (2010) study suggests that students' digital activities "provided students with the opportunity to show what they valued in the specific learning context" of the unit (p. 141). Here, one sees out-of-school digital practices carried into the classroom, while not without "risk" (Dezuanni, 2010, p. 140) in order to help students make more meaningful connections to their in school practices. Yet, even with this information, most schools are not constructed to value these forms of learning and experiences due to curricular demands and state-mandated testing, which suggests why studies like Dezuanni's (2010) are needed to help support learning through digital spaces.

There are also studies that point out how teachers' construct cultural models of their students and their students' online literacies (Chandler-Olcott & Lewis, 2010, p. 163), which are important because they may suggest how and why teachers use students' digital spaces for learning. In their study, Chandler-Olcott and Lewis (2010) worked with secondary English teachers, an academic intervention teacher, and two library media specialists to capture research on the views associated with integrating new literacies in secondary classrooms (p. 166). Almost all teachers in this study reported that they felt students were more engaged when working with online literacies, and one participant felt it helped her relate to her students if she included online spaces within her classroom (p. 171).

While this study was not conducted online, it does suggest that there are positive consequences when educators use digital spaces with their students' learning, which once again emphasizes that there is an increased need to focus on these opportunities for students. After all, youth are still finding ways to engage with "digital landscapes" even though not many are "adequately supported through their school experiences in cultivating this digital imagination" (Vasudevan, 2010, p. 47).

The reality is that youth are often excluded from participating directly in public life; thus, digital spaces are places in which they can become active members and find their voice and develop key literacy skills (Fields et al., 2013, p. 2). Furthermore, specifically focusing on research in these in school, digital spaces, researchers are often presented with more rich, layered data because they use online access to inform their offline interactions, too, which is highlighted in Baker's (2013) study using Facebook to see how students "articulated their authorial selves and performed different aspects of being a student" (p. 139). Now that one sees the need and importance of conducting research in digital spaces through in school learning experiences, one needs to also look to the research literature conducted on students' out-of-school literacy practices that occur within digital spaces, too.

Out-of-School Examples of Research in Digital Spaces

Alvermann (2008) notes that youth are constantly engaging with multimodal content online, "using new tools to show and tell, rewriting their social identities in an effort to become who they say they are" (p.

10). Furthermore, she states that by researching youth's online content and literacy practices, teachers, teacher educators, and researchers will "open themselves to appreciating a wide range of competencies that might otherwise go unmarked in everyday routines of unexamined classroom and research practices" (p. 17). Thus, researching youth's literacy skills in the digital realm is important to gain knowledge that can be applied in face-to-face settings as needed and, perhaps more importantly, change classroom instruction to create more meaningful learning opportunities for students.

Whether through talking "CoffeeTweets" (Zappavinga, 2014) in which adults and students alike can connect through their shared values, creating profiles (Merchant, 2010) in order to make and remake identities, or creating remixes to promote "multimodal explorations of identity, expression, and connection" (Beach et al., 2023, p. 105), digital research is a necessity in today's highly mobile, remixed world. In addition, it is important to consider how young people are now looking up information on their own and engaging in forums or through artificial intelligence to get that information (or help), too.

For example, Harvey et al.'s (2007) study reported on how teenagers interacted with a popular website, *Teenage Health Freak*, to find sexual health information. Youth essentially created their own narrative within the website as they sought out information that they may have felt unable to ask in face-to-face settings with their doctors or parents. Within this study, the researchers noted that the website had its own privacy policy that users must consent to before using the site. Essentially, the consent informs potential participants that their information could be used for research purposes. Thus, these researchers pulled from the exact language used on the site for their analysis. I feel that this is important to note, especially after discussing ethics and the need to obscure data (boyd, 2015) and the editing powers that come into play (Markham, 2005).

Here, one can see there are many implications from the research that show the importance of participants' digital practices and why they matter for literacy researchers. This also further proves why conducting research in digital spaces is becoming an important realm within the qualitative field, too. As more students become mobile and digitally present, researchers cannot ignore the many ways they are actively participating in the world that extends beyond the page of a novel and the classroom walls.

IMPLICATIONS

While the previous examples took a specific approach to show that language use and literacy are increasingly digital today, which is why literacy researchers need to engage within those digital spaces, there is still no arguing the fact that "qualitative research in higher education continues to be dominated by textual research approaches" (Gourlay, 2010, p. 81) and researchers and teachers of adolescents' literacy practices are still sometimes hesitant to engage with the digital due to a variety of factors (Alvermann, 2008; Fields et al., 2013; Vasudevan, 2010). Yet, with access to both on and offline spaces, researchers can find a more rich landscape from which to gather data, such as through DIY-based social networking forums (Fields et al., 2013). Again, access is key when considering how the digital might change the realm of research.

Furthermore, as more research is done in online spaces, researchers need to continue to investigate what it means to conduct inquiry within these spaces. For example, the "challenges for data collection and management such as defining the unit of analysis and how to conceive of this in the often blurry and unbounded landscape of the digital" (Jewitt, 2015, p. 455) need to be taken head on in order to continue the momentum forward and help researchers to come.

With these challenges in mind, a certain transparency stems from being reflexive in one's digital research practices (Paulus, Lester, & Dempster, 2014; Duncan & Watson, 2010). Many studies are including digital research, however, they are not necessarily critically examining the intimate details of *how* that research might vary within the digital space. Yet, whether online or offline, Duncan and Watson (2010) described why being transparent and reflexive matter: "we came to realize that socially responsible research ethics involves consciously adopting stances that would enable us to systematically think through why and how we participated in the world of those we engaged as informants" (p. 50). Thus, when we begin to better understand our own actions within these digital spaces, it is just as important to acknowledge the hows, whys, as well as the discomforts or conflicts "as points of learning about and from research participants [in which] the possibilities for research as social change" can start to be seen (Duncan & Watson, 2010, p. 51).

In addition, materiality and mobility of data and findings (Markham, 2013; Voithofer, 2005) are increasingly imperative to consider when conducting digital research. These notions allow researchers to focus on the creative process of inquiry (Markham, 2013, p. 71) because they are focusing less on methods templates but instead on what is needed within the moment. Furthermore, "as education research evolves to make connections between these [digital] constellations and resources, new methods of conducting research and training researchers will emerge that integrate existing and emerging fields of knowledge" (Voithofer, 2005, p. 11). For this reason, qualitative research is evolving as researchers head into an even more digitized future, and these implications highlight some of the areas researchers will want to consider as they continue down that path.

THE FUTURE IS NOW

We know that there are certain affordances, constraints and ethical considerations that will influence our work in conducting qualitative research in digital spaces. In addition, within any inquiry-based research, there is always uncertainty. However, there is still a lot of uncertainty involving digital research because of its ever-evolving realm, including data, methods, and tools to name a few. With this in mind, Savin-Baden and Major (2010) state: "uncertainty too can be connected with a sense of being stuck. Thus engaging with it can lead to advancement of knowledge and wisdom, as thresholds are crossed and articulation becomes clearer" (p. 172). Pushing through uncertainties and being transparent and reflexive in all researchers do tremendously helps others learn from and navigate the complexities of qualitative research in digital spaces.

In order to make one's work clearer, then, one must remember that data analyses and representations are moving beyond print-based forms (Hagood & Skinner, 2014, p. 7). One must remember to "develop the habit of providing a rationale for the use of digital tools alongside [our] rationale for other design choices made when proposing or reporting a study" (Paulus, Lester, & Dempster, 2014, p. 191). One must be conscious of negotiating space, participants' identities, and the many ways one will see data on-the-go.

Researchers must also not forget the past, but understand that researching in digital spaces does look differently especially as artificial intelligence (AI) starts to be a major focus. For example, ChatGPT is an AI, dialogic program designed to "answer follow-up questions, admit its mistakes, challenge incorrect premises, and reject inappropriate requests." When asked, "What are the ethical considerations, affordances, and constraints of conducting qualitative research in digital spaces?" its response was as follows:

"Conducting qualitative research in digital spaces presents a unique set of ethical considerations, affordances, and constraints. Let's explore each of these aspects in more detail:

Ethical Considerations:

1. Informed Consent: Researchers must ensure that participants in digital spaces are adequately informed about the research purpose, procedures, and potential risks or benefits. Obtaining informed consent becomes more challenging when participants' identities are not easily verifiable.
2. Anonymity and Confidentiality: Protecting the privacy and confidentiality of participants is crucial. In online environments, it can be difficult to ensure anonymity, especially when data may be easily linked to individuals. Researchers should take measures to safeguard participants' identities and ensure their responses cannot be traced back to them.
3. Digital Divide: Researchers need to consider the digital divide, as not everyone has equal access to digital spaces. This divide can introduce biases, limiting the inclusivity of the research and potentially excluding certain demographics.
4. Digital Ethics: Researchers must adhere to ethical guidelines when conducting research online. This includes respecting online community guidelines, avoiding deception or harm, and being mindful of ethical considerations specific to the digital context.

Affordances:

1. Access to Diverse Participants: Digital spaces provide access to a wide range of individuals from diverse backgrounds and geographical locations. This allows researchers to engage with a broader and more varied pool of participants compared to traditional offline research.
2. Rich Data Collection: Digital spaces offer various forms of rich data, including text, images, videos, and social interactions. Researchers can analyze these diverse data sources to gain deeper insights into participants' experiences, behaviors, and perceptions.
3. Real-Time and Longitudinal Research: Digital platforms enable real-time data collection and longitudinal studies, as interactions and conversations can be observed and recorded over an extended period. This allows researchers to capture dynamic processes and changes over time.

Constraints:

1. Lack of Control: Researchers have limited control over the digital environments where data is collected. Online spaces are subject to external influences, such as platform changes, algorithmic biases, or disruptions, which can impact data collection and analysis.
2. Validity and Reliability Concerns: The authenticity and credibility of data collected online may be questioned due to concerns about misrepresentation, trolling, or the presence of bots. Researchers need to critically evaluate the data's validity and reliability to ensure accurate findings.
3. Ethical Dilemmas: Digital spaces may present researchers with ethical dilemmas, such as identifying appropriate boundaries between public and private spaces or navigating conflicts between free speech and participant protection. Researchers must carefully navigate these dilemmas while maintaining ethical integrity.

4. Digital Literacy: Researchers and participants need to possess digital literacy skills to effectively engage in online research. Lack of digital literacy can affect participants' ability to provide informed consent, understand the research purpose, or engage meaningfully in data collection processes.

Considering these ethical considerations, affordances, and constraints is crucial when conducting qualitative research in digital spaces. Researchers should ensure they are well-versed in relevant ethical guidelines, adapt their methodologies to the digital context, and take steps to protect participants' privacy and well-being throughout the research process" (ChatGPT, personal communication, June 2023). As one can see, this breakdown highlights many of the previous sections, and yet, with AI being so new, it's important that we continue our discussions around it and what it means for scholarly research and publication for digital qualitative research.

As Markham (2005) suggests, "one way to meet the future is to learn from but not rely on the past" (p. 815). Thus, an eye on the future is ever important in the work that we do. Just as B3 helped me to understand with digital remix, we may at times, as qualitative researchers in digital spaces "find these energies difficult to direct or assess in standardized ways, [yet] we contend that a recognition of the messy, sometimes cathartic, often beautiful, forever unaccountable traces indeed affect us, take us up, and fundamentally alter our being. In short, we assert that our (inter)connected becoming in the world is something to embrace" (Beach et al., 2023, p. 105). The future is now, and it is up to researchers to continue to learn how digital research *is* different and includes all of its unique affordances, constraints, and ethical considerations.

REFERENCES

Alvermann, D. E. (2008). Why bother theorizing adolescents' online literacies for classroom practice and research? *Journal of Adolescent & Adult Literacy*, *52*(1), 8–19. doi:10.1598/JAAL.52.1.2

Alvermann, D. E., Beach, C. L., & Johnson, J. (2014). *Becoming 3lectric*. Retrieved from http://www.becoming3lectric.com/

Alvermann, D. E., Marshall, J. D., McLean, C. A., Huddleston, A. P., Joaquin, J., & Bishop, J. (2012). Adolescents' web-based literacies, identity construction, and skill development. *Literacy Research and Instruction*, *51*(3), 179–195. doi:10.1080/19388071.2010.523135

Arora, P. (2012). Typology of web 2.0 spheres: Understanding the cultural dimensions of social media spaces. *Current Sociology*, *60*(5), 599–618. doi:10.1177/0011392112440439

Baker, S. (2013). Conceptualising the use of Facebook in ethnographic research: As tool, as data and as context. *Ethnography and Education*, *8*(2), 131–145. doi:10.1080/17457823.2013.792504

Bazeley, P., & Jackson, K. (2013). *Qualitative data analysis with NVIVO*. Sage Publications.

Beach, C. L., Alvermann, D. E., Loomis, S., Hutcherson-Price, L., & Wright, W. (2023). Digital remixing online: Entangled feelings. *Digital Culture & Education*, *14*(4), 92–108.

boyd, d. (2015). Making sense of teen life: Strategies for capturing ethnographic data in a networked era. In E. Hargittai & C. Sandvig (Eds.), *Digital Research Confidential: The Secrets of Studying Behavior Online*. Cambridge, MA: The MIT Press. Retrieved from http://www.danah.org/papers/2012/Methodology-DigitalResearch.pdf

Chandler-Olcott, K., & Lewis, E. (2010). "I think they're being wired differently": Secondary teachers' cultural models of adolescents and their online literacies. In D. E. Alvermann (Ed.), *Adolescents' Online Literacies: Connecting Classrooms, Digital Media, & Popular Culture* (pp. 163–182). Peter Lang Publishing, Inc.

Deumert, A. (2014). The performance of a ludic self on social network(ing) sites. In P. Seargeant & C. Tagg (Eds.), *The language of social media: Identity and community on the internet* (pp. 23–45). Palgrave Macmillan. doi:10.1057/9781137029317_2

Dezuanni, M. (2010). Digital media literacy: Connecting young people's identities, creative production and learning about video games. In D. E. Alvermann (Ed.), *Adolescents' Online Literacies: Connecting Classrooms, Digital Media, & Popular Culture* (pp. 125–143). Peter Lang Publishing, Inc.

Duncan, M., & Watson, R. (2010). Taking a stance: Socially responsible ethics and informed consent. In M. Savin-Baden & C. H. Major (Eds.), *New approaches to qualitative research: Wisdom and uncertainty* (pp. 49–58). Routledge.

Fields, D. A., Grimes, S. M., Magnifico, A., Lammers, J. C., Gomez, K., & Curwood, J. S. (2013). What's next in studying online social networking? Future research directions for creative, DIY-based sites. In *Proceedings of the Annual Conference 9.0 of Games + Learning + Society*. University of Wisconsin-Madison.

Friese, S. (2014). *Qualitative data analysis with ATLAS.ti*. Sage Publications.

Hagood, M. C., & Skinner, E. N. (2014). The vitality of embracing changes in literacies' research, education, and publication. *Journal of Adolescent & Adult Literacy*, *58*(1), 6–8. doi:10.1002/jaal.326

Harvey, K. J., Brown, B., Crawford, P., Macfarlane, A., & McPherson, A. (2007). 'Am I normal?' Teenagers, sexual health and the internet. *Social Science & Medicine*, *65*(4), 771–781. doi:10.1016/j.socscimed.2007.04.005 PMID:17499898

Hine, C. (2015). *Ethnography for the Internet: Embedded, embodied, and everyday*. Bloomsbury Publishing Plc.

James, C., Davis, K., Flores, A., Francis, J. M., Pettingill, L., Rundle, M., & Gardner, H. (2009). *Young people, ethics, and the new digital media: A synthesis of the GoodPlay Project*. The John D. and Catherine T. MacArthur Foundation Reports on Digital Media and Learning. doi:10.7551/mitpress/8520.001.0001

James, N., & Busher, H. (2006). Credibility, authenticity and voice: Dilemmas in online interviewing. *Qualitative Research*, *6*(3), 403–420. doi:10.1177/1468794106065010

Johnson, J. M. (2014). Scholarly innovation for difficult times. *Qualitative Inquiry*, *20*(9), 1119–1121. doi:10.1177/1077800414543695

Kozinets, R. V. (2010). *Netnography: Doing ethnographic research online*. SAGE Publications.

Leander, K., & Vasudevan, L. (2011). Multimodality and mobile culture. In C. Jewitt (Ed.), *The Routledge handbook of multimodal analysis* (pp. 127–139). Routledge.

Macfarlane, B. (2010). Values and virtues in qualitative research. In M. Savin-Baden & C. H. Major (Eds.), *New approaches to qualitative research: Wisdom and uncertainty* (pp. 19–27). Routledge.

Markham, A. (2005). The methods, politics, and ethics of representation in online ethnography. In N. K. Denzin & Y. S. Lincoln (Eds.), *The SAGE Handbook of Qualitative Research* (3rd ed., pp. 793–820). SAGE Publications.

Markham, A. (2013). Remix cultures, remix methods: Reframing qualitative inquiry for social media contexts. In D. K. Denzin & M. D. Giardina (Eds.), *Global Dimensions of Qualitative Inquiry*. Left Coast Press, Inc.

Merchant, G. (2010). View my profiles. In D. E. Alvermann (Ed.), *Adolescents online literacies: Connecting classrooms, digital media, and popular culture* (pp. 51–69). Peter Lang.

Page, R. E. (2013). *Familiar, reconfigured, and emergent dimensions of narrative. In Stories and Social Media: Identities and Interaction*. Routledge.

Paulus, T. M., Lester, J. N., & Dempster, P. G. (2014). *Digital tools for qualitative research*. SAGE Publications. doi:10.4135/9781473957671

Savin-Baden, M., & Major, C. H. (Eds.). (2010). *New approaches to qualitative research: Wisdom and uncertainty*. Routledge. doi:10.4324/9780203849873

Seargeant, P., & Tagg, C. (Eds.). (2014). *The language of social media: Identity and community on the internet*. Palgrave Macmillian. doi:10.1057/9781137029317

Silver, C., & Lewins, A. (2014). *Using software in qualitative research: A step by step guide*. Sage Publications. doi:10.4135/9781473906907

Vasudevan, L. (2010). Literacies in a participatory multimodal world: The arts and aesthetics of Web 2.0. *Language Arts*, *88*(1), 43–50.

Voithofer, R. (2005). Designing new media education research: The materiality of data, representation, and dissemination. *Educational Researcher*, *34*(3), 3–14. doi:10.3102/0013189X034009003

Zappavinga, M. (2014). CoffeeTweets: bonding around the bean on Twitter. In P. Seargeant & C. Tagg (Eds.), *The language of social media: Identity and community on the internet* (pp. 139–166). Palgrave Macmillan. doi:10.1057/9781137029317_7

Chapter 12
Do We Support Ethical Behavior in Digital Tool Use in Early Childhood?

Feyza Aydin Bölükbaş
https://orcid.org/0000-0002-4380-0134
Aksaray University, Turkey

Kadriye Selin Budak
https://orcid.org/0000-0002-8161-7074
Bilecik University, Turkey

Kübra Engin
http://orcid.org/0000-0003-2590-1115
Gazi University, Turkey

Ilkay Ulutaş
https://orcid.org/0000-0002-2234-0773
Gazi University, Turkey

Emine Bozkurt Polat
https://orcid.org/0000-0002-1821-028X
Gazi University, Turkey

ABSTRACT

It is important to be a role model and guide for the child in the process of acquiring the rules and behaviors that should be followed in order to protect from the risks during the use of technology and not to create risks for others. In this context, as a result of the research conducted with preschool teachers who are the guides and role models of children, it was seen that their understanding of digital ethics was based on protection from the harms of technology, correct use, and the application of moral and ethical rules to the digital environment. It was stated that precautions such as restricting posts about children and controlling the channels they share were taken. They stated that they wanted to receive training on digital ethics to improve themselves, guide and protect children, and that they obtained information various resources. They stated that they shared with children, families, and colleagues on issues such as safe internet and digital citizenship and emphasized that teachers and parents should also be trained for digital ethics education to be given to children.

DOI: 10.4018/978-1-6684-8934-5.ch012

INTRODUCTION

Reports based on worldwide data show that 67% of the population uses mobile phones and the number of internet users will reach 4.95 billion in early 2022 (Digital Report, 2022). This rapid growth has a profound impact on children's lives in particular (Bergen, Davis, & Abbitt, 2015; Louv, 2020). Many children are growing up in environments with wide access to different digital tools as part of their daily lives. The use of digital tools such as smartphones, tablets, and computers at home, in schools, and in social spaces is increasing rapidly, and young children use these tools especially for play and entertainment (Arnott & Yelland, 2020; Rideout & Robb, 2020). Recent research shows that access to and use of digital technology by children aged 0-8 is increasing (Ofcom, 2022; Rideout and Robb, 2020). The increase in the use of digital technology has been the subject of debate due to its impact on children's development, and the advantages and disadvantages of its use have been evaluated in various studies.

Technology use also brings online risks. As the time and frequency of children spending time with technology increases, the rate of exposure to risk factors also increases. Although exposure to online risks is directly proportional to technology use, it is a complex process. Because interacting with technology can increase the opportunities to develop strategies to cope with these risks (Livingstone, Mascheroni, & Staksrud, 2015). From a child rights perspective, what is more important than the number and variety of digital devices, platforms and services available to children are the consequences of using these technologies (Gasser & Cortesi, 2017).

In an age of technology that is developing at an incredible pace, parents, teachers, technology producers and policy makers have a great deal of duties and responsibilities to ensure that children's interactions with digital tools are beneficial and of high quality. For example, the concept of digital parenting has emerged to define the responsibilities of parents. Digital parenting includes tasks such as the parent's knowledge of the child's digital rights, protecting the child from online risks, and directing the child to digital literacy trainings to know and protect their own rights. The concept of digital ethics, which appears in the processes related to the protection of their own and others' information and rights in digital environments, deals with issues such as the effects of digital tools on society and the environment, and ethical issues in digital environments. It evaluates issues such as privacy, information overload, addiction, digital divide, surveillance, robotics from an intercultural perspective (Capurro, 2010).

In addition to protecting children from the harms of digital tools, raising them as digital citizens who abide by the rules of digital ethics is also important in terms of maintaining the security of the digital world. In this context, the digital citizenship education process to be offered to children is expected to include respectful, empathetic and responsible behaviors in the digital world, distinguishing right/wrong behaviors and approaches, and implementing the right approach. Digital citizenship education aims to help children respect other people's rights, protect their own rights, understand the consequences of their online behavior, and be aware of online safety, especially in the digital world. It is of great importance to teach children the concept of digital ethics and the principles related to this concept, especially starting from young age groups, and the adults around the child have great responsibilities in this process. Digital ethics for children means that parents, teachers and other adults should be role models, educate about the proper use of digital technologies and help children learn the right digital behaviors. This includes protecting children's personal information, protecting them from online harassment, being aware of their digital footprint and learning how to interact with other people online.

Research has addressed digital ethics in various ways, including the protection and evaluation of digital content (Floridi, 2001); acknowledging the source of an idea and paying attention to copyright

(Bernard, 2012; Freeman & Peace, 2005; Kenton, 2020); and ethical rules to be followed in digital environments (Hansson, 2008).

Digital ethics for children is important to ensure that children exhibit ethical behavior in the digital world and are responsible as digital citizens. This can keep children safe and help them succeed in the digital world by making them aware of the digital world and learning the right digital behaviors.

In this section; the importance of digital ethics in early childhood, studies and policies on digital ethics in the world and in Turkey, teacher as a digital citizen, child as a digital citizen will be examined and teacher views on digital ethics studies will be included.

1. The Importance of Digital Ethics in Early Childhood

With the spread of technology, the issue of digital ethics has gained importance in order to express behaviors in the digital world correctly and appropriately and to adopt these rules in all online technologies. Although there are debates about its definition, it is seen that it has a dynamic structure and expresses the values, norms and rules that people should pay attention to in their digital interactions. Digital ethics, which deals with the impact of Information and Communication Technologies (ICTs) on societies and the environment in general, addresses ethical questions related to communication media such as the internet, online digital tools and navigation services (Capurro, 2009; Özcan, 2021). In the concept of digital ethics, which focuses on how individuals can better manage their lives by using interactive digital media, individuals have the duty to be aware of and take responsibility for this norm.

The widespread use of digital tools and the interaction of young children with these tools is an inevitable reality (Rideout & Robb, 2020), and in this process, the concept of digital ethics in early childhood should be considered as a context that needs to be discussed. In particular, it is emphasized that digital ethics education for everyone should start at an early age and in this way, it is considered necessary to lay the foundations of responsible adulthood in the digital age (Rogerson, 2021).

In the process of promoting the concept of digital ethics for children, parents and teachers have certain responsibilities;

- Knowing and managing online risks
- Encouraging and enabling children to acquire appropriate behaviors
- Ensuring the integration of digital ethics principles into the education process
- The adult himself/herself has digital ethical norms and behaves in an exemplary manner.

2. Risk Factors for Technology Use in Early Childhood

The widespread adoption of the Internet and other digital tools brings about social changes, and risks arise alongside the opportunities offered by these environments (Liu & Hwang, 2021;Yalçın-Irmak & Ardıç, 2018). The digital environment has become one of the areas where children live their daily lives. It is seen that many risks such as making friends in these digital environments, especially meeting strangers in real life, giving personal information online, pedophilia contact, stalking, bullying, racist discourses are among these risks (Livingstone & Haddon, 2009; OECD, 2021). Although it is stated that these risks are concentrated in older age groups, as technology advances, these risks also affect children in early childhood due to the changing nature of the use of digital technologies (Livingstone & Helpser, 2007; OECD, 2021; Shamel et al., 2020). Technological advances, which in themselves have contributed to changes in digital devices and services, have also contributed more broadly to this changing risk landscape. In particular, smartphones have transformed how, when and where the internet can be accessed, and the rise of portable technologies has greatly enhanced online experiences. While the ease of use of portable digital devices has increased usage (Rideout and Robb, 2020), it has also led to the creation of special digital spaces for children and young people, allowing them to share various information and materials away from parental supervision (Shamel et al., 2020). Measures have been taken in various countries against the risks arising in this direction. For example, the German Federal Network Agency (Bundesnetzagentur) has banned smartwatches intended for use by children and has asked parents who have already purchased such devices to destroy them for good measure (ATIF). Not only that, many countries have banned the use of cell phones by children in schools in order to prevent harm (Gao et al., 2014).

Livingstone and Haddon (2009) examined the risks that children may face in digital environments under certain headings: commercial, violent, sexual and cultural risks. In addition, more systematic classifications have been made over time and Livingstone, Mascheroni and Staksrud (2015) 3C online risk classification is widely used as a classic reference point for international stakeholders. However, a new classification has been developed in view of the emerging risks in the digital environment and children's increasing experiences of the risks of online environments (Livingstone & Stoilova, 2021). The classification in their study, which they conducted by starting to investigate the types and contexts of online risk, is presented in Figure 1.

As can be seen in Figure 1, the objective of the classification is elaborated to minimize online risks to children's rights and children's best interests, and to enable them to make the most of online environments. The classification provides the basis for a better understanding of online risks and explains to policy makers, parents, researchers and teachers which risks are important and why, what can be done about different risks and the contexts to be considered, and the need to develop and organize online risk measurements (Livingtsone & Stoilova, 2021). In addition, the "Recommendation of the Council on Children in the Digital Environment" published by the OECD in 2012 was revised and renamed as "Recommendation on Children in the Digital Environment", taking into account the technological, legal and political developments in the following years. The resolution aims to help governments more effectively address technological, legal and policy developments, identify the necessary tools to support children to take advantage of the opportunities of the digital environment and address the new and evolving risks they may face in this environment. Ensuring a safe and beneficial digital environment for children highlights the great importance of the shared responsibility of all stakeholders. It makes clear that governments play a key role in responding to children's needs and that parents need to be supported (OECD, 2021).

In Europe, the EU Kids Online project is a scoping study examining the risks and opportunities that children experience online. EU Kids Online (2020) includes information on 25,101 children between the ages of 9-16 from 19 countries in a survey of online activities and experiences that may have positive or negative effects. The risks and opportunities identified in the report include: general negative experiences online, online aggression and cyberbullying, exposure to potentially harmful content, misuse of data, overuse of the internet, sexting, exposure to sexual content and meeting new people online. When the remarkable results are analyzed, it is seen that the rates indicating that children experience discomfort on the internet are higher in the 2020 report compared to the EU Kids Online 2010 survey. It was also found that when children encounter a problem online, they mostly do not talk to anyone and ignore the problem they are experiencing. Although there are responses indicating that they talk to their parents or peers, this rate is not at the expected level.

Figure 1. The CO:RE classification of online risk to children
Source: Livingstone and Stoilova (2021)

	Agressive	**Sexual**	**Values**	**Cross-Cutting**	Privacy violations (interpersonal, institutional, commercial) Physical and mental health risks (e.g., sedentary lifestyle, excessive screen use, isolation, anxiety) Inequalities and discrimination (in/exclusion, exploiting vulnerability, algorithmic bias/predictive analytics)
Content Child engages with or is exposed to potentially harmful content	Violent, gory, graphic, racist, hateful or extremist information and communication	Pornography (harmful or illegal), sexualization of culture, oppressive body image norms	Mis/disinformation, age-inappropriate marketing or user-generated content		
Contact Child experiences or is targeted by potentially harmful adult contact	Harassment, stalking, hateful **behavior** unwanted or excessive surveillance	Sexual harassment, sexual grooming, sextortion, the generation and sharing of child sexual abuse material	Ideological persuasion or manipulation, **radicalisation** and extremist recruitment conduct		
Conduct Child witnesses, participates in or is a victim of potentially harmful peer conduct	Bullying, hateful or hostile communication or peer activity e.g. trolling, exclusion, shaming	Sexual harassment, non-consensual sexual messaging, adverse sexual pressures	Potentially harmful user communities e.g. self-harm, anti-vaccine, adverse peer pressures		
Contract Child is party to or exploited by potentially harmful contract	Identity theft, fraud, phishing, scams, hacking, blackmail, security risks	Trafficking for purposes of sexual exploitation, streaming (paid-for) child sexual abuse	Gambling, filter bubbles, micro-targeting, dark patterns shaping persuasion or purchase		

Parents and teachers, who are in the position of managing and directing the digital world for children, have important responsibilities such as keeping digital interaction in balance, protecting them from the risks of digital tools, and teaching children the most appropriate use (Chaudron et al., 2018; Bilgin-Ülken & Kılınç, 2018; Livingstone & Helsper, 2008). Within the scope of digital ethics, the adults around the child in this process should make their choices in the digital world in accordance with certain criteria. Although there are many different classifications in the world, the most widely preferred Pan European

Game Information System (PEGI) is defined as "an evaluation and rating system established to promote the responsible and safe use of digital games and to serve as a guide for consumers when purchasing digital games". The PEGI rating consists of five age categories and eight content descriptors that recommend the suitability and content of a game for a specific age range based on the content of the game. The structured system informs the consumer about the qualities of the content and the age group for which it is suitable during the purchase or use of the product with simple and understandable categories and age limits (Digital Game Report, 2019).

Figure 2. PEGI labels
Source: https://pegi.info/what-do-the-labels-mean

As seen in Figure 2, the criteria in the classification are categorized as content and age. When we look at the content classification, there are titles such as "violence, discrimination, sexuality/nudity, drugs, fear/horror, vulgar and sexually explicit language, gambling". Another criterion of the PEGI system is five different age grouping levels (3, 7, 12, 16 and 18). It is stated that the classifications are determined by experts in determining the appropriate age groups, determining the degrees of the criteria and describing the content. For example; "Games with the PEGI 3 label are considered suitable for all age groups. Children cannot associate the characters they see on the screen with real life characters, screen characters are purely fantasy. The game does not contain any sounds or visuals that may frighten young children. Slang (abusive) language is not used." (Digital Game Report, 2019).

As another classification, the Entertainment Software Rating Board (ESRB) is a rating system widely used in the United States, Canada and Mexico and is known as the Entertainment Software Rating Board (ESRB) for digital games. This rating system was legalized by the Entertainment Software Association (ESA) in 1994. At the time, it was aimed at protecting a certain age group from violent digital games, but it is still used effectively today. The main purpose of the ESRB is to help users understand the content and appropriateness of games. This rating system examines game content for various aspects such as alcohol, violence, bad language, drugs, sexuality and gambling, and determines the appropriate age range of users. These descriptors inform game purchasers and/or parents in advance about potential situations they may encounter while playing the game. In this way, users are better informed about the content of the game, enabling them to make more informed choices and choices.

Figure 3. ESRB labels
Source: https://www.esrb.org/

As seen in Figure 3, the rating categories are defined as Early Childhood (EC), Everyone (E), Everyone 10+ (E10+), Teen (T), Mature (M), Adults Only (AO) and Rating Pending (RP). The rating categories indicate age-appropriateness as in the PEGI classification. When the Content Descriptors, which is another criterion, are examined, it is seen that the titles "Substances, Blood/Gore, Humor, Violence, Gambling Language, Nudity Sexuality" are determined. These titles were also examined and detailed in detail depending on different qualities. The "Interactive Elements" rating reveals interactive or online features that do not affect the rating assignment of a product and is expressed as "In-Game Purchases, In-Game Purchases (Includes Random Items), Users Interact, Shares Location, Unrestricted Internet". In addition to informing about the classification, there are also informative webpages with details to raise awareness of parents (https://www.esrb.org/tools-for-parents/). While the commonly preferred classifications are PEGI and ESRB, some countries such as Germany, Australia and Russia prefer their own age-based rating systems (https://www.kaspersky.com/blog/gaming-age-ratings/11647/).

3. Digital Ethics in Early Childhood Digital Learning Environments

"New Information and Communication Technologies (ICTs) affect children's enjoyment of a significant number of fundamental rights guaranteed by the UNCRC, the European Convention on Human Rights and the European Social Charter. According to recommendations issued by the UN Committee on the Rights of the Child, all children should be able to safely access ICTs and digital media, and be empowered to fully participate, express themselves, seek information and enjoy all the rights enshrined in the UNCRC and its Optional Protocols without discrimination of any kind." (Council of Europe Strategy on the Rights of the Child 2016-2021, 20)

Today, concepts such as technology, digital media, applications and digital games are present in most areas where children are present. Adults in the child's environment provide their children with access to digital technologies, sometimes consciously and sometimes unconsciously. Therefore, the most important issue related to children and technology is to determine how digital technologies can contribute to the development of meaningful learning contexts and children's digital habits (Palaiologou, 2016). In the digital age, keeping children safe and protecting their rights in digital environments is also of great importance. For example, publishing children's photos and videos, balance in sharing activity images, protection of privacy and behaviors that respect children's rights are very important (Eckhoff, 2015).

As in all areas, various measures should be taken at national and international level to protect children's rights in digital environments while utilizing technology in education. The OECD Report (2021) recommends the following measures to protect children's rights:

1. ***Strengthening the Legal Framework:*** Countries should establish appropriate legal frameworks to protect children against risks in digital environments. This should cover issues such as combating child abuse, protecting children's privacy and personal information, blocking harmful content and combating cyberbullying.
2. ***Education and Awareness:*** Children, parents, teachers and other relevant stakeholders should be made aware and educated about digital risks. In this way, children will be more aware of how to protect themselves and become more resilient to risks.
3. ***Cooperation with Internet Service Providers:*** Internet service providers should implement effective policies and security measures to protect children. Steps such as blocking harmful content, creating reporting mechanisms and developing technical solutions to ensure children's safety should be taken.
4. ***Child Participation:*** Children should be actively involved in decision-making processes affecting digital environments. Children's participation should be encouraged to advocate for their rights, provide feedback on risks and propose solutions.

Within the scope of Strengthening the Legal Framework recommended by the OECD Report (2021), for example, the Circular No. 2017/12 on the Use of Social Media in Schools sent by the Ministry of National Education in Turkey emphasizes the Constitution of the Republic of Turkey, the Convention on the Rights of the Child adopted by the United Nations General Assembly on 20/11/1989, the Basic Law on National Education No. 1739 and the Turkish Penal Code No. 5237.

Within the scope of Education and Awareness, the following websites are recommended for raising awareness and training for parents, teachers and other stakeholders. The names of the websites, their scope and links for access are detailed below. More detailed content on education and awareness is given under the titles "Teacher as Digital Citizen" and "Child as Digital Citizen".

Common Sense Media: Common Sense Media provides resources and information for parents, educators and children to navigate the digital world safely and responsibly. It offers digital content reviews, advice for parents and educational resources. Website: [https://www.commonsensemedia.org/]

Childnet International: Childnet International is a UK-based organization focused on promoting online safety and digital citizenship among young people. Their website offers resources, videos and teaching materials on topics such as online safety, cyberbullying and digital ethics. Website: [https://www.childnet.com/]

NSPCC Online Safety: The NSPCC (National Society for the Prevention of Cruelty to Children) provides guidance and resources for parents and professionals to keep children safe online. They offer advice on topics such as social media, online gaming and online privacy. Website: [https://www.nspcc.org.uk/keeping-children-safe/online-safety/]

ConnectSafely: ConnectSafely is a non-profit organization dedicated to promoting internet safety, privacy and security. Their website offers resources, guides and tips for parents, educators and young people on topics such as cyberbullying, digital literacy and responsible online behavior. Website: [https://www.connectsafely.org/]

Family Online Safety Institute (FOSI): FOSI is an international non-profit organization that provides resources, research and guidelines to promote a safe and responsible digital environment for children and families. Their website offers educational materials, expert advice and information on online safety and digital citizenship. Website: [https://www.fosi.org/]

GüvenliWeb: Supported by the Informatics Foundation of Turkey, GüvenliWeb provides resources and information on safe internet use for children in Turkey. It offers guidance on digital citizenship, online safety and responsible internet use. Website: [https://www.guvenliweb.org.tr/]

Safe Internet Center: The Safe Internet Center is an initiative supported by the Ministry of Family, Labor and Social Services in Turkey. It offers resources, educational materials and support for parents, children and educators to promote safe internet use and combat online risks. Website: [https://www.guvenliinternet.gov.tr/]

Safer Internet Turkey: Safer Internet Turkey is part of the global Safer Internet Day initiative and aims to promote safe and responsible internet use among children and youth in Turkey. Their website offers resources, tips and awareness campaigns on online safety, cyberbullying and digital citizenship. Website: [https://www.saferinternet.org.tr/]

Bilinçli İnternet: Bilinçli İnternet is an organization focusing on digital citizenship and safe internet use in Turkey. It offers resources, educational materials and workshops for children, parents and educators to promote responsible digital behavior. Website: [https://www.bilincinternet.com/]

Cooperation with Internet Service Providers, one of the topics recommended by the OECD Report (2021), is actually related to Child Participation. This is because the article states that "Children should be actively involved in decision-making processes affecting digital environments. Children's participation should be encouraged in defending their rights, giving feedback on risks and proposing solutions." Safe internet is one of the main points for children to share their thoughts about online environments and for their active participation. The new strategy for a better internet for children (BIK+), adopted on 11 May 2022, will ensure that children are protected, respected and empowered online in the new Digital Decade in line with the European Digital Principles. The children's section of the BIK+ Strategy highlights the importance of safe digital experiences to protect children from harmful and illegal online content, behaviors, contacts and risks, and to improve their online environment by making it safer through a safe, age-appropriate digital environment built to benefit children. Examples of search engines that reduce online risks for children are listed below. The uses, scope and link addresses of safe search engines are given in detail:

Kiddle: Kiddle (www.kiddle.co) is a child-friendly search engine that uses Google's safe search technology to provide child-friendly, filtered search results. It includes large thumbnails for visual search, child-friendly content and blocks obscene or inappropriate material.

KidzSearch: KidzSearch (www.kidzsearch.com) is a safe search engine and online encyclopedia designed for children. It provides filtered search results from trusted sources, including child-friendly educational websites, games, images and videos.

KidRex: KidRex (www.kidrex.org) is a kid-friendly search engine powered by Google Custom Search. It filters out explicit or inappropriate content and displays search results in kid-friendly language and images.

Swiggle Swiggle (www.swiggle.org.uk) is a safe search engine designed for children by the British education organization South West Grid for Learning. It offers filtered search results and provides a child-friendly browsing experience.

Safe Search Kids: Safe Search Kids (www.safesearchkids.com) is a search engine that filters search results to provide kid-friendly content. It aims to provide a safer online search experience for children while blocking inappropriate websites and images.

It is now accepted that the use of digital technologies in all levels of education from early childhood onwards and in administrative structures, educational processes, measurement and evaluation structures related to education is increasing all over the world. In the educational reform that comes with technology, it is very important that teachers and adults around the child, especially in early childhood, have knowledge about children's rights, digital ethics and digital citizenship in order to prevent potential problems that may arise. Although studies, practices and policies have been developed on digital ethics and children's rights, it is stated that teachers need training on ethical issues that may arise regarding the use of technology in educational environments (Buchanan, 2019).

4. Teacher as Digital Citizen

As in every subject, the European Framework for the Digital Competence of Educators (DigCompEdu) has determined a framework consisting of 6 stages for teachers who will be role models for children in terms of acquiring digital ethics. These frameworks consist of "professional engagement" which includes focusing on the professional environment, "digital resources" which includes research and generating new knowledge, "teaching and learning" which includes managing and orchestrating the use of digital tools, "facilitating learners' digital competence", "empowering learners" which includes accessibility and inclusion, personalization, actively engaging learners and finally "assessment" which includes strategies, analysis of evidence and feedback (Redecker, 2017). In addition, Educator Technology Standards in the ISTE (2019) also set technology standards for teachers. These standards consist of "learner" which enables learning practices with others, "collaborator" which involves sharing practices and resources, "designer", "facilitator" which facilitates designing, "leader" which supports children's achievements, "analyst" which makes sense of and protects data, and "citizen" which raises awareness of children to ensure their positive contribution to the digital world. The DigCompEdu framework in the EU includes elements related to "responsible use" safety, security and ethics, while in the case of the ISTE teacher standards, the facilitator and citizen domains are related to responsible use and data security in education.

In digital privacy and data security, it is important that teachers have well-developed digital literacy skills and conscious use of digital tools in the classroom. In addition, it is the teacher's responsibility to ensure the privacy of children's and parents' data and to pay attention to digital privacy (Rodríguez-Triana, Martínez-Monés & Villagrá-Sobrino, 2016). The factors that teachers should pay attention to in this regard can be explained as follows:

- Data Privacy: It is important for the safety of individuals that teachers do not share data on children and parents with third parties. Teachers should store data in a secure environment and take precautions against unauthorized access (Ivanova, Marín, Tur & Buchem, 2019).
- Social Media Privacy: It is seen that the frequency of using social media platforms to share about the educational process has increased, especially after the pandemic (Greenhow, Galvin., Brandon & Askari, 2020). For this reason, teachers should hide children's faces and information in their posts. It is also important for teachers to ensure the privacy of their personal data on social media (Obar & Oeldorf-Hirsch, 2018). In this direction, it is important to provide training to prospective teachers and teachers who will teach in the future on the conscious use of social media (Carpenter & Morrison, 2018).
- Internet Threats and Cyberbullying: With the increasing use of digital tools, the frequency of children's exposure to internet threats and cyberbullying has also increased (Doane, Kelley, Chiang & Padilla, 2013). Due to periodic characteristics in early childhood, media content is often perceived as truthful and children are therefore vulnerable to it (Karakaya, 2021). Therefore, teachers should raise children's awareness about internet threats and cyberbullying and guide them (Kavurucu & Küçük, 2021).
- Educational Technology Security: With the use of digital tools to record and store data in schools, teachers and school administrators have responsibilities such as preventing third parties from accessing data and ensuring its security. Using hard-to-guess passwords on electronic devices used to store data and not sharing data on unsecured platforms with internet access will help teachers to increase data security (Ahmad & Maynard, 2014; Amo-Filva et al., 2021).
- Plagiarism and Copyright: Plagiarism and not paying attention to copyrights is one of the ethical problems encountered (Katıtaş, Seylim & Demirkan, 2022). Especially when teachers, who are role models for children, pay attention to ethical rules and transparently explain the source of the information provided, children's awareness of this issue increases positively. At this point, training pre-service teachers and teachers on plagiarism and copyrights is important for them to comply with ethical principles (Reingold & Baratz, 2011).

The FATIH Project (FP), which has been implemented by the Ministry of National Education (MoNE) with the support of the Ministry of Transportation since 2011 in Turkey, primarily aims to provide access to information technologies for children in disadvantaged regions. One of the components of the project is "Ensuring Conscious and Safe Use of Information Technologies" (MoNE, 2023). In this direction, the project aims to create suitable environmental conditions for teachers' safe internet use by preparing public service announcements and brochures and creating a secure network infrastructure system (Keser & Yayla, 2021).

5. The Child as a Digital Citizen

In Prensky's (2001) definition, digital natives, i.e. children, are born into a digital world. Digital environments offer children many supports in areas such as socialization, education and creativity, but they also bring risks such as cyberbullying and lack of privacy (OECD, 2011). Children need digital citizenship education in the digital world they are born into, firstly to prefer and produce safe digital environments for themselves and to protect themselves from risks, and secondly to exhibit behaviors that respect the rights of others as a part of the digital society.

Digital citizenship involves acting in a way that respects ethical rules and rights in digital environments (Çubukçu & Bayzan, 2013). It requires complying with norms by acting responsibly (Mossberger, Tolbert, & S. McNeal, 2007). Technology literacy education given in the context of digital citizenship; beyond the use of technology, it should include using technology effectively and accurately by considering rights and responsibilities (Çubukçu & Bayzan, 2013).

Digital citizenship is defined with 9 components. Digital access involves participating in society by using digital tools. In this direction, it is evaluated whether all children in the class have equal access to digital tools and equal opportunities to participate in the digital society. It is also important to equalize these opportunities for individuals with special needs and children who do not have access to digital tools outside of school. Digital commerce is the buying and selling of goods and services online. In this context, children need to be informed about the possibility of buying and selling through digital tools and the risks involved. Digital communication is the exchange of information online. Children should be informed about the rules and ethical expectations to be followed when using digital tools for communication. Digital literacy is defined as the ability to decide when and how to use digital tools. Children should be guided in evaluating online resources and using online resources for learning purposes. Digital ethics are the standards of behavior expected by other people using digital tools. In this context, it is aimed for children to be aware of others when using digital tools and to use digital tools in a way that respects others. Digital law includes rights and restrictions on the use of digital tools. Especially with regard to copyrights, it is important to reinforce behaviors that protect the rights of others and their own rights. Digital rights and responsibilities are the freedoms and expected behaviors of everyone who uses digital tools. The aim is to educate children to be aware of their rights and responsibilities in the digital society and to support ethical and responsible behavior. Digital health and well-being are the physical and psychological health components of digital technology. It is important that children who are informed about the physical and psychological effects and risks develop behaviors to take precautions to protect their health. Digital security includes the measures needed to ensure the safety of all users. Children should be given information about the precautions they need to take to ensure the security of their own information, the hardware, software and networks they use, and the software they can use (Ribble & Bailey, 2007, pp.13-37).

Digital citizenship principles are considered as a necessity for being a productive and responsible citizen. Therefore, there is a need for policies that children can implement in schools and that teachers can support. Digital citizenship provides a basis for these policies. While citizenship involves states, societies and borders, digital citizenship is more global in nature and children need to learn how to spend time in this borderless world. In this context, digital tools should be considered as an environment where children socialize and learn to be a part of society rather than a game tool (Ribble & Bailey, 2007, pp. 10-12). In this context, digital ethics education is one of the important dimensions in the process of raising children as digital citizens. Providing children with the principles and appropriate behaviors determined within the scope of digital ethics is important and necessary for children to both prefer a safe internet environment for themselves and to contribute to trust in the internet environment by respecting the rights of others.

A model consisting of 4 steps is proposed for an education process on digital citizenship. In the awareness stage, children are aimed to become technologically literate. For this, they need to be informed about how hardware and software work, their effects on people, possible problems, and accepted rules. Then, the guided practice phase begins and children are expected to experience the appropriate way of use with the support of the teacher. The teacher should allow children to explore and take risks in the process and be understanding of possible problems. Reflecting on the child's actions and recognizing inappropriate uses, examples of misuse or abuse should be worked on. In the modeling and demonstration phase, teachers should act as role models for children regarding appropriate use. Teachers should follow the frequently mentioned principles and rules and fulfill their responsibilities. In the feedback and analysis phase, children should be given opportunities to discuss how to use technology more effectively and appropriately. Analysis of the functions of rules and responsibilities and feedback on the consequences of inappropriate use and ways to avoid or minimize them will allow children to internalize them (Ribble & Bailey, 2007, pp. 75-79). Accordingly, the teacher's acquisition of digital citizenship knowledge, awareness and skills becomes an important component of the process. In this context, the aim of this study is to reveal preschool teachers' conceptions and practices of digital ethics.

METHOD

Research Model

The research was designed in a basic qualitative research design. Basic qualitative research aims to reveal how individuals construct their understanding of the world as a result of their interactions with the world (Merriam, 2013, p. 22). Within the scope of this research, it is aimed to reveal preschool teachers' views on digital ethics and digital ethics education.

Study Group

The study group of the research consists of 20 pre-school teachers determined by maximum diversity sampling method, which is one of the purposeful sampling methods. Maximum diversity sampling method is the creation of common patterns by taking data from different groups in order to determine differences about a subject (Patton, 2005). Attention was paid to the fact that the teachers in the study had different genders, different age groups, and different years of professional experience. In addition, care was taken to include teachers who received and did not receive training on digital ethics in the study. The data about the participants are given in Table 1.

Table 1. Demographic characteristics of participating preschool teachers

Demographic Characteristics of Participating Preschool Teachers		n
Age	30-35 years old	10
	36-40 years old	2
	41 and above	8
Professional Experience	5-10	5
	11-15	9
	16-20	5
	21-25	1
Encountering the concept of digital ethics	Yes	17
	No	3
Receiving training on digital ethics	Yes	3
	No	17

Ten of the teachers are 30-35 years old, 2 are 36-40 years old, and 8 are 41 years old and above. 5 of the teachers have 5-10, 9 of them have 11-15, 5 of them have 16-20 and 1 of them has 21-25 years of professional experience. While 17 teachers stated that they had encountered the concept of digital ethics before, 3 of them stated that they had never encountered this concept before. Only 3 teachers stated that they had received training on digital ethics before.

Data Collection Tools

The data were collected through the semi-structured "Form for Determining Preschool Teachers' Views on the Concept of Digital Ethics" developed by the researchers. The form consisting of 12 open-ended questions was sent to 5 field experts by the researcher. At the end of the pilot application, one more question was added and a final form consisting of 13 questions was created. The form included open-ended questions for the participants to express their views on the use of technology in the classroom, the content of digital ethics, principles and practices related to digital ethics, trainings on digital ethics, sharing with children, families, colleagues in relation to digital ethics, resources they follow on this subject, and digital ethics studies in their institutions.

Data Collection

The participants were informed about the study by the researchers, an appointment was made with the preschool teachers who volunteered to participate in the study, and a convenient time was determined; interviews were conducted via telephone interviews. During the interviews, voice recordings were taken with the permission of the teacher, and then the recordings were transcribed by the researchers. During the transcription, the identities of the participants were concealed by referring to them with codes such as T1, T2.

Data Analysis

The data obtained from the interviews were analyzed by the researchers using the content analysis method. First, the answers given by the teachers were analyzed and codes were created, then the codes were analyzed and categories were created. Inter-coder reliability was calculated as 96% (Miles & Huberman, 1994).

FINDINGS

In this part of the study, findings related to the data obtained from the interviews with preschool teachers are presented.

When the views of preschool teachers on the use of digital tools in the classroom environment are examined, it is seen that all of the teachers (f=20) think that the use of digital tools in the classroom environment is useful. The results of the analysis of the reasons for finding them useful and the factors they pay attention to in the selection of digital tools are shown in Figure 4.

Figure 4. Preschool teachers' views on the use of digital tools in the classroom environment and the factors considered in the selection of digital tools

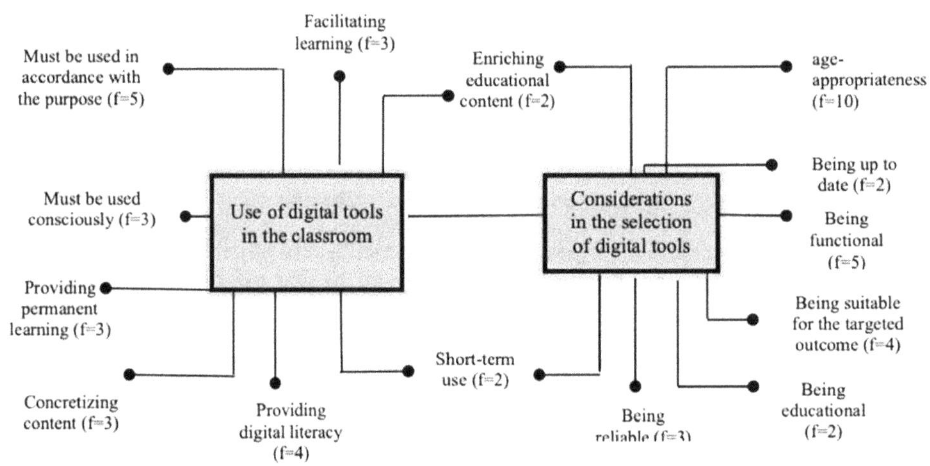

When Figure 4 was analysed, it was seen that the teachers evaluated the use of digital tools in the classroom environment under two different titles as "the process of using digital tools" and "the benefits of digital tools to the educational process". In the process of using digital tools, it was frequently emphasised that they were used in accordance with the purpose (f=5), while in terms of their benefits to the educational process, they mostly emphasised providing digital literacy (f=4) *(T9: I think they should be used as much as possible in the age of digitalisation. It is no longer the right approach to try to keep children away from digital tools.)*, facilitating learning (f=3) *(T11: I find it useful. It increases positive and permanent learning.)* and concretising the content (f=3). In addition to this, when the "factors that teachers pay attention to when choosing digital tools" were analysed, it was determined that they often consider that they are suitable for the age level (f=10) *(T5: I pay attention to choosing content that is suitable for children's age group and that will meet their interests and needs. When using these tools, I give children the opportunity to use them)*. In addition, it is seen that some teachers take into account the elements of being suitable for the targeted outcome (f=4) and being functional (f=5).

When Figure 5 is examined, it is seen that the teachers defined digital ethics as "protecting individuals

Figure 5. Preschool teachers' definitions of the concept of digital ethics

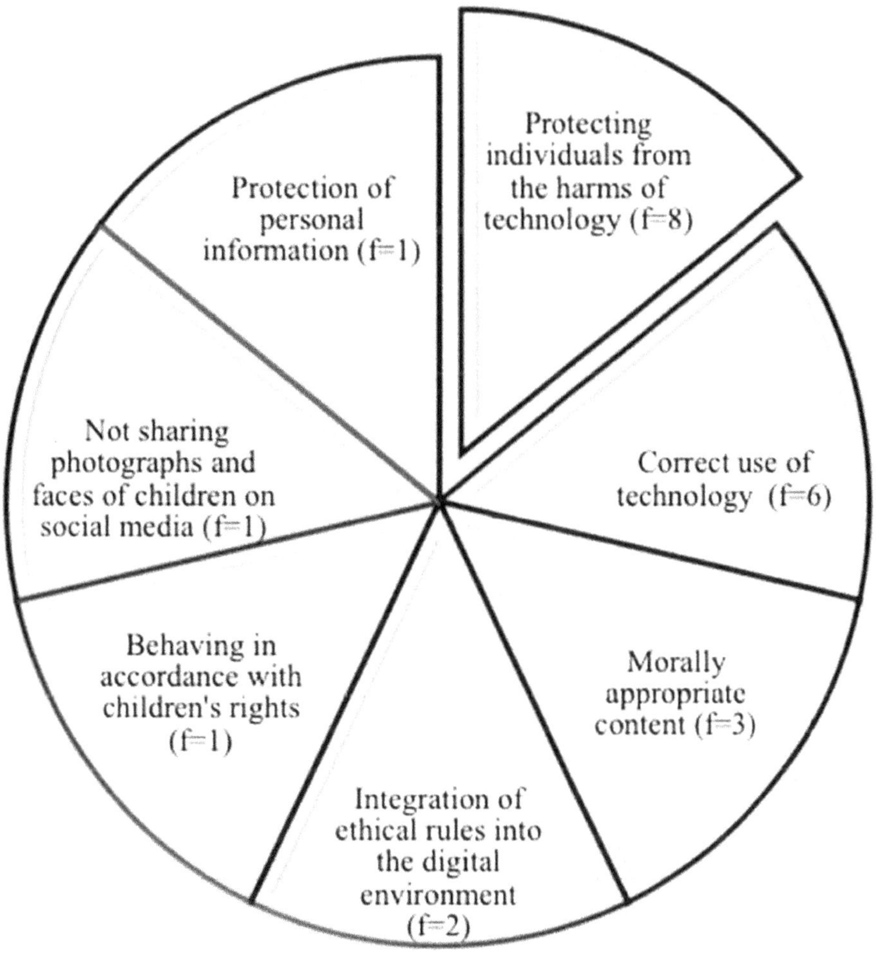

from the harms of technology" the most (f=8) *(T5: In other words, I think that when ethics is considered as a discipline that helps to distinguish between good and bad, right and wrong, digital ethics can be considered as a concept that determines the correct behaviours in the communication process in the digital world. In other words, it can be considered as a concept that helps to prevent harm to individuals through the misuse of technology.)* In addition, some of the teachers (f=6) defined it as "correct use of technology" *(T2: ....digital ethics can be defined as using technology correctly)*, while some teachers (f=6) defined it as "morally appropriate content" *(T16: I think digital ethics are correct behaviours within the framework of our moral values)*. In addition, it can be said that teachers also emphasised data privacy, children's rights and digital citizenship. In addition, it was determined that all of the teachers (f=20) thought that it was necessary to comply with the principles of digital ethics during the training. It was determined that the reason why they considered it necessary to comply with the principles of digital ethics was often "to protect children".

Teachers expressed the principles they paid attention to during social media sharing through three

Figure 6. Principles considered in social media sharing

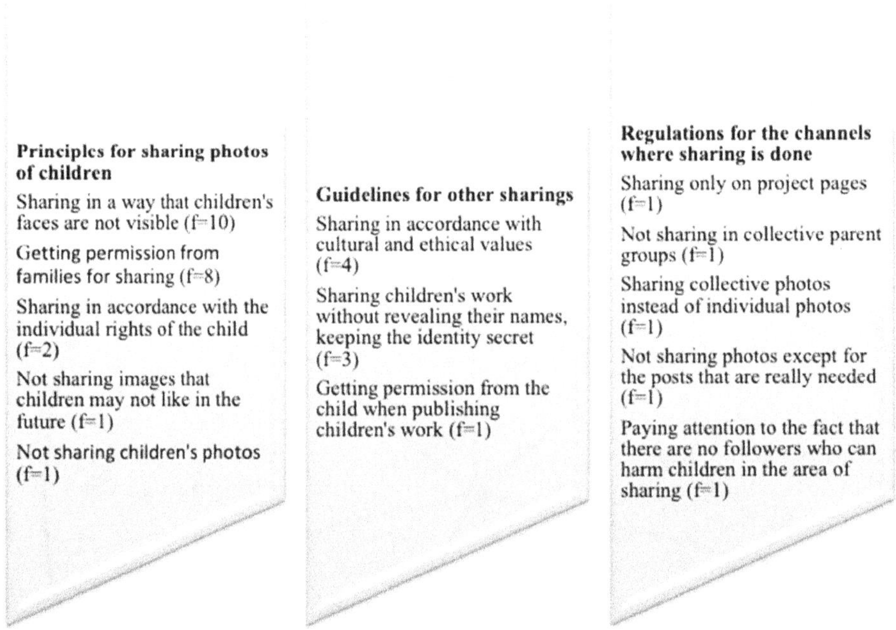

main themes: "principles for sharing photos of children", "principles for other sharings" and "regulations for the channels where sharing is done". The ideas about sharing children's photos were diversified as "not sharing children's photos", "not sharing images that they may not like in the future", "sharing in accordance with individual rights", "getting permission from families for sharing" and "sharing in a way that children's faces are not visible". While T5 stated that he did not share the photographs of children *(T5: I definitely do not share the photographs of children...)*, T3 stated that he avoided sharing images that children would not like in the future *(T3: ....Staying away from posts that may offend them in the future (such as sharing videos, audio recordings).* The most frequently repeated answers were about getting permission from the families and not showing the faces of the children *(T10: We get permission from the families of the children...I make sure that their faces are not visible in videos and photos...; T16: I make sure not to share the children. If there is a child in a photo, I cover his/her face with emoji; T20: I am definitely not in favour of using the images of the faces and voices of children. Now, they can be shared on our school websites or on the websites of the Ministry of National Education for educational purposes. But sharing without permission is extremely harmful from a legal point of view. From the age of 13 onwards, our children have access to social media, but we already receive permission from our parents for every project or activity held at our school at the beginning of each term. But despite this, we still take activity-orientated photos, especially. I think this is the key sentence. "Activity orientated" in quotation marks. Of course, our children will be seen, but without their faces being clearly visible, because the internet has many harms besides its innovations. We need to be careful not to expose them to cyberbullying. I am against it.)*

Regarding their other posts, the teachers stated that they got permission from the children and shared their work without revealing their names *(T5: ...I never share children's photos. Even if I am going to share a work done by them, I get permission from the children and make sure that their names are not visible on the work. In a sense, I do not allow their identities to be seen.)*; they paid attention to sharing posts in accordance with cultural and ethical values *(T6: ...it should be in accordance with cultural and ethical values...; T11: Social morality, clean content, correct values...).*

In addition, regarding the channels they share, they stated that they share only on project pages, they do not share in collective parent groups, they share collective photos instead of individual photos, and they pay attention to the followers in the channels they share *(T10: ...I do not share photos, names, etc. in any social media environment other than children's project pages. I do not share collectively in groups such as WhatsApp parent groups...; T18: ...I pay attention to the fact that the students are not seen individually, the photos are in groups, and the students' faces are blurred. If I can explain the subject I want to explain without uploading photos, I do not upload photos).*

While 19 of the participants stated that they needed to receive training on digital ethics, T17 stated

Figure 7. Willingness to receive in-service training on digital ethics

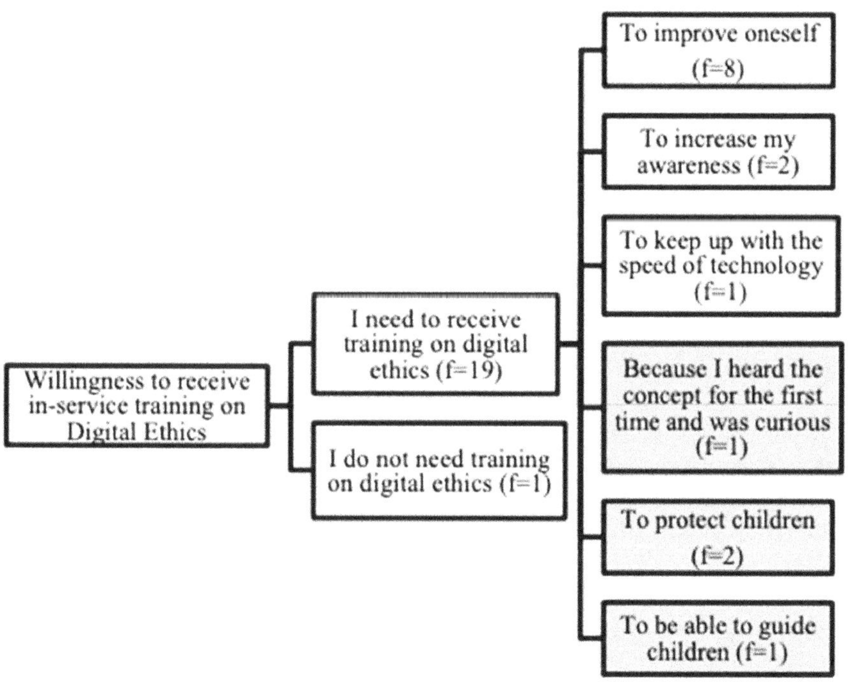

that he did not need training on this subject because he had received a similar training before *(T17: I don't want to. Because I had received it before.)*. The motivations of the teachers who want to receive training on this subject are shaped by their "individual wishes" and "aims related to children". Regarding individual wishes, T15 commented *"In order to be better equipped and to act correctly in this subject"*, while T8 stated *"We should renew ourselves in the rapid change of technology"*. T5 expressed his aims about children as follows: *"I would like to, because we live in a digital world and I do not have competence in this field in order to be able to direct and guide children whose mother tongue is technology more accurately."* T9 expressed his opinion as *"Yes, I would like to take it because we know that something is not right, but we do not investigate how much of it is right and how much is wrong. I would like to learn clearly in order to protect children."*

While 19 of the participants found the teacher trainings on digital ethics insufficient, 1 teacher stated

Figure 8. Opinions on the sufficiency of digital ethics trainings for preschool teachers

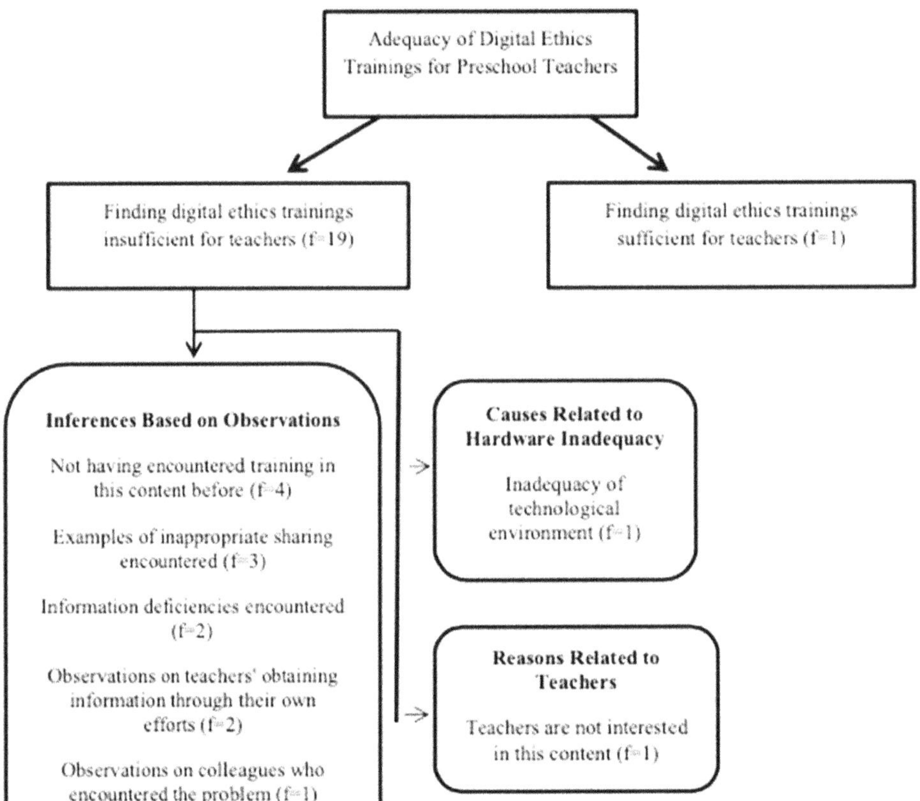

that he/she found it sufficient *(T17: I think it is sufficient.)*. This may be related to his/her access to the training opportunity. The teachers stated that they had not encountered any training on this subject before *(T5: I know that teachers have never been trained on this subject, because as a teacher I have never received any training on this subject; T18: I think it is not enough because although I have received many in-service trainings, I have not received any training on this subject; T20: My observation is that I have been conducting joint projects with the countries in the European Union since two thousand ten. But as far as I observe, only the teachers who do projects in Turkey are more conscious about this issue. Apart from that, our Ministry of National Education is more aware of this issue or some organisations, universities, especially faculties of education, pre-school departments are open to teachers and everyone. They can also offer distance education through webinars.)*, examples of inappropriate sharing encountered *(T1: No. Children's faces are visible on school instagram pages or teachers' personal pages. T12: No, I don't think it is enough because children are used a lot in digital content.)*, information deficiencies *(T16: I think it is not enough, because I even heard the concept for the first time.)*, teachers who access knowledge through their own efforts *(T4: I think that teachers have knowledge in this field through their own efforts. More training should be given in this field.)* and colleagues who encounter problems *(T10: I see that teachers encounter many problems in the news I receive.)* were listed as the reasons why they found the training inadequate. In addition, T6 stated that the trainings were inadequate due to the lack of access to technological facilities, while T19 attributed the inadequacy of the trainings to the teachers' lack of interest in the subject *(T19:...Teachers are not interested in the subject, there are always problems, of course they use it unconsciously).*

When the opinions of the teachers regarding the provision of digital ethics education for children were analysed, it was seen that most of them (f=11) found it necessary to provide digital education, while 4 of the teachers stated that they found the education currently provided on this subject insufficient *(T3: I think that it is not enough, even this awareness raising is not done. Because cyberbullying has increased a lot and most of the children do not know how to use the internet correctly; T18: I did not see any training on this subject, there are trainings given by the teachers with their own efforts, but we cannot provide the students with the necessary training because we do not have enough competence in this subject).* The majority of the teachers stated that it is necessary to provide education in this direction *(T19: ...It has to be pre-school. Because it is the foundation, so the more they learn from the foundation, the greater the effect on the child).* T2 expressed his hesitation on this issue as follows: *I think it is necessary, but I think it will not be very effective in young age groups.* It was expressed that there should also be family and teacher training dimensions for the training to be given *(T5: I think that since the preschool period is small in terms of age group, the trainings to be given on this subject should be planned and carried out in a much more professional way. I think that the teachers who will communicate with children on this subject should be very conscious about how to convey this subject to children and how much of the subject they will convey, as well as having a full knowledge of the subject.; T9: I think that the trainings given to children are in the air. Such trainings should start with families. As long as the family is not involved, the information we provide is insufficient.; T16: Every educator, every parent should definitely receive this training and turn it into behaviour).*

Figure 9. Teachers' views on their sharing with parents about digital ethics

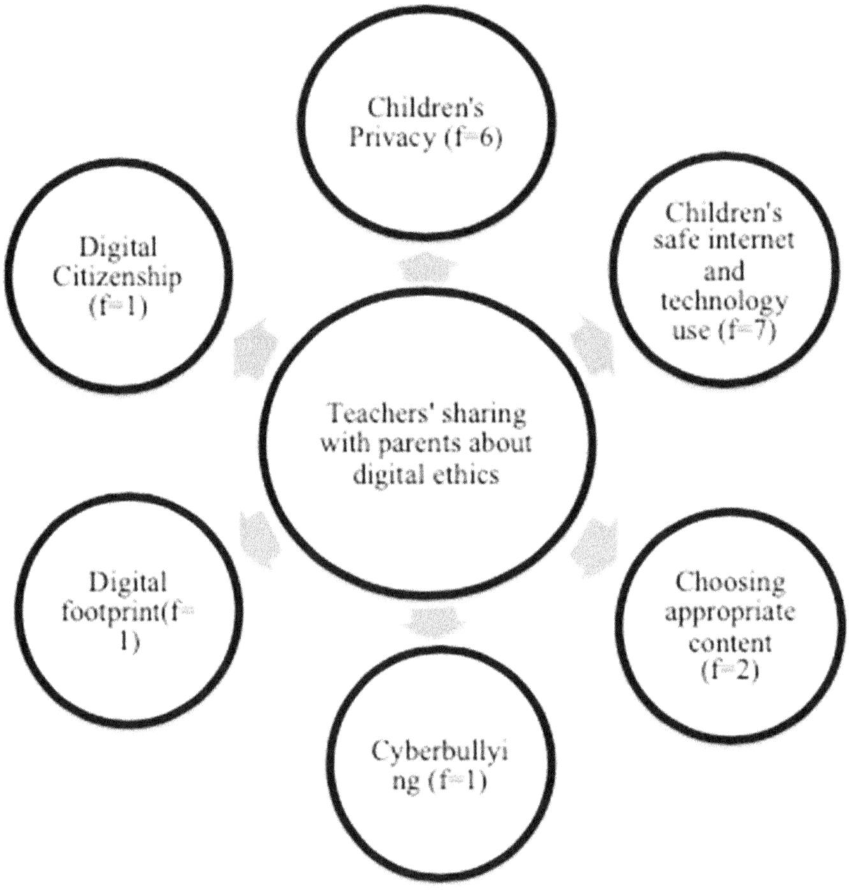

Participants informed parents about "privacy" *(T1: I inform about not giving personal information in social media or virtual programmes and carefully selecting the photos shared)*, "children's safe internet and technology use" *(T17: We informed them in E-safety parent training; T20: There are safe internet sites. We inform about the sites related to them. There are even hours, for example, at certain times, we tell them for other children according to their ages. But by the way, it is not only for our students, we tell them how many hours they can use it from pre-school.)*, "cyberbullying" *(T5: I clearly tell parents about the dangers in the digital world. Cyberbullying, pornographic content, communication with unknown people...)*, "choosing appropriate content" *(T11: If it harms the child mentally, socially and emotionally, if the content is far from education, I convey that I do not consider it appropriate)*, "digital footprint" *(T5: ...I inform them about issues such as digital footprint. For this reason, I remind them to be more careful when sharing)*, "digital citizenship" *(T5: ...I remind them to be respectful to each other in the digital world and to take responsibility for their behaviours and sharing...)*.They also emphasised that they shared resources, prepared videos and brochures, held meetings and cooperated with the guidance service (*T19: I send videos, I send information. We prepare posters together with the directors. We prepare brochures. We ask for help from counselling teachers. Then they are informed. We remind them of this from time to time during the process. They also sent some activities related to children in their own*

posts.; T20: Yes, we usually organise parent-teacher meetings for our parents. In these parent meetings, safe search engines suitable for children, safe search engines that I just mentioned, safe child org of BTK, European Commission, safe in our own country, let me put it this way. Briefly, I give examples about safe platforms. And as I just said, I tell them that they should first use the games that are suitable for children). It can be said that the sharing of teachers with families is related to the negativities that can be encountered during internet use, basic principles and behaviors, and ways that can be followed for protection. A guidance can also be added to the process for children to model behaviors appropriate to digital ethics.

Figure 10. Teachers' views on the shares they share with children and their colleagues about digital ethics

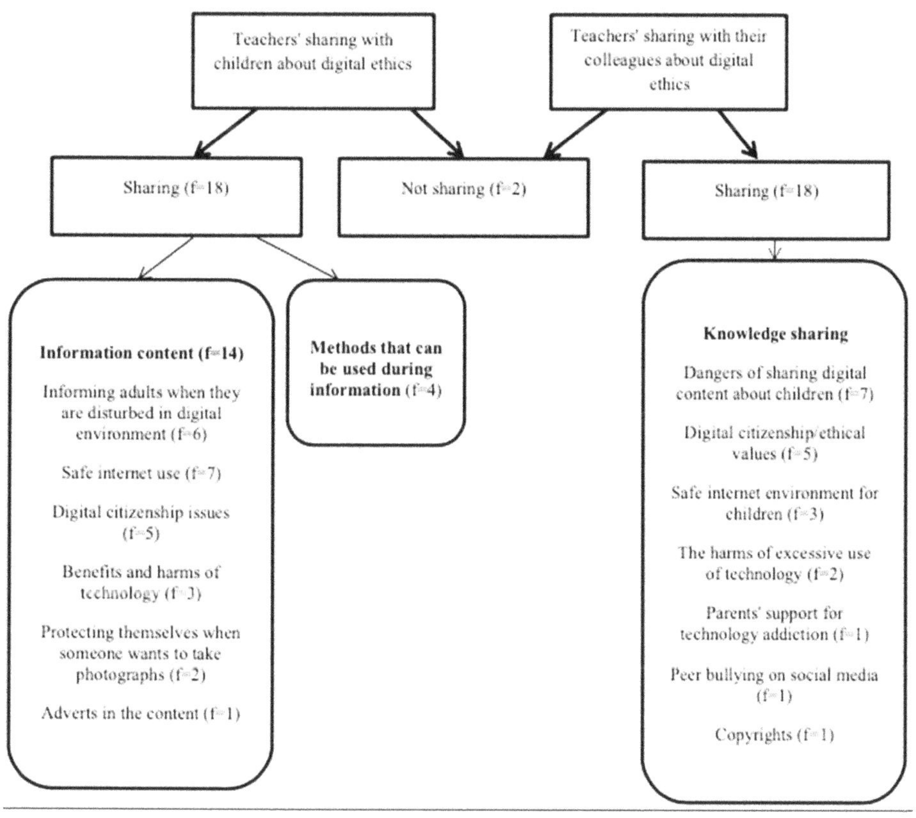

The teachers stated that they informed the children about various issues in their sharing with the children. On the other hand, two teachers stated that they had not shared any information on this subject before (*T3: I have not provided any information until now. I realised this with this interview*). Teachers who shared information about informing adults when they are disturbed in the digital environment (*T5: When they encounter any disrespect or something they don't like, I tell them to share this situation with their families immediately; T20: ...not to give any information about themselves and cyberbullying, of course, how do we say this? When something is said that you don't like, you should immediately inform your family, we can explain cyberbullying more*), safe internet use (*T19: First of all, safe internet use again. We explain what to pay attention to when using technology, which website you can access, how*

long you can use it, what you can do when you see a danger, where you can ask for help, where you should do these things with your family. We do art activities or read stories, then they have learnt exactly), digital citizenship *(T9: I tell the children that we should be as respectful and loving to everyone in our normal life as we are in the digital environment.)*, benefits and harms of technology *(T6: I explain that technology has benefits when used correctly, but it is harmful when used incorrectly.)*, protecting themselves when someone wants to take their photos in daily life *(T1: They cannot be on social media on their own due to their age group. However, I explain that they should intervene when someone wants to take their photos.)*, adverts in the content *(T5: Since we can sometimes encounter adverts that we do not want to see while using digital tools, I remind them that they can watch what they watch using applications such as YouTube Kids without interrupting the advertisements)*.

T2, T10 and T17 stated that they made use of colouring pages, stories, question and answer method and videos while teaching these subjects. *(T2: I show short videos about this subject; T10: We watch videos. We read stories, make colouring and prepare boards. We do question and answer and talk with the children. T17: We made colouring pages about e-safety for children. We watched EBA's educational animations)*. T20 gave the following remarks about the methods that can be used when sharing this subject with children:*...Apart from that, educational stories, educational, cartoons, educational documentaries. We can access these very easily. Apart from puppet plays, we can even produce riddles and rhymes ourselves. This is entirely based on the teacher's own creativity. Every year, there are e-safety days all over Europe that I participate in with my class. We participate in these days. When we participate in these days, we create a class board about e-safety, which we make in areas that children can understand about e-safety. At least all of these are both visual and auditory, appealing to our children from every channel.* It can be said that the content given by teachers to children in the context of digital ethics was limited and focused on children's self-protection and did not reflect the respect for others dimension of digital ethics. It can be stated that the methods used are teacher-oriented and that there is a need for a process enriched with methods such as play-drama that allow learning by experience.

On the other hands two of the participants stated that they did not share with their colleagues about digital ethics. The other participants stated that they shared their work and knowledge on some issues. The dangers of sharing digital content about children *(T18: When they share the pictures of the students one by one on social media, I warn them that this is not right, I inform them that the consent obtained from the parents is not for this. In the projects carried out and the posts that should be made on the school's social media accounts, I state that they should be in accordance with the regulations and guidelines.)*, digital citizenship and digital ethics *(T1: We talk about the need to follow the moral rules in daily life in the virtual environment.)*, safe internet environment for children *(T12: We talk about creating a safe environment for children.)*, excessive use of technology *(T16: We talk about screen time, symptoms such as autism increase in children with increased screen time, children become dull, they start to experience distraction, etc.)*, peer bullying in social media *(T9: We talk about the fact that people are hurtful in social media environments, peer bullying, etc. continues there.)* and copyrights *(T19: ...we talk about both safe internet, technology use, ethical values, copyrights, all of these)*. T20 stated that in the safe internet seminars he gave, he received the most questions from teachers about sharing photographs *(T20: The most frequently asked question at the end of the trainings I usually give to our colleagues is about how much of the children's photographs can be shared and how much cannot be shared)*. Participants' sharing with their colleagues is shaped by predominantly negative content such as dangers, overuse, and peer bullying. It is thought that teachers need training on this subject in order to shape these shares in the direction of developing principles and behaviors appropriate to digital ethics.

Figure 11. Teachers' views on the resources they follow about digital ethics and Studies on digital ethics in the institutions where teachers work

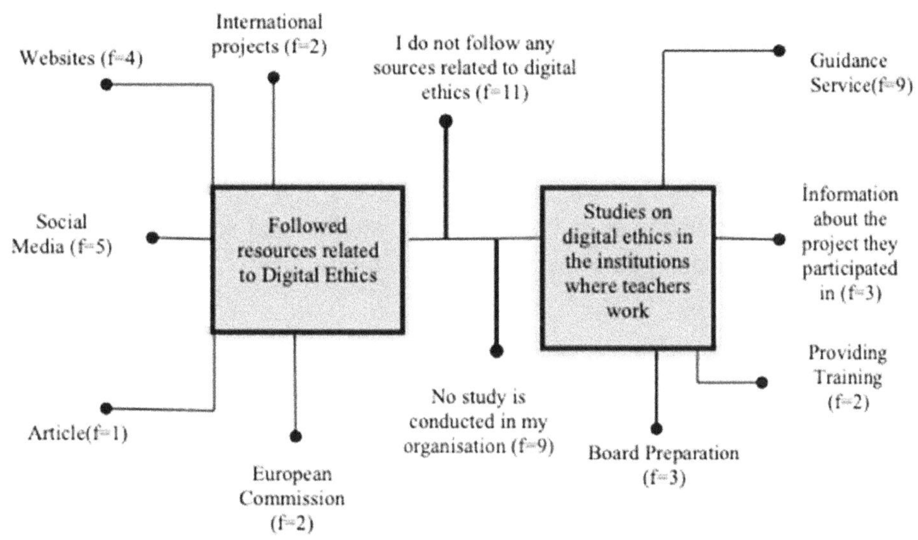

Eleven of the teachers stated that they do not follow any resources related to digital ethics. The teachers who follow the resources related to digital ethics, besides the articles written in this field, are the international projects carried out, resources such as the European Commission, social media accounts *(T3: I follow some social media accounts that do this work. There is clear information given.)* and websites *(T19: There are safe internet sites. There is such a system as safe internet, there are 3 sites, you can get it from there, but I also look at youtube myself).* T20's explanation on this subject is in a way to cover all sources: *I will give a very general answer to your question. Namely, I generally look at Turkish publications on digital ethics, from my Google searches in Turkey. But after that, I translate the English meanings of this and that, and I assign articles to keep myself updated. There are foreign platforms and websites related to preschool. They share information on every topic, but they also share information on digital ethics. And also the European Commission's own site. Have there been Erasmus projects? For example, what has been covered on this topic. As I look at them from time to time, I can also catch up with the current ones. And I definitely include the digital ethics part in the programme under the name of celebrating e-safety day in every project, and since I share it with my colleagues all over Europe, I can see how it enriches their classrooms.* It is a positive aspect that teachers who follow the resources on digital ethics refer to various resources. However, the fact that the majority of the teachers who do not follow the sources points to the need for knowledge and awareness.

9 of the participants stated that there is no study on digital ethics in the institution where they work *(T3: No. It is not a subject that we are sufficiently aware of and informed about yet.)* The teachers who stated that there are studies in this direction in the institution emphasised that these studies are generally carried out by the guidance service (*T11: Our guidance teachers carry out studies for other levels. There are informational activities for students in the classroom and presentations for parents*). T17 stated that the institution where he works provides training in this regard (*T17: Yes, informative training is given twice a year).* In addition, schools are aware of the practices through the projects they participate in

(*T16: In the projects, we are very sensitive when sharing about children, we cover children's faces with emoticons, when a photo is to be shared on social media, their faces are taken on the screen without their faces being clearly visible, in my classroom, I especially open children's songs or instrumental songs that are suitable for them, when watching videos, I prefer very short videos that do not set a negative example*), information is provided by preparing a board *(T19: I prepare and share on the board. Then it attracts the children's attention. As I am in a primary school, I share it in a place where they can all see it)*. The fact that the studies on digital ethics in institutions are limited to guidance service studies can be considered as an indicator that teachers need knowledge and experience in this regard. It can be said that other studies are insufficient and school administrators and senior administrators should be included in the process to encourage these studies and trainings.

DISCUSSION

Conclusion and Recommendations

At the end of the research it was observed that teachers were supportive of the use of digital tools in the classroom. While they emphasised that they mostly choose tools that are appropriate for the age of the child in terms of the issues they pay attention to when using digital tools in the classroom, it can be seen that very few teachers expressed opinions about the way of using the tools and the duration of use. In order to increase teachers' competence in the use of digital tools, training can be provided on the duration of use in the classroom and the selection of tools that are appropriate for children's development. Informative content about age level and duration of use can be added to the digital tools to be used in the classroom.

Teachers defined digital ethics under titles such as protection from the harm of technology, correct use of technology, ethical content, integrating ethical rules into the digital environment, acting in accordance with children's rights, not sharing children's photos and protecting personal information. Teachers are found to have a superficial knowledge of digital ethics. For this reason, elective courses on digital ethics content can be included in undergraduate programmes for pre-service teachers, who are the teachers of the future.

Teachers indicated that they adhere to principles such as not using children's photos or making sure that their faces are not shown, getting permission from parents to share, not sharing images that they might not like in the future, getting permission from the child when sharing their work and making sure that their identity is hidden, not sharing collectively in parent groups, sharing collective photos rather than individual photos, sharing only on project pages, or controlling followers in the sharing medium. However, it can be seen that teachers need to improve in taking measures to ensure data security to prevent third parties from accessing the data. Data security can be ensured by informing teachers and parents about existing data security programmes.

The majority of teachers indicated that they wanted to be trained in digital ethics and that their motivation for doing so was due to factors such as protecting and guiding children, as well as factors such as self-improvement, raising awareness and curiosity. Based on the lack of previous training in this content, sharing that is not in line with the principles, teachers' lack of knowledge, the problems they experienced and their lack of interest in the subject, it is interpreted that teachers' training in digital ethics is insufficient.

Teachers indicated that they include content on safe internet use, telling an adult if they are disturbed, digital citizenship, benefits and harms of technology and advertising in their digital ethics sessions with children, while they include content on safe use of technology, privacy, appropriate content selection, digital footprint, cyberbullying and digital citizenship in their sessions with families. In this context, digital ethics programmes for children, including family involvement, can be prepared and delivered to children to ensure that they have a basic knowledge of digital ethics.

In addition, their sharing with colleagues takes place in the context of the dangers of sharing children's photos as content, digital citizenship, safe internet environment, harms of excessive use of technology, peer bullying on social media and copyright. Social media platforms can be set up for teachers to share their knowledge and experience on this issue with their colleagues and to raise awareness among other teachers.

The participants stated that in the institutions where they work, studies on digital ethics are mostly carried out by the guidance service, they receive information through projects and they raise awareness by preparing panels. Teachers can be encouraged to carry out projects to raise awareness of digital ethics among children and institutional staff.

It was also found that international projects, websites, social media, articles and European Commission studies were used as resources in this area. It can be seen that digital ethics resources in Turkey are prepared in such a way that they address many fields of study and are mostly for adults. It is predicted that the adaptation of these resources according to the level of pre-school education will facilitate the teacher's access to information.

REFERENCES

Ahmad, A., & Maynard, S. (2014). Teaching information security management: Reflections and experiences. *Information Management & Computer Security*, *22*(5), 513–536. doi:10.1108/IMCS-08-2013-0058

Amo Filvá, D., Prinsloo, P., Alier Forment, M., Fonseca Escudero, D., Torres Kompen, R., Canaleta Llampallas, X., & Herrero Martín, J. (2021). Local technology to enhance data privacy and security in educational technology. *International Journal of İnteractive Multimedia and Artificial İntelligence*, *7*(2), 262–273. doi:10.9781/ijimai.2021.11.006

Arnott, L., & Yelland, N. J. (2020). Multimodal lifeworlds: Pedagogies for play inquiries and explorations. *Journal of Early Childhood Education Research*, *9*(1), 124–146.

Bergen, D., Davis, D. R., & Abbitt, J. T. (2015). *Technology play and brain development: Infancy to adolescence and future Implications*. Routledge. doi:10.4324/9781315681436

Bernard, R. (2012). *Moor's Law.* Retrieved from https://blog.soton.ac.uk/comp6044/tag/moors-law/

Bilgin Ülken, F., & Kılınç, F. (2018). Çocukların çevrimiçi ortamda karşılaştıkları risk türleri ile aracılık uygulamaları arasındaki ilişki. *İnsan ve Toplum Bilimleri Araştırmaları Dergisi, 7*(1), 124-145.

Buchanan, R. (2019). Digital ethical dilemmas in teaching. In *Encyclopedia of Teacher Education*. Springer. doi:10.1007/978-981-13-1179-6_150-1

Capurro, R. (2009). Digital ethics. *2009 Global Forum on Civilization and Peace.*

Carpenter, J. P., & Morrison, S. A. (2018). Enhancing teacher education… with Twitter? *Phi Delta Kappan*, *100*(1), 25–28. doi:10.1177/0031721718797118

Chaudron, S., Di Gioia, R., & Gemo, M. (2018). Y*oung children (0-8) and digital technology - A qualitative study across Europe*. JRC Science for Policy Report. doi:10.2760/294383

Çubukcu, A., & Bayzan, Ş. (2013). Türkiye'de dijital vatandaşlık algısı ve bu algıyı internetin bilinçli, güvenli ve etkin kullanımı ile artırma yöntemleri. *Middle Eastern & African Journal of Educational Research*, *5*(1), 148–174.

Doane, A. N., Kelley, M. L., Chiang, E. S., & Padilla, M. A. (2013). Development of the cyberbullying experiences survey. *Emerging Adulthood*, *1*(3), 207–218. doi:10.1177/2167696813479584

Eckhoff, A. (2015). Ethical considerations of children's digital image-making and image-audiancing in early childhood environments. *Early Child Development and Care*, *185*(10), 1617–1628. doi:10.1080/03004430.2015.1013539

Floridi, L. (2001). Ethics in the infosphere. *The Philosophers'. Magazine*, *6*, 18–19.

Freeman, L. A., & Peace, A. G. (2005). *Information Ethics: Privacy and Intellectual Property*. Idea Group. doi:10.4018/978-1-59140-491-0

Gao, Q., Yan, Z., Zhao, C., Pan, Y., & Mo, L. (2014). To ban or not to ban: Differences in mobile phone policies at elementary, middle, and high schools. *Computers in Human Behavior*, *38*, 25–32. doi:10.1016/j.chb.2014.05.011

Gasser, U., & Cortesi, S. (2017). Children's rights and digital technologies: Introduction to the discourse and some meta-observations. In *Handbook of Children's Rights: Global and Multidisciplinary Perspectives*. Taylor and Francis.

Greenhow, C., Galvin, S. M., Brandon, D. L., & Askari, E. (2020). A decade of research on K–12 teaching and teacher learning with social media: Insights on the state of the field. *Teachers College Record*, *122*(6), 1–72. doi:10.1177/016146812012200602

Hansson, T. (2008). Communication and relation building in social systems. In Handbook of Research on Digital Information Technologies: Innovations, Methods and Ethical Issues, Information Science Reference (pp. 367-385). doi:10.4018/978-1-59904-970-0.ch024

International Society for Technology in Education. (2019). *ISTE standards for educators*. Retrieved from https://www.iste.org/standards/for-educators

Ivanova, M., Marín, V. I., Tur, G., & Buchem, I. (2017, June). Towards privacy issues in personal learning environments. A conceptual model of ple privacy. In *EDEN Conference Proceedings* (No. 1, pp. 205-214). Academic Press.

Karakaya, N. (2021). Teknoloji, ailenin rolü ve öğretmenin rolü-çocuk hakları. In Erken Çocuklukta Dijital Eğitim (pp. 197-212). Ankara: Anı.

Katıtaş, S., Seylim, E., & Demirkan, S. (2022). Medya etiği ve medya okuryazarlığı eğitimi üzerine bir inceleme. *Alanyazın*, *3*(1), 153–175. doi:10.22596/ cresjournal.0301.153.175

Kavurucu, Ö., & Küçük, L. (2021). Çağın gelişen sorunu siber zorbalık: Çocuk ve ergenlerde siber zorbalığı önleme. *Medical Research Reports*, *4*(3), 50–56.

Kenton, W. (2020). *Copyright infringement.* İnvestopedia. Retrieved from https://www.investopedia.com/terms/c/copyright.asp

Keser, H., & Yayla, H. G. (2021). Fatih Projesi uygulanan okullardaki öğretmenlerin bilgi güvenliği farkındalık düzeylerinin incelenmesi. *Milli Eğitim Dergisi*, *50*(229), 9–40.

Liu, C., & Hwang, G. J. (2021). Roles and research trends of touchscreen mobile devices in early childhood education: Review of journal publications from 2010 to 2019 based on the technology-enhanced learning model. *Interactive Learning Environments*, 1–20. doi:10.1080/10494820.2020.1855210

Livingstone, S., Haddon, L., Görzig, A., & Ólafsson, K. (2011). *Risks and safety on the internet: The perspective of European children. Full findings.* London: EU Kids Online, LSE.

Livingstone, S., & Helsper, E. (2008). Parental mediation and children's internet use. *Journal of Broadcasting & Electronic Media*, *52*(4), 581–599. doi:10.1080/08838150802437396

Livingstone, S., Mascheroni, G., & Staksrud, E. (2015). *Developing a framework for researching children's online risks and opportunities in Europe*. London: EU Kids Online, LSE.

Louv, R. (2010). *Last child in the woods: Saving our children from nature-deficit disorder*. Algonquin Books.

MEB. (2023). *Milli Eğitim Bakanlığı FATİH Projesi*. Retrieves from http://fatihprojesi.meb.gov.tr

Merriam, S. B. (2013). Nitel araştırma desen ve uygulama için bir rehber. Ankara: Nobel Akademik Yayıncılık.

Mertala, P. (2019). Teachers' beliefs about technology integration in early childhood education: A meta-ethnographical synthesis of qualitative research. *Computers in Human Behavior*, *101*, 334–349. doi:10.1016/j.chb.2019.08.003

Miles, M. B., & Huberman, A. M. (1994). Qualitative data analysis: An expanded sourcebook. *Sage (Atlanta, Ga.).*

Miranda, H. P., & Russell, M. (2012). Understanding factors associated with teacher-directed student use of technology in elementary classrooms: A structural equation modeling approach. *British Journal of Educational Technology*, *43*(4), 652–666. doi:10.1111/j.1467-8535.2011.01228.x

Mossberger, K., Tolbert, C., & McNeal, S. R. (2007). Digital citizenship: The Internet, society and participation. London: The MIT Press.

Obar, J. A., & Oeldorf-Hirsch, A. (2018). The clickwrap: A political economic mechanism for manufacturing consent on social media. *Social Media + Society*, *4*(3), 1–14. doi:10.1177/2056305118784770

OECD. (2021). *Protecting and empowering children in the digital world: OECD recommendations and good practices.* Retrieved from https://www.oecd.org/going-digital/protecting-and-empowering-children-in-the-digital-world.html

OECD. (n.d.) *Children in the digital environment.* Retrieved from https://www.oecd.org/digital/children-digital-environment/

Palaiologou, I. (2016). Children under five and digital technologies: Implications for early years pedagogy. *European Early Childhood Education Research Journal, 24*(1), 5–24. doi:10.1080/1350293X.2014.929876

Patton, M. Q. (2005). *Qualitative Research.* John Wiley & Sons, Ltd.

Prensky, M. (2001). Digital natives, digital immigrants. Do They Really Think Differently? *On the Horizon, 9*(5), 1–6. doi:10.1108/10748120110424816

Redecker, C. (2017). European framework for the digital competence of educators: DigCompEdu. In Y. Punie (Ed.), *EUR 28775 EN.* Publications Office of the European Union.

Reingold, R., & Baratz, L. (2011). An institutional code of ethics—A response to attitude of Israeli teachers' education college students towards academic plagiarism. *Online Submission, 8*(5), 589–598.

Ribble, M., & Bailey, G. (2007). Digital citizenship in schools. Washington, DC: ISTE.

Rodríguez-Triana, M. J., Martínez-Monés, A., & Villagrá-Sobrino, S. (2016). Learning analytics in small-scale teacher-led innovations: Ethical and data privacy issues. *Journal of Learning Analytics, 3*(1), 43–65. doi:10.18608/jla.2016.31.4

Rogerson, S. (2021). *Rebooting ethics education in the digital age.* Retrieved from https://dora.dmu.ac.uk/bitstream/handle/2086/20613/Rebooting%20ethics%20education%20in%20the%20digital%20age%20prepublication%20version.pdf?sequence=2

Sak, R., Erden, F. T., Sak, İ. T. Ş., & Esmeray, H. (2016). Early childhood teachers and computers: Beliefs and self-reported practices. *Journal of Education and Future,* (10), 19–33.

Yalçın-Irmak, A., & Ardıç, A. (2018). Dijital oyunların çocuk ve ergenler üzerindeki etkileri: Literatür inceleme [The effects of digital games on children and adolescents: Literature review]. *Gençlik Araştırmaları Dergisi, 6*(16), 71–84.

Chapter 13
The Politics of Digital Citizenship Education

Vicki Hosek
Illinois State University, USA

ABSTRACT

Digital citizenship has been defined in a multitude of ways in the literature. The definitions promote a range of teaching practices spanning from protectionism to social activism. This chapter examines what the literature and research say about digital citizenship including how it is defined, the theoretical foundations at work in those definitions, and whose responsibility it is for teaching about and supporting digital citizenship. The National Educational Technology Plan (NETP) developed by the U.S. Department of Educational Technology and the International Society for Technology Education (ISTE) has created guidelines for digital citizenship education that incorporate a critical component. This chapter considers the political implications of those definitions and guidelines. Research is presented that examines teacher candidate preparation to provide insight about whether teacher candidates feel prepared to support the digital citizenship goals proposed by both the NETP and the ISTE.

As digital platform use in schools continues to gain prevalence, it is important to consider how educators and students using those platforms are being asked to develop and recruit digital citizenship skills. This chapter gives specific attention to digital citizenship guidelines for educators provided in the National Educational Technology Plan (NETP) developed by the U.S. Department of Educational Technology, and in the International Society for Technology Education (ISTE). Within both sets of guidelines lies a critical digital component that calls for an educator's recruitment of a critical digital lens to help students develop their digital citizenship. Because of this, it is important to look at teacher education programs to determine whether teacher candidates feel prepared to support the digital citizenship goals proposed by both the NETP and the ISTE. To that end, data is presented from a questionnaire completed by 63 teacher candidates (TCs) that asked about their education program experiences with and exposure to critical theory (CL) and critical digital literacy (CDL) in an effort to determine if and how they are prepared to recruit a critical digital lens to their own practices and those of their future students.

DOI: 10.4018/978-1-6684-8934-5.ch013

DEFINING CITIZENSHIP

Before digital classroom environments, citizenship education largely took place in the physical classroom. It is helpful to look at how researchers generally define citizenship as it lays the groundwork for defining it in digital settings. Kerr (2000) and Criddle et al. (2004) describe the ways citizenship is incorporated into curriculum as being on a continuum which ranges from being "passive and historical" to "critical and active" (Kerr, 2000, p. 208; Criddel et al., 2004, p. 31). Westheimer and Kahn (2004) take a critical approach to defining citizenship. They explain that while teaching students about how to follow the rules of society is an important component of citizenship, educators should consider "the proper balance between rule following and thinking about the origins and purpose of those rules" (p. 3). Their findings show that educators believe good citizenship means students listening and following authority, being kind to neighbors, and volunteering in the community (Westheimer & Khan, 2004). This led to their critical analysis of how schools define and teach citizenship. They believe a critical approach to defining citizenship is essential because "politics and the interests of varied groups are often deeply embedded in the ways we conceptualize, implement, and study efforts to educate for democracy" (p. 237). To them, citizenship calls for individuals to "solve problems and improve society by questioning and changing established systems and structures" so as to stop "patterns of injustice" (p. 243).

Johnson and Morris (2010) consider this to be a critical pedagogical view of citizenship. In their research, they analyze distinctions between critical pedagogy and critical thinking when defining citizenship. In their opinion, including a critical component when defining citizenship creates a dilemma within curriculum as educators are asked to "on the one hand to ensure an obedient populace and on the other hand to ensure that citizens are creative and critical" (p. 78). They state that this push for criticality is a push for critical pedagogy (Johnson & Morris, 2010). They explain that if critical pedagogy is central to citizenship, then praxis (reflection combined with action) must occur where students move beyond recognizing injustices in society to working to eliminate them (Fischman and McLaren 2005; Freire, 2021; Kincheloe, 2008; Morrow and Torres, 2002). According to Johnson and Morris (2010), this differs from positioning critical thinking as central to citizenship. Critical thinking includes a "concern for social justice and a desire to improve society," (p. 90) but does not necessarily include taking action to combat social injustices. They identified four distinct elements that are present in a critical pedagogical definition of citizenship: ideology, subjectivity, collectivity, and praxis. To many scholars, citizenship is an active practice of civic engagement that includes critical analysis, dialogue, and debate (Au, 2011; Knoester & Parkinson, 2015; Westheimer & Khan, 2004).

To some researchers, this subjective component of teaching citizenship is essential and the attempt to objectively test citizenship components through the use of national educational standards is problematic. Knoester and Parkison (2015) believe that the adoption by forty-one states of the Common Core State Standards (CCSS) shows a push in education for developing workforce skills rather than critically thinking citizens. They claim that schools are "emphasizing test scores as the primary outcome metric" which has "led to a de-emphasis of social studies within the curriculum" (p. 7; Parkison, 2009). Importantly, social studies is the content area where citizenship typically resides. Knoester and Parkinson (2015) state that schools and education "are valued almost always for the purpose of building a stronger economy, not for the purpose of building a more informed and effective voting public" (p. 10).

From this overview of the literature surrounding defining citizenship, it appears that researchers lean towards the necessity of a critical component ranging from developing critical thinking to a more radical critical pedagogical approach. The research shows that defining citizenship is complex, multifaceted, and often reflects the ideologies of those whose definition it is, making it highly subjective. This complexity of defining and teaching citizenship is also evident when defining digital citizenship.

Defining Digital Citizenship

Digital platforms are ubiquitous in the field of education. In fact, many educators and scholars now consider literacy to include how students learn, create, make meaning, and develop their voices and identities in digital environments (Cope & Kalantzis, 2000; Leu et al., 2004; Street, 1997; Lankshear & Knobel, 2006). As classrooms have grown to include the digital, educators are tasked with guiding students' online behavior in the hope of developing digital citizens. However, defining and measuring digital citizenship has proven difficult (Choi et al., 2018). Researchers have approached the concept of citizenship online in a multitude of ways.

Civic Engagement

It is important to acknowledge that citizenship education in primary and secondary schools in the United States traditionally pertains specifically to developing individuals for participation in our democracy. In that sense, citizenship includes an individual's duties and responsibilities to the country in an effort to support and maintain certain rights (Heater, 2013). However, the nature and reach of digital platforms extends beyond physical borders making defining digital citizenship even more complex.

Bennett et al. (2011) explain that "younger generations are embracing more expressive styles of actualizing citizenship defined around peer content sharing and social media, in contrast to earlier models of dutiful citizenship based on one-way communication managed by authorities" (p. 835). They suggest that digital citizenship is a different "civic style" that will continue to evolve depending on the civic environment. Jones and Mitchell (2016) state that digital citizenship has two components: respectful behavior online and online civic engagement. They believe educators develop digital citizens by "helping youth build and practice specific online social skills versus admonitions against problem behaviors like cyberbullying and sexting" (Jones & Mitchell, 2016, p. 2065). They recommend utilizing the massive community found in online platforms to develop student participation and engagement in community service and civic projects. Jones and Mitchell (2016) suggest practicing respectful communication through guided debates and "perspective engaging activities" (p. 2078). Choi et al. (2018) developed and utilized a five-factor digital citizenship scale to assess an individual's digital citizenship. Their scale includes internet political activism, technical skills, local/global awareness, critical perspective, and networking agency. Similar to how researchers define general citizenship, these different definitions and approaches reflect a widespread spectrum when conceptualizing the critical component of digital citizenship. Another prevalent perspective in the literature centers on a protectionist approach to digital citizenship education.

Protectionism

Most of the studies in this overview of the literature include a focus on security and safety when defining digital citizenship. Mirra (2018) states that "much media education continues to be driven by fear of the world it opens up to students and/or skepticism that it can be used for anything more meaningful than recreation or distraction" (p. 13). Ribble and Bailey (2011) subscribe to this protectionist approach and focus their definition on being responsible, safe, and secure with personal information when using online platforms. In his highly cited work, Ribble (2012) proposes a definition of digital citizenship that focuses on teaching how to be respectful and how to protect yourself when online. Hollandsworth et al. (2011) explain that "Although schools and parents try to ensure the safety of our students, it is imperative that students accept responsibility for using current and future technology in an ethical and legal way" (p. 37). Sturgeon (2008) states that "If we filter out everything in a school, we lose that teaching moment. Filters are not foolproof; we need to educate students on the safe use of the Internet" (p. 5). Buckingham (1998) warns against a protectionist approach stating that policing the ways students talk and/or behave online may lead to superficial conformance but it does not lead to fundamental change.

Digital Citizenship of Teacher Candidates and Practicing Teachers

The U.S. Department of Education developed the National Educational Technology Plan (NETP, 2017) which addresses digital citizenship education and provides direction to both teachers and students about what actions qualify as engaging in digital citizenship. The NEPT (2017) defines digital citizenship according to the standards developed by the International Society for Technology Education (ISTE). ISTE (2018) states that "Digital citizenship shouldn't be a long list of don'ts. It should be about the do's that help create thoughtful, empathetic digital citizens who can wrestle with the important ethical questions at the intersection of technology and humanity." According to the ISTE (2018), digital citizenship education includes teaching students how to: use technology for the betterment of community; engage respectfully online with people with different opinions and perspectives; use technology to reach public leaders in order to shape public policy; and critically analyze online resources to ensure validity. The organization provides the following guidelines for students: be inclusive in online platforms by respectfully engaging with others; evaluate the accuracy, perspective, and validity of online information; use technology for civic engagement and be a "force for good;" balance time between online and offline activities; and know how to be safe online and create safe online spaces for others (ISTE, 2018). The verbiage "shape public policy" and "be a 'force for good'" denotes some level of activism.

While there are many articles about how to define digital citizenship and its value, there are few studies about how teachers learn to incorporate the concept into their methods. Choi et al.'s (2018) study of 348 practicing teachers showed that they had low levels of "internet political activism and critical perspective" (p. 143) both of which the researchers identified as essential to digital citizenship. This means the teachers were "less likely to participate in online political events and tended not to analytically think about bias, prejudice, and injustice" (Choi et al., 2018, p. 159). Pusey and Sadera (2011) surveyed 318 preservice teachers about their knowledge and understanding of cyberethics, cybersafety, and cybersecurity as well as their confidence in teaching those concepts. Their results showed that they felt ill-prepared to teach and model digital citizenship. Martin et al. (2020) used tests, surveys, and semi-structured interviews to study 45 practicing teachers enrolled in a professional development course on digital citizenship. They employed a pre- and post- test to quantitatively examine teachers' initial digital citizenship knowledge and knowledge growth. The topics covered in the tests and course were cyberbullying, digital netiquette, digital footprint, digital privacy, and digital identity. They identified three areas the teachers felt challenged which included: students are not open to learning about digital citizenship; there is no time to teach digital citizenship and it isn't integrated into curriculum; and teachers are not trained in digital citizenship practices.

Responsibility for Digital Citizenship

Because digital platforms are a part of the classroom environment, management of students' use of those online spaces falls to school administrators and teachers. Specifically, citizenship education is most often allocated to teachers in the social studies content area in elementary and middle schools (National Governors' Association, 2010). Regarding high schools, only eight states require a yearlong government or civics class (Viadero et al., 2018). Instead, many states "embed civics, history, and other social studies topics in their required teaching standards" (Viadero et al., 2018). As all classrooms now include digital platforms and teachers across all content areas increasingly rely on technology in their methods, it is important to learn about how all TCs are being prepared to support students' digital citizenship.

The overview of literature for this study shows that the U.S. Department of Education and most scholars include a critical component when defining digital citizenship. While it appears from the research reviewed that little specific attention is given to digital citizenship during teacher education (Pusey & Sadera, 2012; Martin et al., 2020; Walters et al., 2019), one way researchers propose teachers and students can learn to critically navigate digital environments is through the development of critical digital literacy (CDL). Next, CDL is explained as it grounds the study of TCs that is the focus of this chapter.

CONCEPTUAL FRAMEWORK

Many scholars consider a critical lens as essential when teaching and learning in online environments (Avila & Pandya, 2013; Freire, 2021; New London Group, 1996; Pangrazio, 2016; Watulak & Kinzer, 2013). Importantly, there are distinctions in how scholars both describe that criticality and prescribe the purpose of a critical digital lens. Some scholars believe that a critical lens means application of critical theory that serves a specific political purpose. In that sense, educators and students critically analyze power structures located in digital environments and work to dismantle them (Avila & Pandya, 2013; Pangrazio, 2016). This approach is decidedly political and calls for political activism (Giroux, 1988; McLaren, 2017). Avila and Pandya (2013) describe critical digital literacy as an interrogation of digital texts with the purpose of disrupting and disempowering authority.

Other scholars conceptualize critical digital literacy as centered on critical thinking with a focus on seeking, recognizing, analyzing, and understanding diverse perspectives that exist in online environments to locate personal beliefs and perspectives within the broader digital environment. The purpose is to critically analyze the social, cultural, and historical contexts that exist in digital texts and environments to deepen understanding of self and others. While critical thinking may lead to challenge and debate, it does not necessarily serve a specific political agenda (Freire, 2021; Giroux, 1994).

This chapter and the study presented are conceptually grounded on a CDL framework that is rooted in critical thinking rather than critical theory. The findings and discussion of the concepts presented draw heavily on Watulak and Kinzer's (2013) CDL framework. Watulak and Kinzer (2013) define CDL as "encompassing four main elements:

- Understanding cultural, social, and historical contexts of technology use;
- Critical thinking and analysis;
- Reflective practice; and
- Functional skills and tools of digital technology production" (p. 128).

While the political contexts of teaching and learning in digital environments should be considered and critically analyzed by both educators and students, it is the contention of this author that the purpose of that analysis should not be to further any one specific political agenda. The political acts of educators in the classroom, whether implicit or explicit, can contribute to self-censorship of educators and students and the polarization of society (Bar-Tal, 2017; Honeycutt et al., 2023; Zhou & Barabaro, 2023). Rather, the critical digital lens supported here serves to encourage instructors and students to develop a deeper understanding of the concepts studied in courses and a broadening of knowledge of diverse perspectives about such concepts through critical analysis and consideration of digital texts and the online environments and contexts in which they are created.

As seen from the overview of literature presented earlier, scholars often conflate citizenship and politics. Additionally, scholars repeatedly incorporate a criticality when conceptualizing and defining digital citizenship. Because of this, it is important to consider how scholars and educators recruit that criticality in practice. The study presented next provides an example.

METHODS

As noted earlier, there is limited previous research about TCs' knowledge and understanding of digital citizenship. Incorporation of a critical component is one commonality in a large majority of the research reviewed surrounding defining digital citizenship. Because of this, this study's aim is to provide insight about TCs' exposure to and experience with CDL during their undergraduate coursework in the hope of informing discussion about how TCs are prepared to engage critically in learning about and eventually teaching digital citizenship.

The participants in this study were senior level secondary education TCs in the School of Teaching and Learning at a large, public university in the Midwest of the United States. The TCs were enrolled in a course aimed at developing multiliteracies across all content areas. During the course, the students completed a week-long unit focused on development and recruitment of CDL as defined by Watulak and Kinzer (2013). Watulak and Kinzer (2013) utilize the ISTE's National Educational Technology Standards (NETS) to ground their CDL framework. NETS (2008) "encourages teachers to understand local and global societal issues and responsibilities in an evolving digital culture and exhibit legal and ethical behavior in their professional practices" (paragraph 4). Watulak and Kinzer (2013) state that the goal is for teachers and students to "engage with the broader social and cultural contexts of their technology use" and to "move beyond the ability to manipulate a keyboard" (p. 127).

During the unit, the instructor and the TCs also discussed critical literacy (CL) and how it relates to and differs from CDL. At the end of the semester, students were offered the opportunity to take an anonymized survey that contained a multitude of questions including three specifically about personal CL and CDL. Sixty-three TCs consented to take the survey and completed those questions. The 3 questions that are the focus of this study are:

1. Before this course, have you ever learned about CL in your education or other courses in college? If so, when and how was CL made a part of your coursework?
2. Before this course, have you ever learned about CDL in your education or other courses in college? If so, when and how was CDL made a part of your coursework?
3. Do you believe CDL is valuable in your discipline for you and your students?

The quantitative findings from this study were generated from the "yes" and "no" answers to the three questions listed above. In addition, TCs' answers to part two of question one were coded and themed to provide context.

FINDINGS

Table 1 (see below) represents quantitative data generated from surveys taken by the 63 TCs to gain insight about their exposure to CL and CDL in their undergraduate coursework. Specifically, the table contains the quantitative results to part one of questions one and two as well as answers to question 3. The quantitative results in Table 1 provide a discussion point concerning TC preparation to engage in the critical component of digital citizenship that is highly prevalent in most of the research.

Table 1. CL and CDL exposure and value

Content Area	# of TCs	Prior Exposure to CL	Prior Exposure to CDL	Believe CDL Is Valuable
Art/Theatre	5	3	0	3
Consumer Science	8	5	0	8
English	17	17	4	17
History/Social Science	8	6	2	8
Mathematics	7	5	1	4
Music	1	1	0	1
Physical Education	12	7	1	7
Science	4	4	0	4
Technology Education	1	0	0	1
TOTAL	63	48	8	53

To provide additional context for this study, TCs' answers to question one, part two were coded and themed and are presented in Table 2 below.

Table 2. Types of TCs' exposure to CL

Content Area & # of TCs	Total # of TCs	CL Exposure
Art/Theater - 2 Consumer Science - 5 English - 7 History - 5 Math - 5 Music - 1 Physical Education - 3 Science- 3	31	Yes, but did not know it was CL
Art/Theater - 1 English - 10 History - 1 Physical Education - 4 Science -1	17	Yes, explicitly taught about CL
Art/Theater - 2 Consumer Science - 3 History - 2 Math - 2 Physical Education - 5 Technology Ed. - 1	15	Never

A strong diversity of content areas was represented in this study. The largest of those are English at 25%, Physical Education at 19%, and History/Social Science at 17%. Ninety-one percent of the TCs believe they had prior exposure to CL. Of that 91%, 64% of the TCs stated the prior CL exposure was implicit in their coursework. Twenty-seven percent of the students stated they were explicitly taught about CL. Only 12.5% believe they had prior CDL exposure. Overall, 84% of the TCs stated they believe CDL is valuable to their content area, yet most had not been introduced to it until this course which is typically taken during a TC's senior year.

Limitations

This study has limitations. It is not generalizable due to the limited number of participants. The instructor for the course is also the researcher. This potentially introduces bias despite anonymized data collection and analysis of findings occurring after grades were submitted for the course. Finally, because the focus of this study is largely on the quantitative findings, there is limited insight and context concerning the participants' answers.

DISCUSSION

Politics and the interests of varied groups are often deeply embedded in the ways we conceptualize and study efforts to educate for democracy. Politics and the interests associated with the varied conceptions therefore require close attention. (Westheimer & Khan, 2004, p. 266)

According to previous research, two components that appear to be inherent in the definitions of citizenship and digital citizenship are criticality and politics. It may not be possible to remove politics from citizenship education which could result in a one-sided perspective about what it means to be a "good" citizen. One potential way to ensure that all voices and perspectives are fairly and equally represented, analyzed, and debated is to consider what a "critical" lens means and looks like when it is incorporated into digital citizenship education. There is a clear difference between a focus on critical theory versus critical thinking where the former is centered on politics and the latter on critical analysis (Burnett & Merchant, 2011; Giroux, 1991).

As indicated earlier, the U.S. Department of Education subscribes to a definition of digital citizenship that contains a critical component. The research reviewed shows that researchers' and educators' response is to incorporate critical theory and its accompanying political ideology into digital citizenship education. In fact, Westheimer and Kahn (2004) believe the ideological beliefs of educators impact how students are taught to be citizens. They stress that there are political implications when engaging in citizenship education and "suggest that the narrow and often ideologically conservative conception of citizenship embedded in many current efforts at teaching for democracy reflects not arbitrary choices but rather political choices with political consequences" (p. 237). If this is true, then teaching about citizenship, which is required in our schools, is a political act and there is the potential to advance certain political ideologies and agendas during citizenship education both on and offline. According to Pereira (2013), despite decades of research surrounding the importance of incorporating youths' voices in education, they are still considered "as the adults they will become" (p. 175). This fosters a student's passive acceptance of the views of influential educators rather than encouraging a critical questioning and understanding of those views. Recently, there have been large uprisings at schools across the country questioning the embedding of critical theory and related pedagogy, such as critical race theory, in primary and secondary school curriculum (Eden, 2021; Pendharker, 2021). The development of CDL could be a means for finding a middle ground where debate, critical analysis, and discussion could lead to understanding and empathy rather than division and polarization resulting from educators' political agendas.

The results of this small-scale study show that overwhelmingly, CL was taught to TCs whereas CDL was largely absent. Critical pedagogy is grounded in critical theory. The "goal" of critical pedagogy, according to Burnett and Merchant (2011), is "resistance, emancipation or reclaiming power by creating new spaces for dialogue" (p. 43). Burnett and Merchant (2011) state that to apply critical theory/literacy to digital platforms requires acting on political and ideological beliefs. While there is a critical component of CDL, it is focused on promoting critical thinking rather than critical theory. A question that should be asked is why the TCs were largely exposed to CL and only sparingly to CDL? Perhaps by focusing on the development of CDL, the proliferation of CL and its accompanying political agendas (Burnett & Merchant, 2011; Giroux, 1994; McLaren, 2017) would be noticed and challenged, opening the door for more diverse perspectives rather than the promotion of one.

In their answers to the open-ended questions, many of the TCs in this study stated that they believed they were implicitly taught CL. This embedding of CL across all the content areas shows that politics were a factor in these TCs' education. This is concerning in that certain political ideologies could be perpetuated thus potentially limiting diverse perspectives in our schools and country, both of which are supposedly founded on freedom of speech and equality. If instructors continue to implicitly or explicitly push specific political agendas that shut out one side and promote another, we run the risk of jeopardizing the very democracy upon which digital citizenship education is supposedly founded.

As citizenship becomes conceptualized online, educators and students should consider this role that politics is playing in defining digital citizenship. Westheimer and Kahn (2004) suggest that how educators choose to define and teach digital citizenship is reflective of their personal political ideologies particularly when facing issues of social justice. In 2004, they noted a conservative leaning in defining digital citizenship and challenged it. However, when Westheimer and Kahn's (2004) research is considered alongside more recent studies about the political affiliations of university professors, there appears to be a shift. In his study of 5,197 tenure-track PhD tenured college professors in liberal arts colleges/universities, Langbert (2018) found that the ratio of self-identified Democrats to Republicans was 10.4:1, respectively. His results show that there is a "political homogeneity embedded in college culture" (Langbert, 2018, p. 197). If that is the case, and if educators' political ideologies impact how they define and teach about digital citizenship, is there now a liberal leaning?

Future Research

As the classroom environment and citizenship education extends to include digital platforms, questions and concerns grow with it. Consideration of the following questions in future research could provide valuable insight and guidance for educators. If politics and ideologies are embedded in the definitions of citizenship and digital citizenship, then whose perspectives and beliefs are being left out? Do teacher educators knowingly embed their political ideologies into digital citizenship education? Do TCs have space and opportunity to represent their own political beliefs in teacher education courses? Do practicing teachers who teach digital citizenship realize that certain political ideologies are embedded? Is our education system perpetuating a limited and politicized view of being a "good" citizen? As Westheimer and Kahn (2004) state, "the relationship between pedagogical choices and political positions is an important one" (p. 247). Maybe more important is having the ability to recognize when one is being influenced by the other.

It seems one step towards ensuring that digital citizenship is not politicized is to pointedly incorporate CDL into teacher education programs. While there is a strong emphasis on a critical dimension when defining digital citizenship, that emphasis clearly supports the promotion and acceptance of critical theory rather than the questioning of it. The use of a critical digital lens could help students identify which perspectives are being enforced and which are being left out. It is one possible way for TCs to recognize, understand, and perhaps challenge the agendas that are being advanced in digital citizenship definitions and education.

CONCLUSION

Many of us wish to serve politically left-oriented or liberatory goals in our teaching, yet do not want to commit the theoretically totalizing and pedagogically oppressive sins we have inveighed against in the systems we resist. (Giroux, 1988, p. 84)

The above quote problematizes the central issue researchers and educators face when trying to locate personal political and ideological beliefs without embedding biases into citizenship education and curriculum. If educators are hoping to develop digital citizens who work for the "greater good," shouldn't everyone's perspective of what that "greater good" is be represented, considered, and discussed? Presenting one perspective, either overtly or through embedding it in curriculum, does not support an unbiased approach to digital citizenship education. Like Giroux (1988, 1994), Khane and Bower (2017) recognize the problem with self-perpetuating political and ideological beliefs in citizenship education. They state that "in a polarized environment, judgments of truth claims are often shaped more by whether or not individuals' prior perspectives on the issue align with the claims than by how well informed the individuals are or their capacities to reason" (Kahne & Bower, 2017, p. 5). According to Mutz (2002), limited exposure to diverse perspectives could "lead to less tolerance of opposing viewpoints" (p. 61) which produces "a less informed electorate because democratic theory requires that citizens gather and critically evaluate a wide range of views" (p. 61).

The consensus in the research reviewed for this chapter is that politics plays a role in both defining and teaching digital citizenship. This small-scale study examined whether TCs are learning about how to develop and recruit a critical lens to their online teaching and learning. The results show that while TCs see the value of CDL, they have limited exposure and opportunities to develop it. In fact, there was a noted presence of a specific political ideology that is being embedded and incorporated across all content areas in higher education, namely, critical theory. This is a cause for concern as diverse perspectives and political beliefs hold an important place in developing citizens that can consider, empathize with, and debate each other without polarizing the populace. To this point, Druckman (2012) found that when individuals are asked to consider a variety of opinions, provide justification of their own perspectives, and reflect about how their perspective evolved, there appears to be less bias in favor only of their previous views. If this approach to digital citizenship education becomes the goal and not the exception, we stand a stronger chance of finding a middle ground where all citizens are represented and have a voice.

REFERENCES

Au, W. (2011). Teaching under the new Taylorism: High-stakes testing and the standardization of the 21st century curriculum. *Journal of Curriculum Studies*, *43*(1), 25–45. doi:10.1080/00220272.2010.521261

Bar-Tal, D. (2017). Self-Censorship as a socio-political-psychological phenomenon: Conception and research. *Political Psychology*, *38*, 37–65. doi:10.1111/pops.12391

Bennett, W. L., Wells, C., & Freelon, D. (2011). Communicating civic engagement: Contrasting models of citizenship in the youth web sphere. *Journal of Communication, 61*(5), 835-856.

Buckingham, D. (1998). Media education in the UK: Moving beyond protectionism. *Journal of Communication*, *48*(1), 33–43. doi:10.1111/j.1460-2466.1998.tb02735.x

Burnett, C., & Merchant, G. (2011). Is there a space for critical literacy in the context of social media? *English Teaching*, *10*(1), 41–57.

Choi, M., Cristol, D., & Gimbert, B. (2018). Teachers as digital citizens: The influence of individual backgrounds, internet use and psychological characteristics on teachers' levels of digital citizenship. *Computers & Education*, *121*, 143–161. doi:10.1016/j.compedu.2018.03.005

Cope, B., & Kalantzis, M. (Eds.). (2000). *Multiliteracies: Literacy learning and the design of social futures*. Psychology Press.

Criddle, E., Vidovich, L., & O'neill, M. (2004). Discovering democracy: An analysis of curriculum policy for citizenship education. *Westminster Studies in Education*, *27*(1), 27–41. doi:10.1080/0140672040270103

Druckman, J. N. (2012). The politics of motivation. *Critical Review*, *24*(2), 199–216. doi:10.1080/08913811.2012.711022

Eden, M. (2021, June 17). *Parents against critical race theory*. Washington Examiner. https://www.aei.org/op-eds/parents-against-critical-race-theory/

Fischman, G. E., & McLaren, P. (2005). Rethinking critical pedagogy and the Gramscian and Freirean legacies: From organic to committed intellectuals or critical pedagogy, commitment, and praxis. *Cultural Studies? Cultural Studies, Critical Methodologies*, *5*(4), 425–446. doi:10.1177/1532708605279701

Freire, P. (2021). *Pedagogy of hope: Reliving pedagogy of the oppressed*. Bloomsbury Publishing. doi:10.5040/9781350190238

Giroux, H. (1988). *Teachers as intellectuals: Toward a critical pedagogy of learning*. Greenwood Publishing Group.

Giroux, H. (1994). Doing cultural studies: Youth and the challenge of pedagogy. *Harvard Educational Review*, *64*(3), 278–309. doi:10.17763/haer.64.3.u27566k67qq70564

Heater, D. (2013). *What is citizenship?* John Wiley & Sons.

Hollandsworth, R., Dowdy, L., & Donovan, J. (2011). Digital citizenship in K-12: It takes a village. *TechTrends*, *55*(4), 37–47. doi:10.100711528-011-0510-z

Honeycutt, N., Stevens, S. T., & Kaufmann, E. (2023). *The academic mind in 2022: What faculty think about free expression and academic freedom on campus*. The Foundation for Individual Rights and Expression.

International Society for Technology in Education. (2018). *Citizenship in the digital age*. ISTE. https://elearninginfographics.com/citizenship-digital-age-infographic/

James, C., Weinstein, E., & Mendoza, K. (2019). *Teaching digital citizens in today's world: Research and insights behind the common sense K–12 digital citizenship curriculum*. Common Sense Media. https://www.commonsense.org/education/digital-citizenship/curriculum?campaignid

Johnson, L., & Morris, P. (2010). Towards a framework for critical citizenship education. *Curriculum Journal*, *21*(1), 77–96. doi:10.1080/09585170903560444

Jones, L. M., & Mitchell, K. J. (2016). Defining and measuring youth digital citizenship. *New Media & Society*, *18*(9), 2063–2079. doi:10.1177/1461444815577797

Kahne, J., & Bowyer, B. (2017). Educating for democracy in a partisan age: Confronting the challenges of motivated reasoning and misinformation. *American Educational Research Journal*, *54*(1), 3–34. doi:10.3102/0002831216679817

Kerr, D. (2000). Citizenship education: An international comparison. In D. Lawton, J. Cairns, & R. Gardner (Eds.), Education for citizenship (pp. 200 – 227). Continuum.

Kincheloe, J. L. (Ed.). (2008). *Knowledge and critical pedagogy: An introduction*. Springer. doi:10.1007/978-1-4020-8224-5

Knoester, M., & Parkison, P. (2015). Where is citizenship education in the age of common core state standards? *Critical Education, 6*(22).

Langbert, M. (2018). Homogenous: The political affiliations of elite liberal arts college faculty. *Academic Questions, 31*(2), 186–197. doi:10.100712129-018-9700-x

Lankshear, C., & Knobel, M. (2006). *New literacies: Everyday practices and classroom learning*. Open University Press.

Leu, D.J., Kinzer, C. K., Coiro, J. L., & Cammack, D. W. (2004). Toward a theory of new literacies emerging from the Internet and other information and communication technologies. *Theoretical Models and Processes of Reading, 5*(1), 1570-1613.

Martin, F., Gezer, T., Wang, W. C., Petty, T., & Wang, C. (2020). Examining K-12 educator experiences from digital citizenship professional development. *Journal of Research on Technology in Education, 54*(1), 143–160. doi:10.1080/15391523.2020.1815611

McLaren, P. (2017). Critical pedagogy: A look at the major concepts. In A. Darder, R. D. Torres, & M. P. Baltodano (Eds.), *The critical pedagogy reader* (3rd ed., pp. 61–83). Routledge.

Mirra, N. (2018). *Educating for empathy: Literacy learning and civic engagement*. Teachers College Press.

Morrow, R. A., & Torres, C. A. (2002). *Reading Freire and Habermas: Critical pedagogy and transformative social change*. Teachers College Press.

Mutz, D. C. (2002). Cross-cutting social networks: Testing democratic theory in practice. *The American Political Science Review, 96*(1), 111–126. doi:10.1017/S0003055402004264

National Governors Association Center for Best Practices & Council of Chief State School Officers. (2010). *Common Core State Standards*. Authors.

New London Group. (1996). A pedagogy of multiliteracies: Designing social futures. *Harvard Educational Review, 66*(1), 60–92. doi:10.17763/haer.66.1.17370n67v22j160u

Pangrazio, L. (2016). Reconceptualizing critical digital literacy. *Discourse (Abingdon), 37*(2), 163–174. doi:10.1080/01596306.2014.942836

Parkison, P. (2009). Political economy of NCLB: Standards, testing and test scores. *The Educational Forum, 73*(1), 44–57. doi:10.1080/00131720802539606

Pendharker, E. (2023). *Poll: Parents don't want schools to focus on culture wars.* Ed Week. https://www.edweek.org/leadership/poll-parents-dont-want-schools-to-focus-on-culture-wars/2023/01

Pereira, S. (2013). More technology, better childhoods? The case of the Portuguese 'one laptop per child' programme. *Communication Management Quarterly, 29*(29), 171–198. doi:10.5937/comman1329171P

Pusey, P., & Sadera, W. A. (2011). Cyberethics, cybersafety, and cybersecurity: Preservice teacher knowledge, preparedness, and the need for teacher education to make a difference. *Journal of Digital Learning in Teacher Education, 28*(2), 82–85. doi:10.1080/21532974.2011.10784684

Ribble, M. (2012). Digital citizenship for educational change. *Kappa Delta Pi Record, 48*(4), 148–151. doi:10.1080/00228958.2012.734015

Ribble, M., & Bailey, G. (2011). *Digital Citizenship in Schools*. International Society for Technology in Education (ISTE).

Street, B. (1997). The implications of the 'new literacy studies' for literacy education. *English in Education, 31*(3), 45–59. doi:10.1111/j.1754-8845.1997.tb00133.x

Sturgeon, J. (2008, Nov.). Raising good digital citizens. *Cable in the Classroom,* 5-9.

U.S. Department of Education. (2017). *Reimagining the role of technology in education: 2017 national educational technology plan update*. U.S. Department of Education.

Viadero, D., Jones, S., & Baker, L. (2018). *Data: Most states require history but not civics.* Education Week. https://www.edweek.org/teaching-learning/data-most-states-require-history-but-not-civics/2018/10

Walters, M. G., Gee, D., & Mohammed, S. (2019). A literature review: Digital citizenship and the elementary educator. *International Journal of Technology in Education*, *2*(1), 1–21.

Watulak, S. L., & Kinzer, C. K. (2013). Beyond technology skills: Toward a framework for critical digital literacies in pre-service technology education. In J. Avila & J. Z. Pandya (Eds.), Critical digital literacies as social praxis: Intersections and challenges (pp. 127-156). Peter Lang.

Westheimer, J., & Kahne, J. (2004). What kind of citizen? The politics of educating for democracy. *American Educational Research Journal*, *41*(2), 237–269. doi:10.3102/00028312041002237

Zhou, S., & Barbaro, N. (2023). Understanding student expression across higher ed: Heterodox Academy's annual campus expression survey. Heterodox Academy.

Chapter 14
Digital Citizenship and Digital Ethics:
An Educator's Perspective

Niroj Dahal
https://orcid.org/0000-0001-7646-1186
Kathmandu University, Nepal

ABSTRACT

Digital citizenship and digital ethics are closely related concepts regarding how to behave responsibly and ethically in the era of artificial intelligence (AI) in the digital world. The author shared his perspectives on digital citizenship and digital ethics based on his experiences working with PGDs, Master, and MPhil students at Kathmandu University School of Education, Nepal. The author attempted to reflect on how PGDs, Master, and MPhil students understood and acted on these concepts—digital citizenship and digital ethics based on the respecting, educating, and protecting framework of digital citizenship and digital ethics. Further, the author argued that digital citizenship is the responsible and respectful use of technology to engage in online activities: look for trustworthy sources and protect and promote human rights. Similarly, digital ethics are rules and moral principles governing individuals' interpersonal behavior. Overall, the author concluded that digital citizenship and ethics are related concepts that deal with digital technology's moral and social aspects.

INTRODUCTION

Living, working, learning, and interacting have been transformed by digital technology for general people and even for learners globally. The students of today have been raised with digital technologies. They are unable to recall a time without digital technologies. For instance, from the beginning of school education, students are busy watching videos or browsing other resources on the internet, social media, and other platforms, as continue in higher education. Consequently, educators (Choi & Park, 2023; Huffman et al., 2020; Li et al., 2023; Wulandari et al., 2021) concluded that students are fully competent in this domain for handling digital technologies. Even when students are comfortable and proficient with digital

DOI: 10.4018/978-1-6684-8934-5.ch014

technologies, their usage may be in/appropriate. For instance, one of my PGD students shared that *he is careful with his personal and professional life, while posting any post in social media. Even more, he does not believe everything he reads online.* The students may use technology in unethical ways, such as discovering ways to cheat on a test using generative AI such as ChatGPT, Bing Chat and Google Bard Chat. They may use it in socially harmful ways, such as cyberbullying on social media. For instance, to report criminal activity, to assist others, or to conduct oneself properly. In one of the discussions, one MPhil student shared that *he treats others as he would like to be treated by others online. He has not faced real-life consequences but heard a lot about insulting and bullying. He further added that in such cases, he respectfully disagrees.* Thus, Choi (2016) examined the concept of digital citizenship and its relationship to democratic citizenship education in the internet age and argued that digital citizenship is essential for democratic citizenship education in the internet age. In the digital age, citizens need to be able to use technology to access information, participate in civic discourse, and hold their government accountable. For instance, everyone who participates, learns, and works in the digital landscape must also meet certain requirements to handle digital technologies, such as respectful disagreement, treating others as you would like to be treated, taking permission to use the material, thinking carefully about what your goanna post on any platforms, copyright issue, and respect to intellectual property. These requirements define what it means to be a "digital citizen." Isman and Canan Gungoren (2014) added that "digital citizens must have extensive skills, knowledge, internet and technology access, and schools must guide the students to be a digital citizen." (p. 73).

To serve the purpose and fill the gap of handling digital technologies, I offered my perspectives on digital citizenship and ethics in this chapter. Digital literacy is the core concept while handling digital technologies. Moreover, Richardson and Milovidov (2019) argued that digital citizenship education is essential for preparing students—PGDs, Master and MPhil—to be responsible and ethical digital citizens in a world where technology is increasingly pervasive. The slogan of *be aware and be smart* best describes the use of digital technologies. The students must raise questions such who made such technologies?, how do such technologies work? and how do such technologies generate funds? to themselves before using them haphazardly. Richardson and Milovidov (2019) further argued that there are no one-size-fits-all approaches and that the best approach will vary depending on the needs of the students and the context. But digital citizenship education is essential for preparing students to be responsible and ethical digital citizens in the 21st century. Hence, digital citizenship and ethics are closely related concepts regarding how to behave responsibly and ethically in the digital world's era of AI. Digital citizenship is the responsible use of the internet, computers, and digital devices. Teaching digital citizenship in higher education goes beyond having students follow the rules, policies, and procedures. Teaching digital citizenship enables students to think critically, act safely, participate responsibly, and maintain their health and wellness in the digital world. Likewise, digital ethics refers to the ethical considerations and principles that guide behavior in the digital realm in higher education with their rights and responsibilities as a wise user. It involves understanding and addressing the ethical concerns that arise in internet research and digital interactions. The concept of *online bodyguard* as self and *you are your own biggest tread* best described the wise use of digital technologies that give rise to digital ethics that encompasses issues such as privacy, security, intellectual property, online harassment, and the responsible use of technology. Developing an ethical and responsible online culture is important to promote positive digital interactions and protect individuals' rights, responsibilities and well-being. Oxley (2011) discussed the concept and importance of digital citizenship, which is the responsible and ethical use of the internet and digital technologies by examining the issues and challenges that teenagers face online, such as cyberbullying, sexting, privacy, and plagiarism and emphasizes the need for developing an online culture that respects human rights and values diversity among the users.

Digital citizenship and digital ethics deal with the moral and social aspects of using digital technologies. Digital ethics is the study of the values, principles, and norms that guide the design, development, and use of digital technologies. It also examines these technologies' implications, consequences, and challenges for individuals, groups, and society. With the above, Ohler (2011) added that digital citizenship is a form of character education in the digital age and added that it is essential for basic to advance level user as a PGDs, master and MPhil students to be successful in the digital age and can be taught through character and moral education. These key areas of digital citizenship are ethics, responsibility, safety and politeness. Some of the topics that digital ethics covers in higher education are privacy, security, access, inclusion, democracy, accountability, and sustainability. No doubt, digital citizenship is the practice of thinking critically, behaving safely, and participating responsibly and ethically in the digital world. It also involves respecting the rights and responsibilities of others in the online environment. Some of the skills that digital citizenship requires are digital literacy, digital communication, digital etiquette, digital law, digital health and well-being, and digital security. Above all, with this introduction section, this chapter is divided into five sections, digital citizenship: how responsible are you?, what to know about digital ethics, digital citizenship and digital ethics framework: respecting, educating, and protecting, ethics in the digital age: how to be a responsible online citizen and final remarks.

DIGITAL CITIZENSHIP: HOW RESPONSIBLE ARE YOU?

Digital citizenship is the ability to use digital media responsibly and ethically, while understanding the risks involved in protecting information. However, Öztürk (2021) offered that digital citizenship is a complex and multidimensional concept. It is about knowing how to use technology and understanding its ethical and social implications. Further, digital citizenship is about thinking critically, behaving safely, and participating responsibly and ethically in the digital world. It involves the appropriate and responsible use of the internet, computers, and digital devices being aware of communicated writing, backfire of accessibility, private and personal information and security. Indeed, Mulyono et al. (2021) discussed the importance of digital citizenship competence by arguing that digital citizenship competence is essential for individuals to be able to participate effectively in the digital world for increasing the accessibility such as digital citizenship identity, privacy and security management, right and responsibility, digital empathy and active and engaged among others. Teaching digital citizenship goes beyond following rules and policies; it aims to help students think critically, behave safely, participate responsibly, and maintain their health and wellness in the digital world. For instance, one of my master students shared that *he is always aware of information he is sharing in online platform and it real time consequences.* Digital citizenship education focuses on promoting positive and ethical behaviors online. It is important to teach digital citizenship to ensure that individuals are competent in navigating the digital landscape and understand the risks involved. Likewise, when comes to educational practices, Ohler (2011) offered strategies for developing digital citizenship curriculum and assessment, such as using digital stories, creating community guidelines, and involving parents and stakeholders, emphasizing the importance of teaching students to be responsible and respectful digital citizens. Next, Choi (2016) proposed four categories of digital citizenship: ethics, media and information literacy, participation/engagement, and critical resistance. These categories provide a framework for understanding the multidimensional nature of digital citizenship and its relationship to democratic citizenship education.

On the contrary, Öztürk (2021) added that digital citizenship is a complex and multidimensional concept that involves various skills, attitudes, and behaviors, and digital citizenship education is important for preparing students for the challenges and opportunities of the digital age. Further, Öztürk (2021) suggested that digital citizenship education should be integrated into the curriculum, supported by teachers and parents, and evaluated by valid and reliable instruments even in higher education. The focus of digital citizenship education on teaching students should be the ethical use of technology. Critical thinking and problem-solving skills are essential in the digital age, as we are constantly bombarded with information and need to be able to evaluate its accuracy and relevance. We also need to be able to identify and manage online risks, such as cyberbullying, identity theft, and malware. So, digital literacy is the ability to use digital technologies effectively and responsibly, and it is essential for being a responsible and ethical digital citizen. Of particular significance, Wulandari et al. (2021) proposed a framework for implementing digital citizenship education in the era of Society 5.0. The framework includes curriculum, instruction, assessment, and professional development opportunities. Next, Digital citizenship is the use of technology in a way that is responsible, respectful, and ethical. It involves engaging online in a safe and productive way, finding reliable sources of information, and protecting and promoting human rights. In this regard, Wulandari et al. (2021) added that digital citizenship education is essential for preparing students to be responsible and ethical digital citizens in a world where technology is increasingly pervasive in the era of Society 5.0. Digital citizenship teaches communication skills, collaboration, and acting positively on any online platform (Dahal et al., 2022b; Dahal et al., 2022a; Dahal, 2023). It also involves students respecting the rights and responsibilities of others in the online environment. In addition, Li et al. (2023) discussed the importance of teaching digital citizenship in primary school and argued that it is an essential skill for young people in the digital age and should be integrated into existing school curricula. Further, Li et al. (2023) identified three key reasons why digital citizenship should be taught in primary school to higher education. First, they argue that digital citizenship is necessary for young people to be able to use technology safely and responsibly. Second, they argue that digital citizenship is necessary for young people to be able to participate effectively in the digital world. Third, they argue that digital citizenship is necessary for young people to be able to understand the rights and responsibilities of being a digital citizen. Consequently, digital citizenship is important because it helps students to protect themselves and others from online risks, such as cyberbullying, identity theft, phishing, and misinformation. It also helps students participate actively and constructively in the digital society by creating, sharing, and learning from diverse perspectives.

WHAT TO KNOW ABOUT DIGITAL ETHICS

Digital ethics is the study of how to use technology in a way that is ethical, professional, and respectful of others. The branch of ethics deals with the moral guidelines that govern how we interact with each other online, both as individuals and as companies. Dahal and Luitel (2022) remarked that "the ethic of self and others is all about the set of moral principles that govern the overall process" (p. 2672) while using digital tools. Likewise, ethics in South Asia could be looked at from the point of view of Eastern philosophies that are based on morals and faiths. The Vedic and Buddhist philosophies are the basis for these morals and beliefs. People say that ancient ethics gave rise to Vedic philosophy. These old rules have to do with what people do in their social lives. They are called *Dharma* (which means "duty") and *Karma* (which means "action") (Dahal & Luitel, 2022). In Vedic philosophy, *Dharma* and *Karma* change how we act, and others act as wise citizens while using digital tools and/or digital media. In this regard, digital ethics aims to ensure the autonomy and dignity of users is respected on the internet and address the implications, consequences, and challenges of digital technologies for individuals, groups, and society. Some of the issues that digital ethics deals with are privacy, security, access, inclusion, democracy, accountability, and sustainability. For instance, Lyon (2017) explored how surveillance is experienced and practiced in the digital age, where people are constantly exposed to data collection and processing, by discussing the ethical implications of surveillance culture, such as privacy, autonomy, and social justice and provided a critical perspective on the role of surveillance in shaping digital citizenship and modernity. Digital ethics also involves doing the right thing with data, which means being transparent, fair, accountable, and respectful of the rights and interests of data subjects, data owners, data processors, and data users.

Next, the digital age has brought about new challenges and opportunities for educators in higher education. In this changing context, exercising students about digital citizenship and ethics is more important than ever. Digital citizenship refers to the skills and knowledge that we need to participate effectively in the digital world. It includes things like critically evaluating information, creating and sharing content responsibly, and solving problems online (Christou, 2023). Digital ethics, on the other hand, refers to the moral principles that guide our behavior online. It includes things like respecting others' privacy, avoiding cyberbullying, and using technology responsibly. Teaching digital citizenship and ethics are not always easy. The digital world is constantly changing, and new challenges are always emerging. However, it is essential that we equip our students with the skills and knowledge they need to be responsible digital citizens. As an educator, I believe there are three critical perspectives—respecting, educating and protecting that we need to keep in mind when teaching and/or exercising digital ethics and citizenship.

DIGITAL CITIZENSHIP AND ETHICS FRAMEWORK: RESPECTING, EDUCATING, AND PROTECTING

Atif and Chou (2018) discussed the need for innovative approaches to digital citizenship education in order to address the challenges of the digital age by arguing that traditional approaches to digital citizenship education, which focus on teaching students about the dos and don'ts of online behavior, are no longer sufficient. Instead, they call for a more comprehensive approach that emphasizes the development of digital literacy, critical thinking skills, and ethical decision-making skills. Atif and Chou (2018) further called for a renewed focus on digital citizenship education in schools and universities and added that digital citizenship is essential for students to be successful in the digital age, and that innovative approaches to digital citizenship education are essential for preparing students for the challenges of the future. In this chapter, I propose a framework for teaching students about respect, education, and protection in the digital world. This framework is based on the introduction and gap analysis presented above. As an educator, I believe that the framework is important to teach students of PGDs, Master and MPhil how to respect themselves and others online, educate themselves about digital safety, and protect themselves from harm. The respecting, educating, and protecting (REP) framework of digital citizenship and ethics is a model that emphasizes the importance of these three key principles in using technology responsibly. Respect, education, and protection are the three categories of digital citizenship defined by Ribble and Park in 2019 in their book *Digital Citizenship in Schools,* wherein the concept of respect encompasses digital access, digital etiquette, and digital law. This involves using technology in a manner that respects the rights of others, abides by the law, and is sensitive to the feelings of others. The term 'educate' refers to digital communication, digital literacy, and digital commerce. Responsible and ethical use of technology to learn, communicate, and participate in the global economy is required. Included in protect are digital rights and responsibilities, digital safety and security, and digital health and wellness. This entails using technology to protect one from harm, secure one's personal information, and promote a healthy lifestyle (Ribble & Park, 2019).

Instead, the REP framework that I proposed can be used to guide individuals, organizations, and communities in developing and implementing digital citizenship education programs. In my proposal of respect refers to the importance of treating others with dignity and respect online, just as we would in the real world. For instance—a student could practice respect by not sharing personal information about others online without their permission and by avoiding using offensive language or images. This means being mindful of our words and actions and avoiding behaviors that could be harmful or offensive to others. It also means being aware of the potential for cyberbullying and online harassment and taking steps to prevent it. Likewise, in my proposal, educating refers to the importance of learning about the risks and opportunities of using technology and how to use it safely and responsibly. For instance—a teacher could educate students about the risks of cyberbullying by discussing the different types of cyberbullying, how to identify it, and what to do if it happens to them. This includes understanding the different types of online content, protecting personal information, and identifying and avoiding cyberbullying and online harassment. The proposal of protecting refers to the importance of taking steps to protect ourselves and our devices from cybercrime. For instance, a parent could protect their child from online predators by teaching them about online safety and monitoring their online activity. This includes using strong passwords, keeping our software up to date, and being aware of the latest scams and threats. It also means being mindful of the information we share online and ensuring that we only share it with people we trust. For the school and university learners, Ribble and Park (2022) provided a comprehensive guide on how to develop and implement a digital citizenship program in their schools covering a wide range of topics, including the importance of digital citizenship, the different levels of digital citizenship, the role of school leaders in digital citizenship education, and strategies for teaching digital citizenship. Ribble and Park (2022) further provided guidance on creating a shared vision for digital citizenship in the school and university, developing a digital citizenship curriculum, and assessing student learning.

On the whole, the REP framework is a valuable tool for promoting digital citizenship and digital ethics. By following these three principles, we can help to create a more respectful, educated, and protected online environment for everyone. Likewise, the REP framework is a flexible model that can be adapted to different settings and audiences. It promotes digital citizenship and digital ethics and creates a more respectful, educated, and protected online environment for everyone.

ETHICS IN THE DIGITAL AGE: HOW TO BE A RESPONSIBLE ONLINE CITIZEN

Ohler (2011) argued that digital citizenship requires character and/or education for the digital age, which is different from traditional character education by suggesting the educators to address the ethical and moral issues that arise from students' cyber lives, such as cyberbullying, sexting, plagiarism, use of generative AI tool tools such as ChatGPT, Google Bard and Bing Chat and privacy. So, respecting, educating, and protecting self and others are all important parts of digital citizenship and digitial ethics. Respect yourself and others online by not posting anything that could be harmful or offensive to others and not cyberbullying or harassing others. It also means being respectful of people's privacy and not sharing personal information about them without their permission. Educating yourself about digital citizenship and ethics can teach you about the responsibilities of being a digital citizen. Choi (2016) examined how digital citizenship can be defined and practiced for democratic citizenship education in the internet age and identified four dimensions of digital citizenship: digital identity, digital literacy, digital participation, and digital rights and responsibilities. This means learning about copyright laws, privacy settings, communicative practices and how to stay safe online. Protect yourself and others from online dangers, offer yourself tips for making strong passwords, being careful about what information you share online, and being aware of the dangers of malware and viruses. It also means being ready to report any online abuse or harassment you see. Prasetiyo et al. (2021) found that digital citizenship is influenced by three factors: digital readiness, digital citizenship competencies, and educational policies and highlighted the challenges and gaps in the research, such as the lack of empirical studies, the diversity of definitions and frameworks, and the need for more cross-cultural and interdisciplinary perspectives. The ultimate goal of knowing these rules is to make the Internet a more respectful, safe, and ethical place for everyone. Choi and Park (2023) defined digital citizenship as the ability to use technology responsibly and ethically in an online environment. Likewise, Choi and Park (2023) added that critical digital citizenship education should focus on helping adolescents develop the knowledge, skills, and dispositions to use technology in a responsible and ethical way by integrating into all aspects of the curriculum, not just in separate lessons or units as a hidden curriculum. In addition to the above, being aware of your online footprint, knowing the risks of online predators, using privacy settings to protect personal information, knowing the signs of online grooming, and telling a trusted adult about any suspicious activity are some more ways to respect, educate, and be a good digital citizen.

FINAL REMARKS

The digital age is a complex and ever-changing landscape. However, by teaching digital citizenship and digital ethics, we can help our students develop the skills and mindset they need to be responsible digital citizens. This is essential for ensuring that they can thrive in the digital age and contribute positively to society. The purpose of this chapter is to discuss digital citizenship and digital ethics. Concerns on digital citizenship and digital ethics have become increasingly popular in recent years. With the above, this chapter argued that digital citizenship and digital ethics have several ethical and practical considerations that every user and learner needs to consider in general. The development of citizenship skills and knowledge is not automatic. It is taught through both the authorized curriculum and the norms and expectations established in our cultures. Educators and other adults need to lead by example and incorporate digital citizenship into learning plans, extracurricular activities, and non-school-related activities. With so much online communication and learning, students need to have opportunities to explore their digital world safely and engage in discussions about what it means to be a digital citizen. Last but not least, digital citizenship is about thinking critically, acting safely, and participating in the digital world responsibly and ethically.

In conclusion, this chapter has outlined key practical discourse and considerations for ensuring digital citizenship and digital ethics in the academic world. However, further research is needed to explore the research-related outcomes of using digital citizenship and digital ethics and fully understand the best ways to incorporate digital citizenship and digital ethics and execution. In contrast, digital citizenship and digital ethics have profoundly impacted the modern world, academia, and the research community. It is important to ensure that learners that use digital citizenship and ethics adhere to accurate, reliable, justifiable, and ethical considerations. Students can actively involve themselves in the research process by using their cognitive and evaluative skills from conception to conclusions.

ACKNOWLEDGMENT

I am grateful to the PGDs, Master, and MPhil students from the 2021-2022 batches who participated in the discourse, either directly or indirectly, while framing this chapter. Your insights and feedback were invaluable to the development of this chapter. I would also like to thank the anonymous reviewers for their helpful comments and suggestions. Your feedback helped me to improve the chapter significantly. I am deeply indebted to the Kathmandu University School of Education for providing me with a supportive research environment. Last, but not least, I want to acknowledge all who have directly and/or indirectly supported me from drafting to finalizing the chapter.

REFERENCES

Atif, Y., & Chou, C. (2018). Digital citizenship: Innovations in education, practice, and pedagogy. *Journal of Educational Technology & Society*, *21*(1), 152–154. doi:10.48009/4_iis_2019_93-103

Choi, M. (2016). A concept analysis of digital citizenship for democratic citizenship education in the internet age. *Theory and Research in Social Education*, *44*(4), 565–607. doi:10.1080/00933104.2016.1210549

Choi, M., & Park, H. J. (2023). Korean adolescents' profiles of digital citizenship and its relations to Internet ethics: Implications for critical digital citizenship education. *Cambridge Journal of Education*, *53*(4), 1–20. doi:10.1080/0305764X.2023.2191929

Christou, P. A. (2023). How to use artificial intelligence (AI) as a resource, methodological and analysis tool in qualitative research? *Qualitative Report*, *28*(7), 1968–1980. doi:10.46743/2160-3715/2023.6406

Dahal, N. (2023). Integrating collaborative ICT tools in higher education for teaching and learning: A modest proposal for innovation in digital instructions. In *Innovations in Digital Instruction Through Virtual Environments* (pp. 143–156). IGI Global. doi:10.4018/978-1-6684-7015-2.ch008

Dahal, N., & Luitel, B. C. (2022). Understanding and encountering the ethics of self and others in autoethnography: Challenging the extant and exploring possibilities. *Qualitative Report*, *27*(12), 2671–2685. doi:10.46743/2160-3715/2022.5572

Dahal, N., Luitel, B. C., Pant, B. P., & Rajbanshi, R. (2022). Enhancing student-teachers assessment skills: A self-and peer-assessment tool in higher education. *International Journal of Education and Practice*, *10*(4), 313–321. doi:10.18488/61.v10i4.3173

Dahal, N., Luitel, B. C., Pant, B. P., Shrestha, I. M., & Manandhar, N. K. (2020). Emerging ICT tools, techniques and methodologies for online collaborative teaching and learning mathematics. *Mathematics Education Forum Chitwan*, *5*(5), 17–21. doi:10.3126/mefc.v5i5.34753

Huffman, S., Shaw, E., & Loyless, S. (2020). Ensuring ethics and equity: Policy, planning, and digital citizenship. *Education*, *140*(2), 87–99.

Isman, A., & Canan Gungoren, O. (2014). Digital citizenship. *Turkish Online Journal of Educational Technology-TOJET*, *13*(1), 73–77.

Li, Y., Deng, S., Wu, X., Zhao, B., Xie, Y., Luo, X., & Zheng, Y. (2023, July). Integrating Digital Citizenship into a Primary School Course "Ethics and the Rule of Law": Necessity, Strategies and a Pilot Study. In *International Conference on Blended Learning* (pp. 59-70). Springer Nature Switzerland. 10.1007/978-3-031-35731-2_7

Lyon, D. (2017). Digital citizenship and surveillancel Surveillance culture: Engagement, exposure, and ethics in digital modernity. *International Journal of Communication*, *11*, 19.

Mulyono, B., Affandi, I., Suryadi, K., & Darmawan, C. (2021). Digital citizenship competence: Initiating ethical guidelines and responsibilities for digital citizens. In *ICHELSS: International Conference on Humanities, Education, Law, and Social Sciences* (*Vol. 1*, No. 1, pp. 165-175). Academic Press.

Ohler, J. (2011). Digital citizenship means character education for the digital age. *Kappa Delta Pi Record, 47*(sup1), 25-27.

Oxley, C. (2011). Digital citizenship: Developing an ethical and responsible online culture. *Access*, *25*(3), 5–9.

Öztürk, G. (2021). Digital citizenship and its teaching: A literature review. *Journal of Educational Technology and Online Learning*, *4*(1), 31–45.

Prasetiyo, W. H., Naidu, N. B. M., Tan, B. P., & Sumardjoko, B. (2021). Digital Citizenship Trend in Educational Sphere: A Systematic Review. *International Journal of Evaluation and Research in Education*, *10*(4), 1192–1201. doi:10.11591/ijere.v10i4.21767

Ribble, M., & Park, M. (2019). *The digital citizenship handbook for school leaders: Fostering positive interactions online*. International Society for Technology in Education.

Ribble, M., & Park, M. (2022). *The digital citizenship handbook for school leaders: Fostering positive interactions online*. International Society for Technology in Education.

Richardson, J., & Milovidov, E. (2019). *Digital citizenship education handbook: Being online, well-being online, and rights online*. Council of Europe.

Wulandari, E., Winarno, W., & Triyanto, T. (2021). Digital Citizenship Education: Shaping Digital Ethics in Society 5.0. *Universal Journal of Educational Research*, *9*(5), 948–956. doi:10.13189/ujer.2021.090507

Compilation of References

Aaron, L. S., & Lipton, T. (2018). Digital distraction: Shedding light on the 21st-century college classroom. *Journal of Educational Technology Systems*, *46*(3), 363–378. doi:10.1177/0047239517736876

Addas, S., & Pinsonneault, A. (2015). The many faces of information technology interruptions: A taxonomy and preliminary investigation of their performance effects. *Information Systems Journal*, *25*(3), 231–273. doi:10.1111/isj.12064

Agamben, G. (1998). *Image et mémoire*. Éditions Hoëbeke.

Ahmad, A., & Maynard, S. (2014). Teaching information security management: Reflections and experiences. *Information Management & Computer Security*, *22*(5), 513–536. doi:10.1108/IMCS-08-2013-0058

Ahn, S., Kim, J., & Kim, J. (2022). The bifold triadic relationships framework: a theoretical primer for advertising research in the metaverse. *Journal of Advertising*, *51*(5), 592–607. doi:10.1080/00913367.2022.2111729

Albrahim, F. A. (2020). Online teaching skills and competencies. *Turkish Online Journal of Educational Technology-TOJET*, *19*(1), 9–20.

Allam, Z., Bibri, S., Jones, D., Chabaud, D., & Moreno, C. (2022). Unpacking the ‘15-minute city’ via 6g, IoT, and digital twins: towards a new narrative for increasing urban efficiency, resilience, and sustainability. *Sensors (Basel)*, *22*(4), 1369. doi:10.339022041369 PMID:35214271

Allam, Z., Sharifi, A., Bibri, S., Jones, D., & Krogstie, J. (2022). The metaverse as a virtual form of smart cities: opportunities and challenges for environmental, economic, and social sustainability in urban futures. *Smart Cities*, *5*(3), 771–801. doi:10.3390martcities5030040

Allen, I. E., & Seaman, J. (2017). *Digital Compass Learning: Distance Education Enrollment Report 2017*. Babson Survey Research Group.

Alpala, L., Quiroga-Parra, D., Torres, J., & Peluffo-Ordóñez, D. (2022). Smart factory using virtual reality and online multi-user: towards a metaverse for experimental frameworks. *Applied Sciences (Basel, Switzerland)*, *12*(12), 6258. doi:10.3390/app12126258

Alper, M., & Goggin, G. (2017). Digital technology and rights in the lives of children with disabilities. *New Media & Society*, *19*(5), 726–740. doi:10.1177/1461444816686323

Al-Samarraie, H., Bello, K. A., Alzahrani, A. I., Smith, A. P., & Emele, C. (2021). Young users' social media addiction: Causes, consequences and preventions. *Information Technology & People*, *35*(7), 2314–2343. doi:10.1108/ITP-11-2020-0753

Altınay Gazi, Z. (2016). Internalization of digital citizenship for the future of all levels of education. *Education in Science*, *41*(186), 137–148. doi:10.15390/EB.2016.4533

Alvermann, D. E., Beach, C. L., & Johnson, J. (2014). *Becoming 3lectric*. Retrieved from http://www.becoming3lectric.com/

Alvermann, D. E. (2008). Why bother theorizing adolescents' online literacies for classroom practice and research? *Journal of Adolescent & Adult Literacy*, *52*(1), 8–19. doi:10.1598/JAAL.52.1.2

Alvermann, D. E., Marshall, J. D., McLean, C. A., Huddleston, A. P., Joaquin, J., & Bishop, J. (2012). Adolescents' web-based literacies, identity construction, and skill development. *Literacy Research and Instruction*, *51*(3), 179–195. doi:10.1080/19388071.2010.523135

American Academy of Child & Adolescent Psychiatry. (2020). *Screen time and children*. https://www.aacap.org/AACAP/Families_and_Youth/Facts_for_Families/FFF-Guide/Children-And-Watching-TV-054.aspx

American Academy of Pediatrics. (n.d.). *Media family plan*. AAP. https://www.healthychildren.org/English/fmp/Pages/MediaPlan.aspx

Americans with Disabilities Amendments Act of 2008, Public Law 110–325, 122 Stat. 3553 (2008).

Amo Filvá, D., Prinsloo, P., Alier Forment, M., Fonseca Escudero, D., Torres Kompen, R., Canaleta Llampallas, X., & Herrero Martín, J. (2021). Local technology to enhance data privacy and security in educational technology. *International Journal of İnteractive Multimedia and Artificial İntelligence*, *7*(2), 262–273. doi:10.9781/ijimai.2021.11.006

Andrade, M. C. F. (2015). Evidências teóricas para compreensão da inovação aberta (open innovation) nas organizações. *Prespectivas em Gestão & Conhecimento*, *5*(1), 31-42. https://periodicos.ufpb.br/ojs2/index.php/pgc/article/view/19222

Andrews, J. C., Walker, K. L., & Kees, J. (2020). Children and online privacy protection: Empowerment from cognitive defense strategies. *Journal of Public Policy & Marketing*, *39*(2), 205–219. doi:10.1177/0743915619883638

Anthony, W. L., Zhu, Y., & Nower, L. (2021). The relationship of interactive technology use for entertainment and school performance and engagement: Evidence from a longitudinal study in a nationally representative sample of middle school students in China. *Computers in Human Behavior*, *122*, 106846. doi:10.1016/j.chb.2021.106846

Aristotle. (2004). *Nicomachean ethics*. Cambridge University Press.

Arnott, L., Palaiologou, I., & Gray, C. (2018). Digital devices, internet-enabled toys and digital games: The changing nature of young children's learning ecologies, experiences and pedagogies. *British Journal of Educational Technology*, *49*(5), 803–806. doi:10.1111/bjet.12676

Arnott, L., & Yelland, N. J. (2020). Multimodal lifeworlds: Pedagogies for play inquiries and explorations. *Journal of Early Childhood Education Research*, *9*(1), 124–146.

Arora, P. (2012). Typology of web 2.0 spheres: Understanding the cultural dimensions of social media spaces. *Current Sociology*, *60*(5), 599–618. doi:10.1177/0011392112440439

Assante, G. M., Popa, N. L., & Momanu, M. (2022). How personal values and critical dispositions support digital citizenship development in higher education students. *Frontiers in Psychology*, *13*(990518), 1–11. doi:10.3389/fpsyg.2022.990518 PMID:36267083

Ata, R., & Yıldırım, K. (2019). Turkish pre-service teachers' perceptions of digital citizenship in education programs. *Journal of Information Technology Education*, *18*, 419–436. doi:10.28945/4392

Atif, Y., & Chou, C. (2018). Digital citizenship: Innovations in education, practice, and pedagogy. *Journal of Educational Technology & Society*, *21*(1), 152–154. doi:10.48009/4_iis_2019_93-103

Aungst, T. D., & Patel, R. (2020). Integrating digital health into the curriculum—Considerations on the current landscape and future developments. *Journal of Medical Education and Curricular Development*, *7*, 2382120519901275. doi:10.1177/2382120519901275 PMID:32010795

Au, W. (2011). Teaching under the new Taylorism: High-stakes testing and the standardization of the 21st century curriculum. *Journal of Curriculum Studies*, *43*(1), 25–45. doi:10.1080/00220272.2010.521261

Baggio, D., Gavronski, I., & Lima, V. Z. D. (2019). Inovação aberta: Uma vantagem competitiva para pequenas e médias empresas. *Revista Ciências Administrativas*, *25*(3), 3. Advance online publication. doi:10.5020/2318-0722.2019.7507

Baghalzadeh Shishehgarkhaneh, M., Keivani, A., Moehler, R., Jelodari, N., & Roshdi Laleh, S. (2022). Internet of things (IoT), building information modeling (BIM), And Digital Twin (DT) in construction industry: a review, bibliometric, and network analysis. *Buildings*, *12*(10), 1503. doi:10.3390/buildings12101503

Baker, S. (2013). Conceptualising the use of Facebook in ethnographic research: As tool, as data and as context. *Ethnography and Education*, *8*(2), 131–145. doi:10.1080/17457823.2013.792504

Baloch, I., Kaye, T., Koomar, S., & McBurnie, C. (2020). Pakistan topic brief: providing distance learning to hard-to-reach children in urban slums and remote areas. (EdTech Hub Helpdesk Response No 17) DOI:. 3885606. doi:10.5281/zenodo

Barbosa, A. M. A. (2015). *Recomendação de Boas Práticas no Crowdsourcing – um Caso de Estudo* [Dissertação de mestrado, Faculdade de Engenharia da Universidade do Porto]. Repositório Aberto da Universidade do Porto.

Barrera, D., Willner, L. N., & Kukahiko, K. I. (2017). Assessing the development of an emerging critical consciousness through service learning. *Journal of Critical Thought and Praxis*, *6*(3). doi:10.31274/jctp-180810-82

Barron, B., & Darling-Hammond, L. (2008). *Teaching for meaningful learning: a review of research on inquiry-based and cooperative learning*. George Lucas Education Foundation. https://files.eric.ed.gov/fulltext/ED539399.pdf

Barroso, P. M. (2022). From reality to the hyperreality of the simulation. *Texto Livre – Linguagem e Tecnologia ,15*, 1-11.

Barroso, P. M. (2023). Baudrillard on the symbolic violence of the image. *Versus - Quaderni di studi semiotici, 136,* 135-154.

Barrow, R. (2012). *Common sense and the curriculum* (Vol. 13). Routledge. doi:10.4324/9780203815694

Bar-Tal, D. (2017). Self-Censorship as a socio-political-psychological phenomenon: Conception and research. *Political Psychology*, *38*, 37–65. doi:10.1111/pops.12391

Baudrillard, J. (1997). *Simulacra and simulation.* The University of Michigan Press.

Bauman, Z., & Lyon, D. (2013). *Liquid surveillance*. Polity Press.

Bazeley, P., & Jackson, K. (2013). *Qualitative data analysis with NVIVO*. Sage Publications.

Beach, C. L., Alvermann, D. E., Loomis, S., Hutcherson-Price, L., & Wright, W. (2023). Digital remixing online: Entangled feelings. *Digital Culture & Education, 14*(4), 92–108.

Beatty, K. (2019). *Developing student voice through Twitter: Using media literacy education to inspire civic engagement* [Unpublished master's thesis]. University of Rhode Island.

Becker, D. (2019). The digital citizen 2.0. *Arbeiten Aus Anglistik Und Amerikanistik*, *44*(2), 167–194.

Beever, J., McDaniel, R., & Stanlick, N. (2020). *Understanding digital ethics: Cases and contexts.* Routledge.

Bennett, W. L., Wells, C., & Freelon, D. (2011). Communicating civic engagement: Contrasting models of citizenship in the youth web sphere. *Journal of Communication, 61*(5), 835-856.

Bennett, L. B., Aguayo, R. C., & Field, S. L. (2016). At home in the world: Supporting children in human rights, global citizenship, and digital citizenship. *Childhood Education, 92*(3), 189–199. doi:10.1080/00094056.2016.1180892

Bennett, W. L. (2008). Changing citizenship in the digital age. In W. L. Bennett (Ed.), *Civic life online: Learning how digital media can engage youth* (pp. 1–24). MIT Press.

Bergdahl, N., Nouri, J., Fors, U., & Knutsson, O. (2020). Engagement, disengagement and performance when learning with technologies in upper secondary school. *Computers & Education, 149*, 103783. doi:10.1016/j.compedu.2019.103783

Bergen, D., Davis, D. R., & Abbitt, J. T. (2015). *Technology play and brain development: Infancy to adolescence and future Implications*. Routledge. doi:10.4324/9781315681436

Bernard, R. (2012). *Moor's Law.* Retrieved from https://blog.soton.ac.uk/comp6044/tag/moors-law/

Bernardino, S. J. Q., & Santos, J. F. (2020). Crowdsourcing ideas for public investment: the experience of youth participatory budgeting in Portugal. In Crowdasset – Crowdfunding for policy makers (pp. 353–379); Worcropld Scientific Publishing, https://doi.org/ doi:10,1142/11485

Berson, I. R., & Berson, M. J. (2003). Digital literacy for effective citizenship. *Social Education, 67*, 164–167.

Bertrand, M., & Lozenski, B. D. (2023). YPAR dreams deferred? Examining power bases for YPAR to impact policy and practice. *Educational Policy, 37*(2), 437–462. https://doi.org/10.1177/08959048211019975

Beyer, B. (1995). *Critical thinking*. Phi Delta Kappa Educational Foundation.

Bibri, S. (2022). The social shaping of the metaverse as an alternative to the imaginaries of data-driven smart cities: a study in science, technology, and society. *Smart Cities, 5*(3), 832–874. doi:10.3390martcities5030043

Bibri, S., & Allam, Z. (2022). The metaverse as a virtual form of data-driven smart urbanism: on post-pandemic governance through the prism of the logic of surveillance capitalism. *Smart Cities, 5*(2), 715–727. doi:10.3390martcities5020037

Bilgin Ülken, F., & Kılınç, F. (2018). Çocukların çevrimiçi ortamda karşılaştıkları risk türleri ile aracılık uygulamaları arasındaki ilişki. *İnsan ve Toplum Bilimleri Araştırmaları Dergisi, 7*(1), 124-145.

Bilki, Z., Satar, M., & Sak, M. (2023). Critical digital literacy in virtual exchange for ELT teacher education: An interpretivist methodology. *ReCALL, 35*(1), 58–73. doi:10.1017/S095834402200009X

Bishop, A. P. (2003). Using the web to support inquiry-based literacy development. In Bruce, B. C. (Ed.). Literacy in the information age: Inquiries into meaning making with *n*ew technologies. International Reading Association.

Blackburn, S. (2003). *Ethics: A very short introduction*. Oxford University Press. doi:10.1093/actrade/9780192804426.001.0001

Black, R., Walsh, L., Waite, C., Collin, P., Third, A., & Idriss, S. (2022). In their own words: 41 stories of young people's digital citizenship. *Learning, Media and Technology, 47*(4), 524–536. doi:10.1080/17439884.2022.2044848

Bogers, M., Chesbrough, H., & Moedas, C. (2018). Open Innovation: Research, Practices, and Policies. *California Management Review, 60*(2), 5–16. doi:10.1177/0008125617745086

Bolger, R. (2021). Finding Wholes in the Metaverse: Posthuman Mystics as Agents of Evolutionary Contextualization. *Religions, 12*(9), 768. doi:10.3390/rel12090768

Bonanno, K. (2016). Let's talk software: Digital commerce, legalities and copyright awareness, and online wellbeing. *Education Technology Solutions*, 71, 56–58.

boyd, d. (2015). Making sense of teen life: Strategies for capturing ethnographic data in a networked era. In E. Hargittai & C. Sandvig (Eds.), *Digital Research Confidential: The Secrets of Studying Behavior Online*. Cambridge, MA: The MIT Press. Retrieved from http://www.danah.org/papers/2012/Methodology-DigitalResearch.pdf

Brabham, D. C. (2008). Crowdsourcing as a Model for Problem Solving: An Introduction and Cases. *Convergence (London)*, *14*(1), 75–90. doi:10.1177/1354856507084420

Brabham, D. C. (2013). Using crowdsourcing. *Government*. IBM Center for the Business of Government.

Bruce, B. C. (Ed.). (2003). Literacy in the information age: Inquiries into meaning making with *n*ew technologies. International Reading Association.

Buchanan, R. (2019). Digital ethical dilemmas in teaching. In *Encyclopedia of Teacher Education*. Springer. doi:10.1007/978-981-13-1179-6_150-1

Buchholz, B. A., DeHart, J., & Moorman, G. (2020). Digital citizenship during a global pandemic: Moving beyond digital literacy. *Journal of Adolescent & Adult Literacy*, *64*(1), 11–17. doi:10.1002/jaal.1076 PMID:32834710

Buchholz, B. A., & Pyles, D. G. (2018). Scientific literacy in the wild: Using multimodal texts in and out of school. *The Reading Teacher*, *72*(1), 61–70. doi:10.1002/trtr.1678

Buckingham, D. (2015). Defining digital literacy-What do young people need to know about digital media? *Nordic journal of digital literacy, 10* (Jubileumsnummer), 21-35.

Buckingham, D. (1998). Media education in the UK: Moving beyond protectionism. *Journal of Communication*, *48*(1), 33–43. doi:10.1111/j.1460-2466.1998.tb02735.x

Buckingham, D. (2003). *Media education: Literacy, learning and contemporary culture*. Polity Press.

Buckingham, D. (2015). Defining digital literacy: What do young people need to know about digital media? *Nordic Journal of Digital Literacy*, *10*, 21–35. doi:10.18261/ISSN1891-943X-2015-Jubileumsnummer-03

Buckingham, D. (2020). Epilogue: Rethinking digital literacy: Media education in the age of digital capitalism. *Digital Education Review*, (37), 230–239. doi:10.1344/der.2020.37.230-239

Buhalis, D., Lin, M., & Leung, D. (2022). Metaverse as a driver for customer experience and value co-creation: implications for hospitality and tourism management and marketing. *International Journal of Contemporary Hospitality Management*. doi:10.1108/IJCHM-05-2022-0631

Burnett, C., & Merchant, G. (2011). Is there a space for critical literacy in the context of social media? *English Teaching*, *10*(1), 41–57.

Cabannes, Y., & Baierle, S. (2004). Financiamento local e orçamento participativo. Seminário de Lançamento da Rede URB-AL n.o 9. Porto Alegre, Brasil.

Cabannes, Y. (2009). *72 Perguntas frequentes sobre o orçamento participativo*. UN-Habitat.

Cabannes, Y. (2015). The impact of participatory budgeting on basic services: Municipal practices and evidence from the field. *Environment and Urbanization*, *27*(1), 257–284. doi:10.1177/0956247815572297

Cadenas, G. A., Cantú, E. A., Spence, T., & Ruth, A. (2020). Integrating critical consciousness and technology in entrepreneurship career development with diverse community college students. *Journal of Career Development*, *47*(2), 162–176. doi:10.1177/0894845318793968

Caena, F., & Redecker, C. (2019). Aligning teacher competence frameworks to 21st century challenges: The case for the European Digital Competence Framework for Educators (Digcompedu). *European Journal of Education*, *54*(3), 356–369. doi:10.1111/ejed.12345

Cambridge University Press. (n.d.). Digital Citizens. In *Cambridge dictionary*. https://dictionary.cambridge.org/us/dictionary/english/digital-citizenship

Cambridge University Press. (n.d.). Digital Citizenship. In *Cambridge dictionary*. https://dictionary.cambridge.org/us/dictionary/english/digital-citizenship

Cammarota, J. (2017). Youth participatory action research: A pedagogy of transformational resistance for critical youth studies. *The Journal for Critical Education Policy Studies*, *15*(2), 188–213.

Cammarota, J., & Fine, M. (2008). *Revolutionizing education: Youth participatory action research in motion*. Routledge.

Campbell, C., & Haines, C. (2016). *Becoming a media mentor: A guide for working with children and families*. American Library Association.

Campbell, R., Martin, C. R., & Fabos, B. (2014). *Media & culture: Mass communication in a digital age*. Bedford / St. Martin's.

Cappannari, L., & Vitillo, A. (2022). XR and metaverse software platforms. In Roadmapping Extended Reality (pp. 135-156). doi:10.1002/9781119865810.ch6

Capurro, R. (2009). Digital ethics. *2009 Global Forum on Civilization and Peace*.

Carpenter, J. P., & Morrison, S. A. (2018). Enhancing teacher education… with Twitter? *Phi Delta Kappan*, *100*(1), 25–28. doi:10.1177/0031721718797118

Chandler-Olcott, K., & Lewis, E. (2010). "I think they're being wired differently": Secondary teachers' cultural models of adolescents and their online literacies. In D. E. Alvermann (Ed.), *Adolescents' Online Literacies: Connecting Classrooms, Digital Media, & Popular Culture* (pp. 163–182). Peter Lang Publishing, Inc.

Chassiakos, Y. L. R., & Stager, M. (2020). Current trends in digital media: How and why teens use technology. In *Technology and adolescent health* (pp. 25–56). Academic Press. doi:10.1016/B978-0-12-817319-0.00002-5

Chaudron, S., Di Gioia, R., & Gemo, M. (2018). Y*oung children (0-8) and digital technology - A qualitative study across Europe*. JRC Science for Policy Report. doi:10.2760/294383

Chen, L., Nath, R., & Tang, Z. (2020). Understanding the determinants of digital distraction: An automatic thinking behavior perspective. *Computers in Human Behavior*, *104*, 106195. doi:10.1016/j.chb.2019.106195

Chen, Q., & Lee, S. (2021). Research status and trend of digital twin: visual knowledge mapping analysis. *International Journal of Advanced Smart Convergence*, *10*(4), 84–97. doi:10.7236/IJASC.2021.10.4.84

Chen, Y., Rodgers, J., & Mcconachie, H. (2009). Restricted and repetitive behaviours, sensory processing and cognitive style in children with autism spectrum disorders. *Journal of Autism and Developmental Disorders*, *39*(4), 635–642. doi:10.100710803-008-0663-6 PMID:19015969

Cheong, P. H., Shuter, R., & Suwinyattichaiporn, T. (2016). Managing student digital distractions and hyperconnectivity: Communication strategies and challenges for professorial authority. *Communication Education*, *65*(3), 272–289. doi:10.1080/03634523.2016.1159317

Chesbrough, H. (2011). Everything you need to know about open innovation. *Forbes*. https://www.forbes.com/sites/henrychesbrough/2011/03/21/everything-you-need-to-know-about-open-innovation/

Chesbrough, H. (2003). The era of open innovation. *MIT Sloan Management Review*, *44*(3), 35–41.

Chesbrough, H. W., & Crowther, A. K. (2006). Beyond high tech: Early adopters o open innovation in other industries. *R & D Management*, *36*(3), 229–236. doi:10.1111/j.1467-9310.2006.00428.x

Choi, M. (2016). A concept analysis of digital citizenship for democratic citizenship education in the internet age. *Theory and Research in Social Education*, *44*(4), 565–607. doi:10.1080/00933104.2016.1210549

Choi, M., & Cristol, D. (2021). Digital citizenship with intersectionality lens: Towards participatory democracy driven digital citizenship education. *Theory into Practice*, *60*(4), 361–370. doi:10.1080/00405841.2021.1987094

Choi, M., Cristol, D., & Gimbert, B. (2018). Teachers as digital citizens: The influence of individual backgrounds, internet use and psychological characteristics on teachers' levels of digital citizenship. *Computers & Education*, *121*, 143–161. doi:10.1016/j.compedu.2018.03.005

Choi, M., Glassman, M., & Cristol, D. (2017). What it means to be a citizen in the internet age: Development of a reliable and valid digital citizenship scale. *Computers & Education*, *107*, 100–112. doi:10.1016/j.compedu.2017.01.002

Choi, M., & Park, H. J. (2023). Korean adolescents' profiles of digital citizenship and its relations to internet ethics: Implications for critical digital citizenship education. *Cambridge Journal of Education*, *53*(4), 567–586. doi:10.1080/0305764X.2023.2191929

Christians, C. (2011). Journalism ethics in theory and practice. In G. Cheney, S. May, & D. Munshi (Eds.), *The Handbook of Communication Ethics* (pp. 190–203). Routledge. doi:10.4324/9780203890400.ch12

Christou, P. A. (2023). How to use artificial intelligence (AI) as a resource, methodological and analysis tool in qualitative research? *Qualitative Report*, *28*(7), 1968–1980. doi:10.46743/2160-3715/2023.6406

Cladis, A. E. (2020). A shifting paradigm: An evaluation of the pervasive effects of digital technologies on language expression, creativity, critical thinking, political discourse, and interactive processes of human communications. *E-Learning and Digital Media*, *17*(5), 341–364. doi:10.1177/2042753017752583

Clarke, K., & Lhamon, C. E. (2023). *Dear Colleague: Online Accessibility.* Washington DC: United States Department of Justice Civil Rights Division. United States Department of Education Office of Civil Rights. https://www2.ed.gov/about/offices/list/ocr/docs/postsec-online-access-051923.pdf

Collins, A., & Halverson, R. (2018). *Rethinking education in the age of technology: The digital revolution and schooling in America.* Teachers College Press.

Connected Learning Alliance. (n.d.) What is connected learning? *About Connected Learning.* Connected Learning Alliance. Accessed July, 21, 2023. https://clalliance.org/about-connected-learning/

Cope, B., & Kalantzis, M. (Eds.). (2000). *Multiliteracies: Literacy learning and the design of social futures.* Psychology Press.

Cortesi, S., Hasse, A., Lombana-Bermudez, A., Kim, S., & Gasser, U. (2020). Youth and digital citizenship+ (plus): Understanding skills for a digital world. *Berkman Klein Center Research Publication*, (2020-2).

Costa, C. A. S. (2014). *Mecanismos de participação cívica: orçamento participativo.* [Dissertação de Mestrado, Escola de Economia e Gestão da Universidade do Minho]. Repositorium.

Council of Europe. (2019). *Digital Citizenship Education Handbook.* European Commission. https://ec.europa.eu/newsroom/just/items/672450/en

Criddle, E., Vidovich, L., & O'neill, M. (2004). Discovering democracy: An analysis of curriculum policy for citizenship education. *Westminster Studies in Education, 27*(1), 27–41. doi:10.1080/0140672040270103

Csikszentmihalyi, M. (1997). Flow and the psychology of discovery and invention. *HarperPerennial, New York, 39*, 1–16.

Çubukcu, A., & Bayzan, Ş. (2013). Türkiye'de dijital vatandaşlık algısı ve bu algıyı internetin bilinçli, güvenli ve etkin kullanımı ile artırma yöntemleri. *Middle Eastern & African Journal of Educational Research, 5*(1), 148–174.

D'Amico, M., Denov, M., Khan, F., Linds, W., & Akesson, B. (2016). Research as intervention? Exploring the health and well-being of children and youth facing global adversity through participatory visual methods. *Global Public Health: An International Journal for Research, Policy and Practice, 11*(5-6), 528–545. doi:10.1080/17441692.2016.1165719 PMID:27043374

Dahal, N. (2023). Integrating collaborative ICT tools in higher education for teaching and learning: A modest proposal for innovation in digital instructions. In *Innovations in Digital Instruction Through Virtual Environments* (pp. 143–156). IGI Global. doi:10.4018/978-1-6684-7015-2.ch008

Dahal, N., & Luitel, B. C. (2022). Understanding and encountering the ethics of self and others in autoethnography: Challenging the extant and exploring possibilities. *Qualitative Report, 27*(12), 2671–2685. doi:10.46743/2160-3715/2022.5572

Dahal, N., Luitel, B. C., Pant, B. P., & Rajbanshi, R. (2022). Enhancing student-teachers assessment skills: A self-and peer-assessment tool in higher education. *International Journal of Education and Practice, 10*(4), 313–321. doi:10.18488/61.v10i4.3173

Dahal, N., Luitel, B. C., Pant, B. P., Shrestha, I. M., & Manandhar, N. K. (2020). Emerging ICT tools, techniques and methodologies for online collaborative teaching and learning mathematics. *Mathematics Education Forum Chitwan, 5*(5), 17–21. doi:10.3126/mefc.v5i5.34753

Dahan, N., Al-Razgan, M., Al-Laith, A., Alsoufi, M., Al-Asaly, M., & Alfakih, T. (2022). Metaverse framework: a case study on e-learning environment (ELEM). *Electronics (Basel), 11*(10), 1616. doi:10.3390/electronics11101616

Dahlander, L., & Gann, D. M. (2010). How open is innovation? *Research Policy, 39*(6), 699–709. doi:10.1016/j.respol.2010.01.013

Dai, Y., Wang, J., & Gao, S. (2022). Advanced electronics and artificial intelligence: must-have technologies toward human body digital twins. *Advanced Intelligent Systems, 4*(7), 2100263. doi:10.1002/aisy.202100263

Davenport, T. H., & Ronanki, R. (2018). Artificial intelligence for the real world. *Harvard Business Review, 96*(1), 108–116.

Davidson, A., & Silkane, V. (2019). The relationship between media literacy and civic participation among young adults in Latvia. In S. Kurbanoğlu, S. Spiranec, Y. Ünal, J. Boustany, M. L. Huotari, E. Grassian, D. Mizrachi, & L. Roy (Eds.), *Information literacy in everyday life* (pp. 103–112). Springer. doi:10.1007/978-3-030-13472-3_10

Davis, K., Katz, S., Santo, R., & James, C. (2010). Fostering Cross-Generational Dialogues About the Ethics of Online Life. *Davis, K., Katz, SL, Santo, R., & James, C.(2010). Fostering cross-generational dialogues about the ethics of online life. The Journal of Media Literacy Education, 2*(2), 124–150.

Dawkins, A. (2020). Digital Citizenship, Digital Legacy, and School Librarians. *Library Technology Reports, 56*(5), 17+. https://link.gale.com/apps/doc/A632141257/ITOF?u=gale15691&sid=bookmark-ITOF&xid=753c1ab5

De Abreu, B. (2010). Changing Technology= Empowering Students through Media Literacy Education. *New Horizons in Education, 58*(3), 26–33.

De Ketelaere, B., Smeets, B., Verboven, P., Nicolaï, B., & Saeys, W. (2022). Digital twins in quality engineering. *Quality Engineering, 34*(3), 404–408. doi:10.1080/08982112.2022.2052731

De Renzio, P., & Wehner, J. (2017). The impacts of fiscal openness. *The World Bank Research Observer, 32*(2), 185–210. doi:10.1093/wbro/lkx004

DeHart, J. D. (Ed.). (2023). *Innovations in digital instruction through virtual environments*. IGI Global. doi:10.4018/978-1-6684-7015-2

Deigh, J. (2010). *An introduction to ethics*. Cambridge University Press. doi:10.1017/CBO9780511750519

Deumert, A. (2014). The performance of a ludic self on social network(ing) sites. In P. Seargeant & C. Tagg (Eds.), *The language of social media: Identity and community on the internet* (pp. 23–45). Palgrave Macmillan. doi:10.1057/9781137029317_2

Dezuanni, M. (2010). Digital media literacy: Connecting young people's identities, creative production and learning about video games. In D. E. Alvermann (Ed.), *Adolescents' Online Literacies: Connecting Classrooms, Digital Media, & Popular Culture* (pp. 125–143). Peter Lang Publishing, Inc.

Dias, J. C. V., Sant'Anna, L. S., & Portilho, R. M. R. (2018). Open Innovation e o Papel Instrumentalizador dos Contratos. *RFD-Revista da Faculdade de Direito da UERJ., 0*(34), 77–100. doi:10.12957/rfd.2018.18126

Diaz, J. R. D., Ucharima, I. H., & Egas, M. M. S. (2022). Digital competencies as pedagogical predictors for learning to learn. *Revista De La Universidad Del Zulia, 13*(38), 643–656.

Doane, A. N., Kelley, M. L., Chiang, E. S., & Padilla, M. A. (2013). Development of the cyberbullying experiences survey. *Emerging Adulthood, 1*(3), 207–218. doi:10.1177/2167696813479584

Dow, B. J., Johnson, A. L., Wang, C. S., Whitson, J., & Menon, T. (2021). The COVID-19 pandemic and the search for structure: Social media and conspiracy theories. *Social and Personality Psychology Compass, 15*(9), 1–22. doi:10.1111pc3.12636 PMID:34512798

Drucker, P. F. (1961). The technological revolution: Notes on the relationship of technology, science and culture. *Technology and Culture, 2*(4), 342–351. doi:10.2307/3100889

Druckman, J. N. (2012). The politics of motivation. *Critical Review, 24*(2), 199–216. doi:10.1080/08913811.2012.711022

Duan, H., Li, J., Fan, S., Lin, Z., Wu, X., & Cai, W. (2021). *Metaverse for Social Good: A University Campus Prototype*. Paper presented at the Proceedings of the 29th ACM International Conference on Multimedia, Virtual Event, China. 10.1145/3474085.3479238

Duncan, M., & Watson, R. (2010). Taking a stance: Socially responsible ethics and informed consent. In M. Savin-Baden & C. H. Major (Eds.), *New approaches to qualitative research: Wisdom and uncertainty* (pp. 49–58). Routledge.

Dunn, S. (2023). Cultural connections based upon technology availability. In V. Brown (Ed.), *Global societal impact of digital literacy*. Kendall Hunt.

Eckhoff, A. (2015). Ethical considerations of children's digital image-making and image-audiancing in early childhood environments. *Early Child Development and Care, 185*(10), 1617–1628. doi:10.1080/03004430.2015.1013539

Eden, M. (2021, June 17). *Parents against critical race theory*. Washington Examiner. https://www.aei.org/op-eds/parents-against-critical-race-theory/

Education Law Center. (2020, August). *Fact sheet: Resolving special education disagreements.* Education Law Center. https://www.elc-pa.org/wp-content/uploads/2019/08/Resolving-Special-Ed-Disagreements-Rev-Aug-2019.pdf

Elbow, P. (1973). *Writing without teachers.* Oxford University Press.

Elliott, D., & Spence, E. H. (2018). *Ethics for a digital era.* Wiley Blackwell.

Elmali, F, Tekin, A., & Polat, E. (2020). A study on digital citizenship: Preschool Teacher Candidates vs. Computer Education and instructional technology teacher candidates. *Turkish Online Journal of Distance Education, 21*(4)-15

Elmali, F., Tekin, A., & Polat, E. (2020). A study on digital citizenship: Preschool teacher candidates vs. computer education and instructional technology teacher candidates. *Turkish Online Journal of Distance Education, 21*(4), 251–269. doi:10.17718/tojde.803423

Emig, J. (1968). *The composing process of twelfth graders.* NCTE.

Ennis, R. H. (1993). Critical thinking assessment. *Theory into Practice, 32*(3), 179–186. doi:10.1080/00405849309543594

Eom, S. (2022). The emerging digital twin bureaucracy in the 21st century. *Perspectives on Public Management and Governance, 5*(2), 174–186. doi:10.1093/ppmgov/gvac005

Epstein, R. L., & Kernberger, C. (2012). *Critical thinking.* Advanced Reasoning Forum.

Erdei, T., Krakó, R., & Husi, G. (2022). Design of a digital twin training centre for an industrial robot arm. *Applied Sciences (Basel, Switzerland), 12*(17), 8862. doi:10.3390/app12178862

Erdem, C., & Koçyiğit, M. (2019). Exploring undergraduates' digital citizenship levels: Adaptation of the digital citizenship scale to Turkish. *Malaysian Online Journal of Educational Technology, 7*(3), 22–38. doi:10.17220/mojet.2019.03.003

Erdem, C., Oruç, E., Atar, C., & Bağcı, H. (2023). The mediating effect of digital literacy in the relationship between media literacy and digital citizenship. *Education and Information Technologies, 28*(5), 4875–4891. doi:10.100710639-022-11354-4

Eshet, Y. (2012). Thinking in the digital era: A revised model for digital literacy. *Issues in Informing Science and Information Technology, 9*, 267–276. doi:10.28945/1621

Ess, C. (2011). Ethical dimensions of new technology/media. In G. Cheney, S. May, & D. Munshi (Eds.), *The Handbook of Communication Ethics* (pp. 204–220). Routledge. doi:10.4324/9780203890400.ch13

Estellés-Arolas, E., & González-Ladrón-de-Guevara, F. (2012). Towards an integrated crowdsourcing definition. *Journal of Information Science, 38*(2), 189–200. doi:10.1177/0165551512437638

European Commission, Joint Research Centre, Vuorikari, R., Kluzer, S., & Punie, Y. (2022). *DigComp 2.2: The Digital Competence framework for citizens.* Publications Office of the European Union.

Facione, P. A. (1990). *Critical thinking: A statement of expert consensus for purposes of educational assessment and instruction - executive summary - the Delphi report.* The California Academic Press.

Faraboschi, P., Frachtenberg, E., Laplante, P., Milojicic, D., & Saracco, R. (2022). Virtual worlds (Metaverse): from skepticism, to fear, to immersive opportunities. *Computer, 55*(10), 100–106. doi:10.1109/MC.2022.3192702

Farivar, F., Esmaeelinezhad, O., & Richardson, J. (2022). Digital intrusions or distraction at work and work-Life conflict. *New Technology, Work and Employment, 37*(3), 363–380. doi:10.1111/ntwe.12235

Faul, F., Erdfelder, E., Lang, A. G., & Buchner, A. (2007). G* power 3: A flexible statistical power analysis program for the social, behavioral, and biomedical sciences. *Behavior Research Methods, 39*(2), 175–191. doi:10.3758/BF03193146 PMID:17695343

Fernandes, D. B. G. (2012). *O Crowdsourcing: Quando? Como? E porquê?* [Projeto de Mestrado em Gestão. ISCTE Business School. Instituto Universitário de Lisboa]. Repositório ISCTE.

Ferreira, M. I. R. (2018). *A Inovação Aberta em Portugal – O nível de recetividade das PME* [Dissertação de mestrado]. Católica Porto Business School.

Fetterhoff, T. J., & Voelkel, D. (2006). Managing open innovation in biotechnology. *Research Technology Management, 49*(3), 14–18. doi:10.1080/08956308.2006.11657373

Fields, D. A., Grimes, S. M., Magnifico, A., Lammers, J. C., Gomez, K., & Curwood, J. S. (2013). What's next in studying online social networking? Future research directions for creative, DIY-based sites. In *Proceedings of the Annual Conference 9.0 of Games + Learning + Society.* University of Wisconsin-Madison.

Fischman, G. E., & McLaren, P. (2005). Rethinking critical pedagogy and the Gramscian and Freirean legacies: From organic to committed intellectuals or critical pedagogy, commitment, and praxis. *Cultural Studies? Cultural Studies, Critical Methodologies, 5*(4), 425–446. doi:10.1177/1532708605279701

Fisher, K., Williamson, H., Guerra, N., & Kupferman, S. (2021). Digital citizenship: Technology access and use for youth with and without intellectual and developmental disabilities. *Inclusion (Washington, D.C.), 9*(4), 263–275. doi:10.1352/2326-6988-9.4.263

Flanigan, A. E., & Babchuk, W. A. (2022). Digital distraction in the classroom: Exploring instructor perceptions and reactions. *Teaching in Higher Education, 27*(3), 352–370. doi:10.1080/13562517.2020.1724937

Flayelle, M., Maurage, P., & Billieux, J. (2017). Toward a qualitative understanding of binge-watching behaviors: A focus group approach. *Journal of Behavioral Addictions, 6*(4), 457–471. doi:10.1556/2006.6.2017.060 PMID:29025269

Flew, T., & McElhinney, S. (2006). Globalization and the structure of new media industries. In L. Lievrouw & S. Livingstone (Eds.), *The Handbook of New Media* (pp. 287–306). Sage.

Floridi, L. (2001). Ethics in the infosphere. *The Philosophers'. Magazine, 6*, 18–19.

Fontcuberta, J. (2014). *A câmera de Pandora: A fotografí@ depois da fotografia* [translation: *Pandora's Camera: Photography After Photography*]. Gustavo Gili.

Forzani, E., & Leu, D. J. (2017). Multiple perspectives on literacy as it continuously changes: Reflections on opportunities and challenges when literacy is deictic. *Journal of Education, 197*(2), 19–24. doi:10.1177/002205741719700203

Franklin, A., Ho, A., & Ebdon, C. (2009). Participatory budgeting in midwestern states: democratic connection or citizen disconnection? *Public Budgeting & Finance, 29*(3), 52–73. doi:10.1111/j.1540-5850.2009.00936.x

Frau-Meigs, D., Brian, O. N., Soriani, A., & Tome, V. (2017). Digital citizenship education: Overview and new perspectives. Council of Europe Publishing.

Freeman, L. A., & Peace, A. G. (2005). *Information Ethics: Privacy and Intellectual Property.* Idea Group. doi:10.4018/978-1-59140-491-0

Freire, P. (2021). *Pedagogy of hope: Reliving pedagogy of the oppressed.* Bloomsbury Publishing. doi:10.5040/9781350190238

Freire, R. S., Falcão, F. S., & Rocha, A. C. B. (2022). Conceitos históricos e técnicas da inovação aberta – uma revisão sistemática. *DAT Journal*, *7*(1), 322–340. doi:10.29147/datjournal.v7i1.546

Friese, S. (2014). *Qualitative data analysis with ATLAS.ti*. Sage Publications.

Fritz, S., McCallum, I., Schill, C., Perger, C., Grillmayer, R., Achard, F., Kraxner, F., & Obersteiner, M. (2009). GeoWiki. Org: The use ofcrowdsourcing to improve global land cover. *Remote Sensing (Basel)*, *1*(3), 345–354. doi:10.3390/rs1030345

Gallagher, F. (2014). Media literacy education: A requirement for today's digital citizens. In D. B. S. Abreu & P. Mihailidis (Eds.), *Media literacy education in action: Theoretical and pedagogical perspectives*. Routledge.

Gallagher, J. R. (2020). A pedagogy of ethical interface production based on virtue ethics. In J. Reyman & E. Sparby (Eds.), *Digital ethics: Rhetoric and responsibility in online aggression* (pp. 69–84). Routledge.

Gao, Q., Yan, Z., Zhao, C., Pan, Y., & Mo, L. (2014). To ban or not to ban: Differences in mobile phone policies at elementary, middle, and high schools. *Computers in Human Behavior*, *38*, 25–32. doi:10.1016/j.chb.2014.05.011

Garcia, A. (2016, May 26). How remix culture informs student writing & creativity. *School Library Journal*. https://www.slj.com/story/how-remix-culture-informs-student-writing-creativity

Garcia, A. G., McGrew, S., Mirra, N., Tynes, B., & Kahne, J. (2021). Rethinking digital citizenship: Learning about media, literacy, and race in turbulent times. *Educating for civic reasoning and discourse*, 319-352.

Garcia, A., & Mirra, N. (2021). Writing toward justice: Youth speculative civic literacies in online policy discourse. *Urban Education*, *56*(4), 640–669. doi:10.1177/0042085920953881

Gasser, U., & Cortesi, S. (2017). Children's rights and digital technologies: Introduction to the discourse and some meta-observations. In *Handbook of Children's Rights: Global and Multidisciplinary Perspectives*. Taylor and Francis.

Gassmann, O., & Enkel, E. (2004). Towards a theory of open innovation: three core process archetypes. *R&D Management Conference*, 1–18.

Gee, J. (1996). *Social linguistics and literacies: Ideology in discourses* (2nd ed.). Taylor & Francis.

Giangrande, N., White, R. M., East, M., Jackson, R., Clarke, T., Saloff Coste, M., & Penha Lopes, G. (2019). A competency framework to assess and activate education for sustainable development: Addressing the UN sustainable development goals 4.7 challenge. *Sustainability (Basel)*, *11*(10), 2832. doi:10.3390/su11102832

Gibaldi, J. (1999). *MLA handbook for writers of research papers* (5th ed.). Modern Language Association of America.

Giroux, H. (1988). *Teachers as intellectuals: Toward a critical pedagogy of learning*. Greenwood Publishing Group.

Giroux, H. (1994). Doing cultural studies: Youth and the challenge of pedagogy. *Harvard Educational Review*, *64*(3), 278–309. doi:10.17763/haer.64.3.u27566k67qq70564

Gleason, B., & Von Gillern, S. (2018). Digital citizenship with social media. *Journal of Educational Technology & Society*, *21*(1), 200–212.

Goggin, G. (2016). Reimagining digital citizenship via disability. In A. McCosker, S. Vivienne, & A. Johns (Eds.), *Negotiating digital citizenship: Control, contest, culture* (pp. 61–80). Rowman and Littlefield.

Gonçalves, A. T., 2009. *Orçamento Participativo, Um Contributo para a Construção da Cidadania e da Democracia nos Níveis Locais*. Instituto da Segurança Social, IP.

Greenhow, C., Galvin, S. M., Brandon, D. L., & Askari, E. (2020). A decade of research on K–12 teaching and teacher learning with social media: Insights on the state of the field. *Teachers College Record*, *122*(6), 1–72. doi:10.1177/016146812012200602

Greenhow, C., Robelia, B., & Hughes, J. E. (2009). Learning, teaching, and scholarship in a digital age Web 2.0 and classroom research: What path should we take now? *Educational Researcher*, *38*(4), 246–259. doi:10.3102/0013189X09336671

Gui, M., Fasoli, M., & Carradore, R. (2017). Digital well-being. Developing a new theoretical tool for media literacy research. *Italian Journal of Sociology of Education*, *9*(1), 155–173.

Gulbahar, Y., & Guven, I. (2008). A survey on ICT usage and the perceptions of social studies teachers in Turkey. *Journal of Educational Technology & Society*, *11*(3), 37–51.

Gutek, G. L. (2022). *A history of the western educational experience*. Waveland Press.

Haase, A. (2022). *Teacher Perceptions and Implementation of Digital Citizenship in Elementary Classrooms: A Mixed Methods Study*. ProQuest Dissertations.

Hagel, J., & Brown, J. S. (2008). Creation Nets: Harnessing the Potential of Open Innovation. *Journal of Service Science*, *1*(2), 27–40. doi:10.19030/jss.v1i2.4293

Hagood, M. C., & Skinner, E. N. (2014). The vitality of embracing changes in literacies' research, education, and publication. *Journal of Adolescent & Adult Literacy*, *58*(1), 6–8. doi:10.1002/jaal.326

Haleem, A., Javaid, M., Qadri, M. A., & Suman, R. (2022). Understanding the role of digital technologies in education: A review. *Sustainable Operations and Computers*, *3*, 275–285. doi:10.1016/j.susoc.2022.05.004

Halstead, M., & Pike, M. (2006). *Citizenship and moral education: Values in action*. Routledge. doi:10.4324/9780203088593

Hämäläinen, E. K., Kiili, C., Marttunen, M., Räikkönen, E., González-Ibáñez, R., & Leppänen, P. H. (2020). Promoting sixth graders' credibility evaluation of Web pages: An intervention study. *Computers in Human Behavior*, *110*, 106372. doi:10.1016/j.chb.2020.106372

Hammon, L., & Hippner, H. (2012). Crowdsourcing. *Business & Information Systems Engineering*, *4*(3), 163–166. doi:10.100712599-012-0215-7

Han, B.-C. (2017). *In the swarm: Digital prospects*. Massachusetts Institute of Technology. doi:10.7551/mitpress/11055.001.0001

Hanson, J. (2011). Authors, authority, ownership, and ethics in digital media and news. In R. S. Fortner & P. M. Fackler (Eds.), *The Handbook of Global Communication and Media Ethics* (pp. 803–822). Wiley-Blackwell. doi:10.1002/9781444390629.ch41

Hanson, K., & Carlson, B. (2005). *Effective Access: Teachers' Use of Digital Resources in STEM Teaching*. Education Development Center, Inc.

Hansson, T. (2008). Communication and relation building in social systems. In Handbook of Research on Digital Information Technologies: Innovations, Methods and Ethical Issues, Information Science Reference (pp. 367-385). doi:10.4018/978-1-59904-970-0.ch024

Harris, B. N., McCarthy, P. C., Wright, A. M., Schutz, H., Boersma, K. S., Shepherd, S. L., Manning, L. A., Malisch, J. L., & Ellington, R. M. (2020). From panic to pedagogy: Using online active learning to promote inclusive instruction in ecology and evolutionary biology courses and beyond. *Ecology and Evolution*, *10*(22), 12581–12612. doi:10.1002/ece3.6915 PMID:33250996

Harvey, K. J., Brown, B., Crawford, P., Macfarlane, A., & McPherson, A. (2007). 'Am I normal?' Teenagers, sexual health and the internet. *Social Science & Medicine*, *65*(4), 771–781. doi:10.1016/j.socscimed.2007.04.005 PMID:17499898

Haskey-Valerius, B., Gould, H., & von Gillern, S. (2022). Preservice teachers' conceptualizations of media literacy, digital citizenship, and their relationship. In E. Langran (Ed.), *Proceedings of society for information technology & teacher education international conference* (pp. 1378–1383). Association for the Advancement of Computing in Education.

Hassani, H., Huang, X., & MacFeely, S. (2022). Impactful digital twin in the healthcare revolution. *Big Data and Cognitive Computing*, *6*(3), 83. doi:10.3390/bdcc6030083

Heater, D. (2013). *What is citizenship?* John Wiley & Sons.

Heath, S. B. (1983). *Ways with words: Language, life, and work in communities and classrooms*. Cambridge University Press.

Heberle, A. E., Rapa, L. J., & Farago, F. (2020). Critical consciousness in children and adolescents: A systematic review, critical assessment, and recommendations for future research. *Psychological Bulletin*, *146*(6), 525–551. doi:10.1037/bul0000230 PMID:32271028

Hermes, J. (2006). Citizenship in the age of the internet. *European Journal of Communication*, *21*(3), 295–309. doi:10.1177/0267323106066634

Hicks, T., & Turner, K. H. (2011). "That's not writing": Exploring the intersection of digital writing, community literacy, and social justice. [Doi: 10.1353/clj.2012.0000]. *Community Literacy Journal*, *6*(1), 55–78.

Hillocks, G. Jr. (1982). Inquiry and the composing process: Theory and research. *College English*, *44*(7), 659–673. doi:10.2307/376805

Hillocks, G. Jr. (1995). *Teaching writing as reflective practice*. TC Press.

Hine, C. (2015). *Ethnography for the Internet: Embedded, embodied, and everyday*. Bloomsbury Publishing Plc.

Hobbs, R. (2019). Meditation: Ethical practices. In Turner, K. H. (Ed.). The ethics of digital literacy: Developing knowledge and skills across grade levels. 155-158. Rowman & Littlefield.

Hobbs, R. (2010). *Copyright clarity: How fair use supports digital learning*. Corwin.

Hobbs, R. (2010). *Digital and media literacy: a plan of action*. The Aspen Institute.

Hobbs, R. (2017). Teaching and learning in a post-truth world. *Educational Leadership*, *75*(3), 26–31.

Hobbs, R. (2021). *Media literacy in action: Questioning the media*. Rowman & Littlefield Publishers.

Hobbs, R., Donnelly, K., Friesem, J., & Moen, M. (2013). Learning to engage: How positive attitudes about the news, media literacy, and video production contribute to adolescent civic engagement. *Educational Media International*, *50*(4), 231–246. doi:10.1080/09523987.2013.862364

Hobbs, R., & Jensen, A. (2009). The past, present and future of media literacy education. *The Journal of Media Literacy Education*, *1*(1), 1–11. doi:10.23860/jmle-1-1-1

Hollandsworth, R., Dowdy, L., & Donovan, J. (2011). Digital citizenship in K-12: It takes a village. *TechTrends*, *55*(4), 37–47. doi:10.100711528-011-0510-z

Honeycutt, N., Stevens, S. T., & Kaufmann, E. (2023). *The academic mind in 2022: What faculty think about free expression and academic freedom on campus*. The Foundation for Individual Rights and Expression.

Höttecke, D., & Allchin, D. (2020). Reconceptualizing nature-of-science education in the age of social media. *Science Education, 104*(4), 641–666. doi:10.1002ce.21575

Howe, J. (2006). *The Rise of Crowdsourcing.* Wired. https://www.wired.com/2006/06/crowds/

Hsiao, K. L., Ku, Y. Y., & Lee, Y. T. (2023). Online first). Development and validation of a scale for assessing university students' new online media and reporting literacy. *Education and Information Technologies*. Advance online publication. doi:10.100710639-023-11730-8

Huffman, S., Shaw, E., & Loyless, S. (2020). Ensuring ethics and equity: Policy, planning, and digital citizenship. *Education, 140*(2), 87–99.

Huizingh, E. K. R. E. (2011). Open innovation: State of the art and future perspectives. *Technovation, 31*(1), 2–9. doi:10.1016/j.technovation.2010.10.002

Hunt, M. G., Marx, R., Lipson, C., & Young, J. (2018). No more FOMO: Limiting social media ' decreases loneliness and depression. *Journal of Social and Clinical Psychology, 37*(10), 751–768. doi:10.1521/jscp.2018.37.10.751

Individuals with Disabilities Education Act (IDEA). 20 U.S.C. § 1400 (2004). International Society for Technology in Education. Digital Citizens in Education. https://www.iste.org/areas-of-focus/digital-citizenship

International Society for Technology in Education. (2018). *Citizenship in the digital age.* ISTE. https://elearninginfographics.com/citizenship-digital-age-infographic/

International Society for Technology in Education. (2019). *ISTE standards for educators.* Retrieved from https://www.iste.org/standards/for-educators

Işıklı, S. (2015). Digital citizenship: An actual contribution to theory of participatory democracy. *Online Academic Journal of Information Technology, 6*(18), 21–38. doi:10.5824/1309-1581.2015.1.002.x

Isin, E., & Ruppert, E. (2015). Being Digital Citizens. Rowman & Littlefield.

Isman, A., & Canan Gungoren, O. (2014). Digital citizenship. *Turkish Online Journal of Educational Technology-TOJET, 13*(1), 73–77.

ISTE (International Society for Technology in Education). (2023). *ISTE standards for students*. Retrieved from https://www.iste.org/standards/iste-standards-for-students

Ito, M., Antin, J., Finn, M., Law, A., Manion, A., Mitnick, S., Schlossberg, D., Yardi, S., & Horst, H. A. (2013). *Hanging out, messing around, and geeking out: Kids living and learning with new media*. MIT Press.

Ivanova, M., Marín, V. I., Tur, G., & Buchem, I. (2017, June). Towards privacy issues in personal learning environments. A conceptual model of ple privacy. In *EDEN Conference Proceedings* (No. 1, pp. 205-214). Academic Press.

Jackson, L. (2022). Civic identity, citizenship and foreign language education. In C. Lütge, T. Merse, & P. Rauschert (Eds.). Global citizenship in foreign language education: Concepts, practices, connection (pp. 41-61). Routledge. https://doi.org/10.4324/9781003183839

James, C., Weinstein, E., & Mendoza, K. (2019). *Teaching digital citizens in today's world: Research and insights behind the common sense K–12 digital citizenship curriculum.* Common Sense Media. https://www.commonsense.org/education/digital-citizenship/curriculum?campaignid

James, C., Davis, K., Flores, A., Francis, J. M., Pettingill, L., Rundle, M., & Gardner, H. (2009). *Young people, ethics, and the new digital media: A synthesis of the GoodPlay Project*. The John D. and Catherine T. MacArthur Foundation Reports on Digital Media and Learning. doi:10.7551/mitpress/8520.001.0001

James, N., & Busher, H. (2006). Credibility, authenticity and voice: Dilemmas in online interviewing. *Qualitative Research*, *6*(3), 403–420. doi:10.1177/1468794106065010

Jamil, S., Rahman, M., & Fawad, M. (2022). A comprehensive survey of digital twins and federated learning for industrial internet of things (IIoT), internet of vehicles (IoV) and internet of drones (IoD). *Applied System Innovation*, *5*(3), 56. doi:10.3390/asi5030056

Jemal, A. (2017). Critical consciousness: A critique and critical analysis of the literature. *The Urban Review*, *49*(4), 602–626. doi:10.100711256-017-0411-3 PMID:29657340

Jenkins, H. (2016). Youth voice, media, and political engagement. *By any media necessary: The new youth activism, 3*(1).

Jenkins, H., Clinton, K., Purushotma, R., Robison, A. J., & Weigel, M. (2006). *Confronting the challenges of participatory culture: Media education for the 21st century*. The MacArthur Foundation.

Jennings, L. B., Parra-Medina, D. M., Hilfinger-Messias, D. K., & McLoughlin, K. (2006). Toward a critical social theory of youth empowerment. *Journal of Community Practice*, *14*(1-2), 31–55. doi:10.1300/J125v14n01_03

Jin, J., Xu, H., & Leng, B. (2022). Adaptive points sampling for implicit field reconstruction of industrial digital twin. *Sensors (Basel)*, *22*(17), 6630. doi:10.339022176630 PMID:36081088

Jocius, R., O'Byrne, I., Blanton, M., Albert, J., Joshi, D., Robinson, R., & Andrews, A. (2021). Leveraging virtual professional development to build computational thinking literacies in English language arts classrooms. *Contemporary Issues in Technology & Teacher Education*, *21*(4), 626–654.

Johnson-Eilola, J. (2004). The database and the essay: Understanding composition as articulation. In *Wysocki, A. F., et al. (2004). Writing new media: Theory and applications for expanding the teaching of composition*. Utah State University Press.

Johnson, J. M. (2014). Scholarly innovation for difficult times. *Qualitative Inquiry*, *20*(9), 1119–1121. doi:10.1177/1077800414543695

Johnson, L., & Morris, P. (2010). Towards a framework for critical citizenship education. *Curriculum Journal*, *21*(1), 77–96. doi:10.1080/09585170903560444

Jones, L. M., & Mitchell, K. J. (2016). Defining and measuring youth digital citizenship. *New Media & Society*, *18*(9), 2063–2079. doi:10.1177/1461444815577797

Jungwirth, B. (2003). Information overload: Threat of opportunity? In B. C. Bruce (Ed.), *Literacy in the information age: Inquiries into meaning making with new technologies*. International Reading Association.

Junior, D. J. K., Ireland, R. D., & Snow, C. C. (2007). Strategic Entrepreneurship, Collaborative Innovation, and Wealth Creation. *Strategic Entrepreneurship Journal*, *01*(3-4), 371–385. doi:10.1002ej.20

Kahne, J. E., & Sporte, S. E. (2008). Developing citizens: The impact of civic learning opportunities on students' commitment to civic participation. *American Educational Research Journal*, *45*(3), 738–766. doi:10.3102/0002831208316951

Kahne, J., & Bowyer, B. (2017). Educating for democracy in a partisan age: Confronting the challenges of motivated reasoning and misinformation. *American Educational Research Journal*, *54*(1), 3–34. doi:10.3102/0002831216679817

Kahne, J., Lee, N. J., & Feezel, J. T. (2012). Digital media literacy education and online civic and political participation. *International Journal of Communication*, *6*, 1–14.

Karakaya, N. (2021). Teknoloji, ailenin rolü ve öğretmenin rolü-çocuk hakları. In Erken Çocuklukta Dijital Eğitim (pp. 197-212). Ankara: Anı.

Katıtaş, S., Seylim, E., & Demirkan, S. (2022). Medya etiği ve medya okuryazarlığı eğitimi üzerine bir inceleme. *Alanyazın*, *3*(1), 153–175. doi:10.22596/ cresjournal.0301.153.175

Kaun, A., & Uldam, J. (2018). Digital activism: After the hype. *New Media & Society*, *20*(6), 2099–2106. doi:10.1177/1461444817731924

Kavurucu, Ö., & Küçük, L. (2021). Çağın gelişen sorunu siber zorbalık: Çocuk ve ergenlerde siber zorbalığı önleme. *Medical Research Reports*, *4*(3), 50–56.

Kearney, S., & Maakrun, J. (2020). Let's get engaged: The nexus between digital technologies, engagement and learning. *Education Sciences*, *10*(12), 357. doi:10.3390/educsci10120357

Kellner, D., & Share, J. (2005). Toward critical media literacy: Core concepts, debates, organizations, and policy. *Discourse (Abingdon)*, *26*(3), 369–386. doi:10.1080/01596300500200169

Kellner, D., & Share, J. (2007). Critical media literacy, democracy, and the reconstruction of education. In D. Macedo & S. R. Steinberg (Eds.), *Media literacy: A reader* (pp. 3–23). Peter Lang Publishing.

Kenton, W. (2020). *Copyright infringement*. İnvestopedia. Retrieved from https://www.investopedia.com/terms/c/copyright.asp

Kerr, D. (2000). Citizenship education: An international comparison. In D. Lawton, J. Cairns, & R. Gardner (Eds.), Education for citizenship (pp. 200 – 227). Continuum.

Keser, H., & Yayla, H. G. (2021). Fatih Projesi uygulanan okullardaki öğretmenlerin bilgi güvenliği farkındalık düzeylerinin incelenmesi. *Milli Eğitim Dergisi*, *50*(229), 9–40.

Keshmiri Neghab, H., Jamshidi, M., & Keshmiri Neghab, H. (2022). Digital twin of a magnetic medical microrobot with stochastic model predictive controller boosted by machine learning in cyber-physical healthcare systems. *Information (Basel)*, *13*(7), 321. doi:10.3390/info13070321

Ketchen, D. J. Jr, Snow, C. C., & Street, V. L. (2004). Improving firm performance by matching strategic decision-making processes to competitive dynamics. *The Academy of Management Perspectives*, *18*(4), 29–43. doi:10.5465/ame.2004.15268671

Khalifa, M. A., Gooden, M. A., & Davis, J. E. (2016). Culturally responsive school leadership: A synthesis of the literature. *Review of Educational Research*, *86*(4), 1272–1311. doi:10.3102/0034654316630383

Kincheloe, J. L. (Ed.). (2008). *Knowledge and critical pedagogy: An introduction*. Springer. doi:10.1007/978-1-4020-8224-5

King, J., & South, J. (2017). *Reimagining the role of technology in higher education: A supplement to the national education technology plan*. US Department of Education, Office of Educational Technology.

Kirkgöz, Y. (2009). Globalization and English language policy in Turkey. *Educational Policy*, *23*(5), 663–684. doi:10.1177/0895904808316319

Kirlappos, I., & Sasse, M. A. (2011). Security education against phishing: A modest proposal for a major rethink. *IEEE Security and Privacy*, *10*(2), 24–32. doi:10.1109/MSP.2011.179

Klikauer, T. (2008). *Management communication: Communicative ethics and action*. Palgrave MacMillan. doi:10.1057/9780230583238

Knoester, M., & Parkison, P. (2015). Where is citizenship education in the age of common core state standards? *Critical Education, 6*(22).

Koç, M., & Barut, E. (2016). Development and validation of New Media Literacy Scale (NMLS) for university students. *Computers in Human Behavior*, *63*, 834–843. doi:10.1016/j.chb.2016.06.035

Kogut, B., & Metiu, A. (2001). Open-source software development and distributed innovation. *Oxford Review of Economic Policy*, *17*(2), 248–264. doi:10.1093/oxrep/17.2.248

Kozinets, R. V. (2010). *Netnography: Doing ethnographic research online*. SAGE Publications.

Kozyreva, A., Lewandowsky, S., & Hertwig, R. (2020). Citizens versus the internet: Confronting digital challenges with cognitive tools. *Psychological Science in the Public Interest*, *21*(3), 103–156. doi:10.1177/1529100620946707 PMID:33325331

Krutka, D. G., & Carpenter, J. P. (2017). Digital Citizenship in the Curriculum. *Educational Leadership*, *75*(3), 50–55.

Kunnath, J. P., & Jackson, A. (2019). Developing student critical consciousness: Twitter as a tool to apply critical literacy in the English classroom. *The Journal of Media Literacy Education*, *11*(1), 52–74. doi:10.23860/JMLE-2019-11-1-3

Kutsyuruba, B., Klinger, D. A., & Hussain, A. (2015). Relationships among school climate, school safety, and student achievement and well-being: A review of the literature. *Review of Education*, *3*(2), 103–135. doi:10.1002/rev3.3043

Langbert, M. (2018). Homogenous: The political affiliations of elite liberal arts college faculty. *Academic Questions*, *31*(2), 186–197. doi:10.100712129-018-9700-x

Lankshear, C., & Knobel, M. (2003). *New literacies: Changing knowledge and classroom culture*. Open University Press.

Lankshear, C., & Knobel, M. (2006). *New literacies: Everyday practices and classroom learning*. Open University Press.

Lauricella, A. R., Herdzina, J., & Robb, M. (2020). Early childhood educators' teaching of digital citizenship competencies. *Computers & Education*, *158*(103989), 1–9. doi:10.1016/j.compedu.2020.103989

Lawless, B., & Chen, Y. W. (2021). *Teaching social justice: Critical tools for the intercultural communication classroom*. Rowman & Littlefield.

Leander, K., & Vasudevan, L. (2011). Multimodality and mobile culture. In C. Jewitt (Ed.), *The Routledge handbook of multimodal analysis* (pp. 127 139). Routledge.

Leaning, M. (2017). *Media and information literacy: An integrated approach for the 21st century*. Elsevier Science & Technology.

Lee, A. Y., & Hancock, J. T. (2023). Developing Digital Resilience: An Educational Intervention Improves Elementary Students' Response to Digital Challenges. *Computers and Education Open*, 100144.

Lee, D. (2018). *Design thinking in the classroom: Easy-to-use teaching tools to foster creativity, encourage innovation and unleash potential in every student*. Simon and Schuster.

Lee, L., Chen, D. T., Li, J. Y., & Lin, T. B. (2015). Understanding new media literacy: The development of a measuring instrument. *Computers & Education*, *85*, 84–93. doi:10.1016/j.compedu.2015.02.006

Lee, S., Park, G., Yoon, B., & Park, J. (2010). Open innovation in SMEs—An intermediated network model. *Research Policy*, *39*(2), 290–300. doi:10.1016/j.respol.2009.12.009

Lenhart, A., Arafeh, S., & Smith, A. (2008). *Writing, technology and teens*. Pew Internet and American Life. https://www.pewresearch.org/internet/2008/04/24/writing-technology-and-teens/

Leu, D.J., Kinzer, C. K., Coiro, J. L., & Cammack, D. W. (2004). Toward a theory of new literacies emerging from the Internet and other information and communication technologies. *Theoretical Models and Processes of Reading, 5*(1), 1570-1613.

Leu, D. J. Jr, Kinzer, C. K., Coiro, J., & Cammack, D. W. (2004). Toward a theory of new literacies emerging from the Internet and other information and communication technologies. In R. B. Ruddell & N. Unrau (Eds.), *Theoretical models and processes of reading* (5th ed., pp. 1570–1613). International Reading Association.

Levine, P. (2007). *The future of democracy: Developing the next generation of American citizens*. University Press of New England.

Lewallen, T. C., Hunt, H., Potts-Datema, W., Zaza, S., & Giles, W. (2015). The whole school, whole community, whole child model: A new approach for improving educational attainment and healthy development for students. *The Journal of School Health*, *85*(11), 729–739. doi:10.1111/josh.12310 PMID:26440815

Li, Y., Deng, S., Wu, X., Zhao, B., Xie, Y., Luo, X., & Zheng, Y. (2023, July). Integrating Digital Citizenship into a Primary School Course "Ethics and the Rule of Law": Necessity, Strategies and a Pilot Study. In *International Conference on Blended Learning* (pp. 59-70). Springer Nature Switzerland. 10.1007/978-3-031-35731-2_7

Liao, C. H., & Wu, J. Y. (2022). Deploying multimodal learning analytics models to explore the impact of digital distraction and peer learning on student performance. *Computers & Education*, *190*, 104599. doi:10.1016/j.compedu.2022.104599

Li, M., & Yu, Z. (2022). Teachers' satisfaction, role, and digital literacy during the COVID-19 pandemic. *Sustainability (Basel)*, *14*(3), 1121. doi:10.3390u14031121

Liu, C., & Hwang, G. J. (2021). Roles and research trends of touchscreen mobile devices in early childhood education: Review of journal publications from 2010 to 2019 based on the technology-enhanced learning model. *Interactive Learning Environments*, 1–20. doi:10.1080/10494820.2020.1855210

Livingstone, S., & Sefton-Green, J. (2016) *The class: Living and learning in the digital age.* NYU Press. muse.jhu.edu/book/49200

Livingstone, S., Haddon, L., Görzig, A., & Ólafsson, K. (2011). *Risks and safety on the internet: The perspective of European children. Full findings.* London: EU Kids Online, LSE.

Livingstone, S., Mascheroni, G., & Staksrud, E. (2015). *Developing a framework for researching children's online risks and opportunities in Europe*. London: EU Kids Online, LSE.

Livingstone, S., & Helsper, E. (2008). Parental mediation and children's internet use. *Journal of Broadcasting & Electronic Media*, *52*(4), 581–599. doi:10.1080/08838150802437396

Livingstone, S., Mascheroni, G., & Stoilova, M. (2021). The outcomes of gaining digital skills for young people's lives and wellbeing: A systematic evidence review. *New Media & Society*. doi:10.1177/14614448211043189

Logic, J. (1998). Champing at the bits: Computers, copyright, and the composition classroom. *Computers and Composition*, *15*(2), 201–214. doi:10.1016/S8755-4615(98)90054-8

Louv, R. (2010). *Last child in the woods: Saving our children from nature-deficit disorder*. Algonquin Books.

Luke, A. (2018). Digital ethics now. *Language and Literature*, *20*(3), 185–198.

Lyon, D. (2017). Digital citizenship and surveillance| Surveillance culture: Engagement, exposure, and ethics in digital modernity. *International Journal of Communication*, *11*, 19.

Macfarlane, B. (2010). Values and virtues in qualitative research. In M. Savin-Baden & C. H. Major (Eds.), *New approaches to qualitative research: Wisdom and uncertainty* (pp. 19–27). Routledge.

MacIntyre, G., Cogan, N., Stewart, A., Quinn, N., O'Connell, M., & Rowe, M. (2022). Citizens defining citizenship: A model grounded in lived experience and its implications for research, policy and practice. *Health & Social Care in the Community*, *30*(3), 695–705. doi:10.1111/hsc.13440 PMID:34155710

Maddahi, Y., & Chen, S. (2022). Applications of digital twins in the healthcare industry: case review of an IoT-enabled remote technology in dentistry. *Virtual Worlds*, *1*(1), 20–41. doi:10.3390/virtualworlds1010003

Magalhães, L., Magalhães, L., Ramos, J., Moura, L., de Moraes, R., Gonçalves, J., Hisatugu, W. H., Souza, M. T., de Lacalle, L. N. L., & Ferreira, J. (2022). Conceiving a digital twin for a flexible manufacturing system. *Applied Sciences (Basel, Switzerland)*, *12*(19), 9864. doi:10.3390/app12199864

Mahmoud, A. H. (2021). *Cultivating Digital Citizenship By Creating An Effective Roadmap For New Generation Of Distance Learning.*

Maker Castro, E., Wray-Lake, L., & Cohen, A. K. (2022). Critical consciousness and wellbeing in adolescents and young adults: A systematic review. *Adolescent Research Review*, *7*(4), 499–522. doi:10.100740894-022-00188-3 PMID:35582551

Marchetti, A., & O'Dell, R. (2018). *Beyond literary analysis: Teaching students to write with passion and authority about any text*. Heinemann.

Markham, A. (2005). The methods, politics, and ethics of representation in online ethnography. In N. K. Denzin & Y. S. Lincoln (Eds.), *The SAGE Handbook of Qualitative Research* (3rd ed., pp. 793–820). SAGE Publications.

Markham, A. (2013). Remix cultures, remix methods: Reframing qualitative inquiry for social media contexts. In D. K. Denzin & M. D. Giardina (Eds.), *Global Dimensions of Qualitative Inquiry*. Left Coast Press, Inc.

Martin, F., Gezer, T., Wang, W. C., Petty, T., & Wang, C. (2020). Examining K-12 educator experiences from digital citizenship professional development. *Journal of Research on Technology in Education*, *54*(1), 143–160. doi:10.1080/15391523.2020.1815611

Masi, A., Mendoza Diaz, A., Tully, L., Azim, S. I., Woolfenden, S., Efron, D., & Eapen, V. (2021). Impact of the COVID-19 pandemic on the well-being of children with neurodevelopmental disabilities and their parents. *Journal of Paediatrics and Child Health*, *57*(5), 631–636. doi:10.1111/jpc.15285 PMID:33426739

Masiero, S., & Bailur, S. (2021). Digital identity for development: The quest for justice and a research agenda. *Information Technology for Development*, *27*(1), 1–12. doi:10.1080/02681102.2021.1859669

Mason, M. (2008). *Critical thinking and learning*. Blackwell Publishing. doi:10.1002/9781444306774

Mattson, K. (2017). *Digital citizenship in action empowering students to engage in online communities*. International Society for Technology in Education.

May, K. E., & Elder, A. D. (2018). Efficient, helpful, or distracting? A literature review of media multitasking in relation to academic performance. *International Journal of Educational Technology in Higher Education*, *15*(1), 1–17. doi:10.118641239-018-0096-z

McDaniel, J. W. (2020). Implementing Personal Devices in Math. Networks: *An Online Journal for Teacher Research, 22*(3).

McGillivray, D., McPherson, G., Jones, J., & McCandlish, A. (2016). Young people, digital media making and critical digital citizenship. *Leisure Studies*, *35*(6), 724–738. doi:10.1080/02614367.2015.1062041

McGrail, E., Turner, K. H., Piotrowski, A., Caprino, K., Zucker, L., & Greenwood, M. E. (2021). An interconnected framework for assessment of digital multimodal composition. *English Education*, *53*(4), 277–302.

McGrew, S. (2020). Learning to evaluate: An intervention in civic online reasoning. *Computers & Education*, *145*, 1–13. doi:10.1016/j.compedu.2019.103711

McGrew, S., Ortega, T., Breakstone, J., & Wineburg, S. (2017). The challenge that's bigger than fake news: Civic reasoning in a social media environment. *American Educator*, *41*(3), 4–39.

McLaren, P. (2017). Critical pedagogy: A look at the major concepts. In A. Darder, R. D. Torres, & M. P. Baltodano (Eds.), *The critical pedagogy reader* (3rd ed., pp. 61–83). Routledge.

McLuhan, M. (1994). *Understanding media: The extensions of man*. The MIT Press.

Mead, J. F., & Green, P. C. (2012). Special education and school choice: a special leadership challenge. In J. Crockett, B. Billingsley, & M. L. Boscardin (Eds.), *Handbook of leadership in special education* (pp. 115–128). Routledge.

MEB. (2023). *Milli Eğitim Bakanlığı FATİH Projesi*. Retrieves from http://fatihprojesi.meb.gov.tr

Melo, C. B. B., Nascimento, J. C. H. B., Melo, M. A., Bernardes, J. R., & Sousa, W. D. (2015). Crowdsourcing como uma ferramenta à inovação estratégica empresarial: Uma revisão de literatura. *Revista de Empreendedorismo. Inovação e Tecnologia*, *1*(1), 13–24. doi:10.18256/2359-3539/reit-imed.v1n1p13-24

Merchant, G. (2010). View my profiles. In D. E. Alvermann (Ed.), *Adolescents online literacies: Connecting classrooms, digital media, and popular culture* (pp. 51–69). Peter Lang.

Merriam, S. B. (2013). Nitel araştırma desen ve uygulama için bir rehber. Ankara: Nobel Akademik Yayıncılık.

Mertala, P. (2019). Teachers' beliefs about technology integration in early childhood education: A meta-ethnographical synthesis of qualitative research. *Computers in Human Behavior*, *101*, 334–349. doi:10.1016/j.chb.2019.08.003

Middleton, S. (2022). *Towards digital citizenship: a digital literacy curriculum to support teachers in the classroom*. Northumbria University.

Mihailidis, P., & Thevenin, B. (2013). Media literacy as a core competency for engaged citizenship in participatory democracy. *The American Behavioral Scientist*, *57*(11), 1611–1622. doi:10.1177/0002764213489015

Milenkova, V., & Lendzhova, V. (2021). *Digital Citizenship and Digital Literacy in the Conditions of Social Crisis. Computers, 10(4), 40*. MDPI AG., doi:10.3390/computers10040040

Miles, M. B., & Huberman, A. M. (1994). Qualitative data analysis: An expanded sourcebook. *Sage (Atlanta, Ga.)*.

Miller, A. (2000). *The crucible*. Penguin Classics.

Mills, K. A. (2010). A Review of the digital turn in the new literacy studies. *Review of Educational Research*, *80*(2), 246–271. doi:10.3102/0034654310364401

Miranda, H. P., & Russell, M. (2012). Understanding factors associated with teacher-directed student use of technology in elementary classrooms: A structural equation modeling approach. *British Journal of Educational Technology*, *43*(4), 652–666. doi:10.1111/j.1467-8535.2011.01228.x

Mirra, N. (2018). *Educating for empathy: Literacy learning and civic engagement*. Teachers College Press.

Mirra, N., Garcia, A., & Morrell, E. (2015). *Doing youth participatory action research: Transforming inquiry with researchers, educators, and students*. Routledge. doi:10.4324/9781315748047

Mitchell, W. J. T. (1992). *The reconfigured eye: Visual truth in the post-photographic era*. The MIT Press.

Mitchell, W. J. T. (1994). *Picture theory: Essays on verbal and visual representation*. Chicago University Press.

MLA. (2023a). How do I cite generative AI in MLA style? MLA Style Center. Accessed July 19, 2023. https://style.mla.org/citing-generative-ai/

MLA. (2023b). MLA Style [timeline]. MLA Style Center. Accessed July 19, 2023. https://www.mla.org/About-Us/About-the-MLA/MLA-Archives/Time-Lines/MLA-Style

MLA. (2023c) Submitting manuscripts to PMLA. Modern Language Association. Retrieved July 19, 2023, from https://www.mla.org/Publications/Journals/PMLA/Submitting-Manuscripts-to-PMLA

MLA-CCCC Joint Task Force. (2023). MLA-CCCC joint task force on writing and AI: Overview of the issues, statement of principles, and recommendations. Modern Language Association and Conference on College Communication and Communication. Accessed July 20, 2023. https://aiandwriting.hcommons.org/working-paper-1/

Moreira, P., Dantas, F., & Martin, C. (2016). Algumas contribuições de Milton Santos para a compreensão do plano diretor participativo das cidades. *Caminhos de Geografia*, *16*(55), 174-185

Morrow, R. A., & Torres, C. A. (2002). *Reading Freire and Habermas: Critical pedagogy and transformative social change*. Teachers College Press.

Mossberger, K., Tolbert, C., & McNeal, S. R. (2007). Digital citizenship: The Internet, society and participation. London: The MIT Press.

Mossberger, K., Tolbert, C. J., & Hamilton, A. (2012). Broadband adoption| measuring digital citizenship: Mobile access and broadband. *International Journal of Communication*, *6*(37), 2492–2528.

Mossberger, K., Tolbert, C. J., & McNeal, R. S. (2007). *Digital citizenship: The Internet, society, and participation*. MIT Press. doi:10.7551/mitpress/7428.001.0001

Mossberger, K., Tolbert, C., & McNeal, R. (2007). *The internet, society, and participation*. The MIT Press.

Moss, T., Short, J., Payne, T., & Lumpkin, G. (2011). Dual identities in social ventures: An exploratory study. *Entrepreneurship Theory and Practice*, *35*(4), 805–830. doi:10.1111/j.1540-6520.2010.00372.x

Mulyono, B., Affandi, I., Suryadi, K., & Darmawan, C. (2021). Digital citizenship competence: Initiating ethical guidelines and responsibilities for digital citizens. In *ICHELSS: International Conference on Humanities, Education, Law, and Social Sciences* (*Vol. 1*, No. 1, pp. 165-175). Academic Press.

Murray, D. M. (1968). *A writer teaches writing: A practical method of teaching composition*. Houghton Mifflin.

Mutz, D. C. (2002). Cross-cutting social networks: Testing democratic theory in practice. *The American Political Science Review*, *96*(1), 111–126. doi:10.1017/S0003055402004264

National Association for Media Literacy Education. (2023). Core principles of media literacy education [Fact sheet]. Accessed July 14, 2023. https://namle.net/resources/core-principles/

National Association for Media LIteracy Education. Student Television Network, Media Commission of the National Council of Teachers of English, Action Coalition for Media Education, & Visual Communication Division of the International Communication Association. (200802018). *Code of best practices in fair use for media literacy education*. Accessed July 20, 2023. https://ncte.org/statement/fairusemedialiteracy/

National Council of Teachers of English. (2005). *Multimodal literacies*. NCTE Position Statements. Retrieved July 19, 2023. https://ncte.org/statement/multimodalliteracies/

National Council of Teachers of English. (2019). *Definition of Literacy in a Digital Age*. NCTE Position Statements. Accessed July 19, 2023. https://ncte.org/statement/nctes-definition-literacy-digital-age/

National Governors Association Center for Best Practices & Council of Chief State School Officers. (2010). *Common Core State Standards*. Authors.

National Writing Project & Nagin. C. (2003). Because writing matters: Improving student writing in our schools. Jossey-Bass.

National Writing Project. Devoss, D. N., Eidman-Aadahl, E., & Hicks, T. (2010). Because digital writing matters: Improving student writing in online and multimedia environments. Jossey-Bass.

Neto, J. C. A. C., & Silva, M. E. R. (2013). Orçamento participativo: decisão de muitos, benefícios de todos. *SANARE – Revista De Políticas Públicas*, *2*(3). https://sanare.emnuvens.com.br/sanare/article/view/70

New London Group. (1996). A pedagogy of multiliteracies: Designing social futures. *Harvard Educational Review*, *66*(1), 60–92. doi:10.17763/haer.66.1.17370n67v22j160u

Nguyen, M. H., Gruber, J., Fuchs, J., Marler, W., Hunsaker, A., & Hargittai, E. (2020). <? covid19?> Changes in Digital Communication During the COVID-19 Global Pandemic: Implications for Digital Inequality and Future Research. *Social Media + Society*, *6*(3), 2056305120948255. doi:10.1177/2056305120948255 PMID:34192039

Nguyen, T. (2022). Toward human digital twins for cybersecurity simulations on the metaverse: ontological and network science approach. *JMIRx Med*, *3*(2), e33502. doi:10.2196/33502

Njoku, J., Nwakanma, C., Amaizu, G., & Kim, D. (2022). Prospects and challenges of Metaverse application in data-driven intelligent transportation systems. *IET Intelligent Transport Systems*. doi:10.1049/itr2.12252

Northwestern Medicine. (n.d.). *Technology overuse and addiction*. Northwestern Medicine. https://www.nm.org/conditions-and-care-areas/behavioral-health/technology-overuse-and-addiction

Nöth, W. (2011). Visual semiotics: Key features and an application to picture ads. In E. Margolis & L. Pauwels (Eds.), *The SAGE Handbook of Visual Research Methods* (pp. 298–316). Sage. doi:10.4135/9781446268278.n16

O'Brien, M. (2023, July 12). Sarah Silverman and novelists sue ChatGPT-maker OpenAI for ingesting their books. AP-News. Accessed July 20, 2023. https://apnews.com/article/sarah-silverman-suing-chatgpt-openai-ai-8927025139a8151e-26053249d1aeec20

O'Bryan, V. (2023, May 7). Using ChatGPT to get started with the writing process. NCTE. Accessed July 20, 2023. https://ncte.org/blog/2023/05/chatgpt-writing-process/

O'Byrne, W. I., Turner, K., Paciga, K. A., & Stevens, E. (2021). Co-constructing digital futures: Parents and children becoming thoughtful, connected, and critical users of digital technologies. *Algorithmic rights and protections for children*.

O'Byrne, W. I. (2018). Empowering students as critical readers and writers in online spaces. In *Best practices in teaching digital literacies* (pp. 233–250). Emerald Publishing Limited. doi:10.1108/S2048-045820180000009018

Obar, J. A., & Oeldorf-Hirsch, A. (2018). The clickwrap: A political economic mechanism for manufacturing consent on social media. *Social Media + Society*, *4*(3), 1–14. doi:10.1177/2056305118784770

OECD. (2021). *Protecting and empowering children in the digital world: OECD recommendations and good practices*. Retrieved from https://www.oecd.org/going-digital/protecting-and-empowering-children-in-the-digital-world.html

OECD. (n.d.) *Children in the digital environment.* Retrieved from https://www.oecd.org/digital/children-digital-environment/

OFS Health Care. (2022). *Kids' screen time: How much is too much?* OFS Health Care. https://www.osfhealthcare.org/blog/kids-screen-time-how-much-is-too-much/

Ohler, J. (2011). Digital citizenship means character education for the digital age. *Kappa Delta Pi Record, 47*(sup1), 25-27.

Ohler, J. (2010). *Digital community, digital citizen.* Corwin Press. doi:10.4135/9781452219448

Ong, W. J. (2002). *Orality and literacy: The technologizing of the word.* Routledge.

Orhan, A. (2023a). Fake news detection on social media: The predictive role of university students' critical thinking dispositions and new media literacy. *Smart Learning Environments, 10*(29), 1–14. doi:10.118640561-023-00248-8

Orhan, A. (2023b). Investigating psychometric properties of the Turkish version of Sosu Critical Thinking Disposition Scale: Evidence from two independent samples. *International Journal of Psychology and Educational Studies, 10*(2), 348–359. doi:10.52380/ijpes.2023.10.2.1017

Orlowski, S. K., Lawn, S., Venning, A., Winsall, M., Jones, G. M., Wyld, K., Damarell, R. A., Antezana, G., Schrader, G., Smith, D., Collin, P., & Bidargaddi, N. (2015). Participatory research as one piece of the puzzle: A systematic review of consumer involvement in design of technology-based youth mental health and well-being interventions. *JMIR Human Factors, 2*(2), e4361. doi:10.2196/humanfactors.4361 PMID:27025279

Ortiz, K., Rice, M., Curry, T., Mellard, D., & Kennedy, K. (2021). Parent perceptions of virtual school support for their children with disabilities. *American Journal of Distance Education, 35*(4), 276–292. doi:10.1080/08923647.2021.1979343

Osamuyimen, S., Juan, M. H., & Melissa, S. (2009). Designing crowdsourcing community for the enterprise. *Proceedings of the ACM SIGKDD,* 50-53. 10.1145/1600150.1600168

Oudeweetering, K., & Voogt, J. (2018). Teachers' conceptualization and enactment of twenty-first century competences: Exploring dimensions for new curricula. *Curriculum Journal, 29*(1), 116–133. doi:10.1080/09585176.2017.1369136

Oxley, C. (2010). *Digital citizenship: Developing an ethical and responsible online culture.* International Association of School Librarianship.

Oxley, C. (2011). Digital citizenship: Developing an ethical and responsible online culture. *Access, 25*(3), 5–9.

Öztürk, G. (2021). Digital citizenship and its teaching: A literature review. *Journal of Educational Technology and Online Learning, 4*(1), 31–45.

Page, R. E. (2013). *Familiar, reconfigured, and emergent dimensions of narrative. In Stories and Social Media: Identities and Interaction.* Routledge.

Palaiologou, I. (2016). Children under five and digital technologies: Implications for early years pedagogy. *European Early Childhood Education Research Journal, 24*(1), 5–24. doi:10.1080/1350293X.2014.929876

Palmeri, J. (2012). *Remixing composition: A history of multimodal writing pedagogy.* Southern Illinois University Press.

Pamucar, D., Deveci, M., Gokasar, I., Tavana, M., & Köppen, M. (2022). A metaverse assessment model for sustainable transportation using ordinal priority approach and Aczel-Alsina norms. *Technological Forecasting and Social Change, 182*, 121778. doi:10.1016/j.techfore.2022.121778

Pangrazio, L. (2016). Reconceptualizing critical digital literacy. *Discourse (Abingdon), 37*(2), 163–174. doi:10.1080/01596306.2014.942836

Pangrazio, L., & Sefton-Green, J. (2021). Digital rights, digital citizenship and digital literacy: What's the difference? NAER. *Journal of New Approaches in Educational Research*, *10*(1), 15–27. doi:10.7821/naer.2021.1.616

Parkison, P. (2009). Political economy of NCLB: Standards, testing and test scores. *The Educational Forum*, *73*(1), 44–57. doi:10.1080/00131720802539606

Park, S., & Kim, Y. (2022). A metaverse: taxonomy, components, applications, and open challenges. *IEEE Access: Practical Innovations, Open Solutions*, *10*, 4209–4251. doi:10.1109/ACCESS.2021.3140175

Park, Y. M., Chae, J. H., Kim, S. K., & Kwon, H. S. (2021). The effects of media literacy education and its influence on digital citizenship: Focusing on CMF education programs in Korea. *Journal of Information Systems*, *30*(3), 113–135.

Passey, D., Shonfeld, M., Appleby, L., Judge, M., Saito, T., & Smits, A. (2018). Digital agency: Empowering equity in and through education. *Technology. Knowledge and Learning*, *23*(3), 425–439. doi:10.100710758-018-9384-x

Patton, M. Q. (2005). *Qualitative Research*. John Wiley & Sons, Ltd.

Paul, R. (1990). *Critical thinking: What every person needs to survive in a rapidly changing world.* Center for Critical Thinking and Moral Critique.

Paulus, T. M., Lester, J. N., & Dempster, P. G. (2014). *Digital tools for qualitative research.* SAGE Publications. doi:10.4135/9781473957671

Pavlik, J. V. (2001). *Journalism and new media*. Columbia University Press. doi:10.7312/pavl11482

Pedersen, J., Kocsis, D., Tripathi, S., Tarrell, A., Weerakoon, A., Tahmasbi, N., Xiong, J., & Deng, W. (2013). Conceptual foundations of crowdsourcing: a review of is research. *HICCS-46 Hawaii International Conference on System Sciences.* 10.1109/HICSS.2013.143

Pendharker, E. (2023). *Poll: Parents don't want schools to focus on culture wars.* Ed Week. https://www.edweek.org/leadership/poll-parents-dont-want-schools-to-focus-on-culture-wars/2023/01

Pereira, S. (2013). More technology, better childhoods? The case of the Portuguese 'one laptop per child' programme. *Communication Management Quarterly*, *29*(29), 171–198. doi:10.5937/comman1329171P

Pérez-Juárez, M. Á., González-Ortega, D., & Aguiar-Pérez, J. M. (2023). Digital Distractions from the Point of View of Higher Education Students. *Sustainability (Basel)*, *15*(7), 6044. doi:10.3390u15076044

Pinheiro, L. (2009). *Open Innovation e a redescoberta da importância das relações Universidade-Empresa: o caso paradigmático da Brisa* [Dissertação de mestrado. Faculdade de Economia da Universidade do Porto]. Repositório Aberto da Universidade do Porto.

Pinkard, N., Erete, S., Martin, C. K., & McKinney de Royston, M. (2017). Digital youth divas: Exploring narrative-driven curriculum to spark middle school girls' interest in computational activities. *Journal of the Learning Sciences*, *26*(3), 477–516. doi:10.1080/10508406.2017.1307199

Popescu, D., Dragomir, M., Popescu, S., & Dragomir, D. (2022). Building better digital twins for production systems by incorporating environmental related functions—literature analysis and determining alternatives. *Applied Sciences (Basel, Switzerland)*, *12*(17), 8657. doi:10.3390/app12178657

Prasetiyo, W. H., Naidu, N. B. M., Tan, B. P., & Sumardjoko, B. (2021). Digital Citizenship Trend in Educational Sphere: A Systematic Review. *International Journal of Evaluation and Research in Education*, *10*(4), 1192–1201. doi:10.11591/ijere.v10i4.21767

Prasetiyo, W. H., Sumardjoko, B., Muhibbin, A., Naidu, N. B. M., & Muthali'in, A. (2023). Promoting Digital Citizenship among Student-Teachers: The Role of Project-Based Learning in Improving Appropriate Online Behaviors. [PER]. *Participatory Educational Research*, *10*(1), 389–407. doi:10.17275/per.23.21.10.1

Prensky, M. (2001). Digital natives, digital immigrants. Do They Really Think Differently? *On the Horizon*, *9*(5), 1–6. doi:10.1108/10748120110424816

Puolimatka, T. (1995). *Democracy and education: The critical citizen as an educational aim*. Suomalaisen Tiedeakatemia.

Pusey, P., & Sadera, W. A. (2011). Cyberethics, cybersafety, and cybersecurity: Preservice teacher knowledge, preparedness, and the need for teacher education to make a difference. *Journal of Digital Learning in Teacher Education*, *28*(2), 82–85. doi:10.1080/21532974.2011.10784684

Pyles, D. G., Rish, R. M., & Warner, J. (Eds.). (2019). *Negotiating place and space through digital literacies: Research and practice*. IAP.

Ravitch, D. (2016). *The death and life of the great American school system: How testing and choice are undermining education*. Basic books.

Redecker, C. (2017). European framework for the digital competence of educators: DigCompEdu. In Y. Punie (Ed.), *EUR 28775 EN*. Publications Office of the European Union.

Redmond, T. (2015). Media literacy is common sense: Bridging Common Core Standards with the media experiences of digital learners: Findings from a case study highlight the benefits of an integrated model of literacy, thereby illustrating the relevance and accessibility of media literacy education. *Middle School Journal*, *46*(3), 10–17. doi:10.1080/00940771.2015.11461910

Reeves, J., & Crowther, T. (2019). Teacher feedback on the use of innovative social media simulations to enhance critical thinking in young people on radicalisation, extremism, sexual exploitation and grooming. *Pastoral Care in Education*, *37*(4), 280–296. doi:10.1080/02643944.2019.1618377

Rehabilitation Act of 1973. Section 504 Regulations, 34 C.F.R. § 104.1 et seq.

Reingold, R., & Baratz, L. (2011). An institutional code of ethics—A response to attitude of Israeli teachers' education college students towards academic plagiarism. *Online Submission*, *8*(5), 589–598.

Resnick, M. (2017). *Lifelong kindergarten: Cultivating creativity through projects, passion, peers, and play*. MIT press. doi:10.7551/mitpress/11017.001.0001

Rhodes, J., & Alexander, J. (2014). Reimagining the social turn: New work from the field. *College English*, *76*(6), 481–487.

Ribble, M. (2015). *Digital citizenship in schools: Nine elements all students should know*. International Society for Technology in Education.

Ribble, M. (2017). *Nine elements: Nine themes of digital citizenship*. Retrieved from https://www.digitalcitizenship.net/nine-elements.html

Ribble, M. (2017). *Nine Themes of Digital Citizenship*. Digital Citizenship. https://www.digitalcitizenship.net/nine-elements.html

Ribble, M., & Bailey, G. (2007). Digital citizenship in schools. Washington, DC: ISTE.

Ribble, M., & Bailey, G. (2011). *Digital Citizenship in Schools*. International Society for Technology in Education (ISTE).

Ribble, M. (2008). Passport to digital citizenship. *Learning and Leading with Technology*, *36*(4), 14–17.

Ribble, M. (2008). Passport to Digital Citizenship: Journey toward Appropriate Technology Use at School and Home. *Learning and Leading with Technology*, *36*, 14–17.

Ribble, M. (2012). Digital citizenship for educational change. *Kappa Delta Pi Record*, *48*(4), 148–151. doi:10.1080/00228958.2012.734015

Ribble, M. (2015). *Digital citizenship in schools: Nine elements all students should know*. International Society for Technology in Education.

Ribble, M. S., & Bailey, G. D. (2004). Digital citizenship focus questions for implementation. *Learning and Leading with Technology*, *32*(2), 12–15.

Ribble, M., & Bailey, G. (2007). *Digital Citizenship in Schools*. ISTE.

Ribble, M., & Bailey, G. D. (2011). *Digital citizenship in schools*. International Society for Technology in Education.

Ribble, M., Bailey, G. D., & Ross, T. (2004). Digital citizenship: addressing appropriate technology behavior. Learning and Leading with Technology, 32(1). 6-9. https://eric.ed.gov/?id=EJ695788

Ribble, M., Bailey, G. D., & Ross, T. W. (2004). Digital citizenship: Addressing appropriate technology behavior. *Learning and Leading with Technology*, *32*(1), 6–9.

Ribble, M., & Miller, T. (2013). Educational leadership in an online world: Connecting students to technology responsibly, safely, and ethically. *Online Learning : the Official Journal of the Online Learning Consortium*, *17*(1), 137–145. doi:10.24059/olj.v17i1.310

Ribble, M., & Park, M. (2019). *The digital citizenship handbook for school leaders: Fostering positive interactions online*. International Society for Technology in Education.

Rice, M. (2021). Reconceptualizing teacher professional learning about technology integration as intra-active entanglements. *Professional Development in Education*, *47*(3), 524–537. doi:10.1080/19415257.2021.1891953

Rice, M. (2022). Special education teachers' use of technologies during the COVID-19 era (Spring 2020—Fall 2021). *TechTrends*, *66*(2), 310–326. doi:10.100711528-022-00700-5 PMID:35098256

Rice, M. (in press). Using Narrative Inquiry research methodology in online and distance education environments. *Educational Technology Research and Development*.

Rice, M., & Bailon, M. (2023). A qualitative analysis of publicly available standards and guidance about digital literacies in U.S. states. *Education and Information Technologies*, *28*(6), 6927–6948. doi:10.100710639-022-11482-x PMID:36465420

Rice, M., & Deschaine, M. (2020). Orienting towards teacher education in online environments for all students. *The Educational Forum*, *84*(2), 114–125. doi:10.1080/00131725.2020.1702747

Rice, M., & Ortiz, K. (2022). Reframing parental involvement in online learning as shared work. *Journal of Research on Technology in Education*. doi:10.1080/15391523.2022.2030269

Richardson, J., & Milovidov, E. (2019). *Digital citizenship education handbook: Being online, well-being online, and rights online*. Council of Europe.

Richards, R. (2010). Digital citizenship and web 2.0 tools. *Journal of Online Learning and Teaching*, *6*(2), 516–522.

RobbGrieco. M., & Hobbs, R. (2013). A field guide to media literacy education in the United States. Kingston, RI: Media Education Lab, University of Rhode Island.

Rodríguez-Triana, M. J., Martínez-Monés, A., & Villagrá-Sobrino, S. (2016). Learning analytics in small-scale teacher-led innovations: Ethical and data privacy issues. *Journal of Learning Analytics*, *3*(1), 43–65. doi:10.18608/jla.2016.31.4

Rogerson, S. (2021). *Rebooting ethics education in the digital age*. Retrieved from https://dora.dmu.ac.uk/bitstream/handle/2086/20613/Rebooting%20ethics%20education%20in%20the%20digital%20age%20prepublication%20version.pdf?sequence=2

Rogers, R. (2016). The harm of video games: The ethics behind regulating minors' access to violent video games in light of the supreme court ruling. In A. Davisson & P. Booth (Eds.), *Controversies in digital ethics* (pp. 216–230). Bloomsbury.

Rozgonjuk, D., Täht, K., & Vassil, K. (2021). Internet Use at and outside of School in Relation to Low- and High-Stakes Mathematics Test Scores across 3 Years. *International Journal of STEM Education*, *8*(1), 8. doi:10.118640594-021-00287-y

Ruggerio, V. R. (1988). *Teaching thinking across the curriculum*. Harper & Row.

Ruggerio, V. R. (1990). *Beyond feelings: A guide to critical thinking*. Mayfield Publishing Company.

Ruiz Mejia, J., & Rawat, D. (2022). Recent advances in a medical domain metaverse: status, challenges, and perspective. Paper presented at the *2022 Thirteenth International Conference on Ubiquitous and Future Networks (ICUFN)*. IEEE. 10.1109/ICUFN55119.2022.9829645

Sadaf, A., Martin, F., & Ahlgrim-Delzell, L. (2019). Student perceptions of the impact of quality matters—Certified online courses on their learning and engagement. *Online Learning : the Official Journal of the Online Learning Consortium*, *23*(4), 214–233. doi:10.24059/olj.v23i4.2009

Sakorn, W. (2022). Assessing the digital citizenship self-efficacy of pre-service teachers. *Journal of Educational Issues.*, *8*(2), 513. doi:10.5296/jei.v8i2.20246

Sak, R., Erden, F. T., Sak, İ. T. Ş., & Esmeray, H. (2016). Early childhood teachers and computers: Beliefs and self-reported practices. *Journal of Education and Future*, (10), 19–33.

Salem, T., & Dragomir, M. (2022). Options for and challenges of employing digital twins in construction management. *Applied Sciences (Basel, Switzerland)*, *12*(6), 2928. doi:10.3390/app12062928

Sampaio, R. (2016). e-Participatory Budgeting as an initiative of e-requests: Prospecting for leading cases and reflections on e-Participation. *Revista de Administração Pública*, *50*(6), 937–958. doi:10.1590/0034-7612152210

Sartori, G. (2011). *Homo videns: Televisione e post-pensiero*. Editori Laterza.

Savin-Baden, M., & Major, C. H. (Eds.). (2010). *New approaches to qualitative research: Wisdom and uncertainty*. Routledge. doi:10.4324/9780203849873

Schellekens, E. (2007). Aesthetics & morality. *Continuum*.

Schumpeter, J. (1934). *The theory of economic development: An inquiry into profits, capital, credit, interest, and the business cycle*. Transaction Publishers.

Schwartz, D. (2003). Professional oversight: Policing the credibility of photojournalism. In L. Gross, J. S. Katz, & J. Ruby (Eds.), *Image ethics in the digital age* (pp. 27–51). University of Minnesota Press.

Schwartz, M. (2023). *Adaptive ethics for digital transformation: A new approach for enterprise leadrs*. IT Revolution Press.

Schwartz, S. E., Rhodes, J. E., Liang, B., Sánchez, B., Spencer, R., Kremer, S., & Kanchewa, S. (2014). Mentoring in the digital age: Social media use in adult–youth relationships. *Children and Youth Services Review*, *47*, 205–213. doi:10.1016/j.childyouth.2014.09.004

Scott, L. (2017). Use of technology for children with Autism: Academic success in the Least Restrictive Environment. *Journal of Educational Leadership in Action*, *5*(1), 5. https://digitalcommons.lindenwood.edu/cgi/viewcontent.cgi?article=1051&context=ela

Seale, J., & Schoenberger, N. (2018). Be internet awesome: A critical analysis of google's child-focused internet safety program. *Emerging Library & Information Perspectives*, *1*, 34–58. doi:10.5206/elip.v1i1.366

Seargeant, P., & Tagg, C. (Eds.). (2014). *The language of social media: Identity and community on the internet*. Palgrave Macmillian. doi:10.1057/9781137029317

Selfe, C. (2004). Students who teach us. In *Wysocki, A. F., et al. (2004). Writing new media: Theory and applications for expanding the teaching of composition*. Utah State University Press.

Serdyukov, P. (2017). Innovation in education: what works, what doesn't, and what to do about it? *Journal of research in innovative teaching & learning, 10*(1), 4-33.

Sharp, R., Green, A., & Lewis, J. (2017). *Education and social control: A study in progressive primary education* (Vol. 49). Routledge. doi:10.4324/9781315210179

Shellgren, M., & Gunder, A. (2022). *Advancing universal access to quality digital learning through global coalitions and narrative practices*. Online Learning Consortium.

Shen, B., Tan, W., Guo, J., Zhao, L., & Qin, P. (2021). How to promote user purchase in metaverse? a systematic literature review on consumer behavior research and virtual commerce application design. *Applied Sciences (Basel, Switzerland)*, *11*(23), 11087. doi:10.3390/app112311087

Silva, M. G. V. S. (2017). *Implementação do orçamento participativo nos municípios portugueses* [Tese de Mestrado. Universidade do Minho]. Repositorium.

Silver, C., & Lewins, A. (2014). *Using software in qualitative research: A step by step guide*. Sage Publications. doi:10.4135/9781473906907

Simsek, E., & Simsek, A. (2013). New literacies for digital citizenship. *Contemporary Educational Technology*, *4*(2), 126–137. doi:10.30935/cedtech/6097

Sintomer, Y., Herzberg, C., & Röcke, A. (2008). Participatory budgeting in europe: potentials and challenges. *International Journal of Urban and Regional Research*, *32*(1), 164–178. doi:10.1111/j.1468-2427.2008.00777.x

Sintomer, Y., Herzberg, C., Röcke, A., & Allegretti, G. (2012). Transnational models of citizen participation: the case of participatory budgeting. *Journal of Public Deliberation*, *8*(2), 9. doi:10.16997/jdd.141

Siyaev, A., & Jo, G. (2021). Towards Aircraft Maintenance Metaverse Using Speech Interactions with Virtual Objects in Mixed Reality. *Sensors (Basel)*, *21*(6), 2066. doi:10.339021062066 PMID:33804253

Skulmowski, A., & Xu, K. M. (2021). Understanding cognitive load in digital and online learning: A new perspective on extraneous cognitive load. *Educational Psychology Review*, 1–26.

Smith, E. E., & Storrs, H. (2023). Digital literacies, social media, and undergraduate learning: What do students think they need to know? *International Journal of Educational Technology in Higher Education*, *20*(1), 1–33. doi:10.118641239-023-00398-2 PMID:37220550

Sosu, E. M. (2013). The development and psychometric validation of a Critical Thinking Disposition Scale. *Thinking Skills and Creativity*, *9*, 107–119. doi:10.1016/j.tsc.2012.09.002

Souers, K., & Hall, P. (2016). *Fostering resilient learners: Strategies for creating a trauma-sensitive classroom*. ASCD.

Souza, C. (2001). Construção e consolidação de instituições democráticas: Papel do orçamento participativo. *São Paulo em Perspectiva, 15*(4), 84–97. doi:10.1590/S0102-88392001000400010

Spinoza, B. (2009). *Ethics*. The Floating Press.

Spithoven, A., Clarysse, B., & Knockaert, M. (2010). Building absorptive capacity to organise inbound open innovation in traditional industries. *Technovation, 30*(2), 130–141. doi:10.1016/j.technovation.2009.08.004

Stiegler, B. (1998). *Technics and time 1: The Fault of Epimetheu*. Stanford University Press. doi:10.1515/9781503616738

Stiegler, B. (2009). *Technics and time 2: Disorientation*. Stanford University Press.

Street, B. (1997). The implications of the 'new literacy studies' for literacy education. *English in Education, 31*(3), 45–59. doi:10.1111/j.1754-8845.1997.tb00133.x

Stubbs, C. F., & Hullinger, H. (2022). Developing Critical Consciousness in Students for Cultural Transformation. *Journal of the Scholarship of Teaching and Learning for Christians in Higher Education, 12*(1), 5. doi:10.31380otlched.12.1.9

Sturgeon, J. (2008, Nov.). Raising good digital citizens. *Cable in the Classroom,* 5-9.

Suppo, C. (2018). *Digital citizenship instruction in Pennsylvania public schools: School leaders expressed beliefs and current practices*. ProQuest Dissertations.

Swenson, S., & Ryder, R. E. (2016). *Dear Colleague: Virtual Schools*. Washington DC: Department of Education Office of Special Education and Rehabilitative Services. https://sites.ed.gov/idea/files/dcl-virtual-schools-08-05-2016.pdf

Switaj, E. (2015). Web Writing and Citation: The Authority of Communities. In J. Dougherty & T. O'Donnell (Eds.), *Web Writing: Why and How for Liberal Arts Teaching and Learning* (pp. 223–232). University of Michigan Press., doi:10.2307/j.ctv65sxgk.24

Taber, K. (2018). The use of Cronbach's Alpha when developing and reporting research instruments in science education. *Research in Science Education, 48*(6), 1273–1296. doi:10.100711165-016-9602-2

Ten Dam, G., & Volman, M. (2004). Critical thinking as a citizenship competence: Teaching strategies. *Learning and Instruction, 14*(4), 359–379. doi:10.1016/j.learninstruc.2004.01.005

The Neglected "R": The Need for Writing Revolution. (2003). College Entrance Examination Board. https://archive.nwp.org/cs/public/download/nwp_file/21478/the-neglected-r-college-board-nwp-report.pdf?x-r=pcfile_d

Thoman, E., & Jolls, T. (2004). Media literacy: A national priority for a changing world. *The American Behavioral Scientist, 48*(1), 18–29. doi:10.1177/0002764204267246

Thomas, S. (2016). *Future Ready Learning: Reimagining the Role of Technology in Education. 2016 National Education Technology Plan*. Office of Educational Technology, US Department of Education.

Thorson, K. (2012). What does it mean to be a good citizen? Citizenship vocabularies as resources for action. *The Annals of the American Academy of Political and Social Science, 644*(1), 70–85. doi:10.1177/0002716212453264

Toffler, A. (1981). *The third wave*. Morrow.

Tomczyk, Ł. (2019). What do teachers know about digital safety? *Computers in the Schools, 36*(3), 167–187. doi:10.1080/07380569.2019.1642728

Tromp, J. (2022). Extended Reality & The Backbone: Towards a 3D Mirrorworld. In Roadmapping Extended Reality (pp. 193-227).

Türel, Y. K., & Dokumaci, Ö. (2022). Use of media and technology, academic procrastination, and academic achievement in adolescence. *Participatory Educational Research*, *9*(2), 481–497. doi:10.17275/per.22.50.9.2

Turner, K. H. (2013). Error or strength? Competencies developed in adolescent digitalk. In Pytash, K. E. & Ferdig, R. E., Eds. *Exploring Technology for Writing and Writing Instruction*, 114-134. IGI Global.

Turner, K. H. (2019). Where did I find that?: Helping students develop ethical practices in digital writing. In Turner, K. H. (Ed.). The ethics of digital literacy: Developing knowledge and skills across grade levels. 159-164. Rowman & Littlefield.

Turner, K. H. (Ed.). (2019). *The ethics of digital literacy: Developing knowledge and skills across grade levels*. Rowman & Littlefield.

Turner, K. H., Abrams, S., Katic, E., & Donovan, M. J. (2014). Digitalk: The what and the why of adolescent digital language. *Journal of Literacy Research*, *46*(2), 157–193. Doi.org/10.1177/1086296X14534061

Turner, K. H., & Hicks, T. (2017). *Argument in the real world: teaching adolescents to read and write digital texts*. Heinemann.

Turner, K. H., Hicks, T., & Zucker, L. (2020). Connected reading: A framework for understanding how adolescents encounter, evaluate, and engage with texts in the digital age. *Reading Research Quarterly*, *55*(2), 291–309. doi:10.1002/rrq.271

U.S. Department of Education. (2017). *Reimagining the role of technology in education: 2017 national educational technology plan update*. U.S. Department of Education.

Ukko, J., Saunila, M., Nasiri, M., Rantala, T., & Holopainen, M. (2022). Digital twins' impact on organizational control: Perspectives on formal vs social control. *Information Technology & People*, *35*(8), 253–272. doi:10.1108/ITP-09-2020-0608

Umble, E. J., Haft, R. R., & Umble, M. M. (2003). Enterprise resource planning: Implementation procedurs and critical success factors. *European Journal of Operational Research*, *146*(2), 241–257. doi:10.1016/S0377-2217(02)00547-7

UNESCO. (2017). *Outcome document of the Conference on Digital Citizenship Education in Asia-Pacific*. UNESCO.

UNESCO. (2018). *Skills for a connected world Programme*. UNESCO.

UNESCO. (2019). *Digital kids Asia-Pacific: insights into children's digital citizenship*. UNESCO. https://unesdoc.unesco.org/ark:/48223/pf0000367985

Unsworth, N., & McMillan, B. D. (2017). Attentional disengagements in educational contexts: A diary investigation of everyday mind-wandering and distraction. *Cognitive Research: Principles and Implications*, *2*(1), 1–20. doi:10.118641235-017-0070-7 PMID:28890917

Valente, D. (2022). Scaffolding in-depth learning: picturebooks for intercultural citizenship in primary English teacher education. In D. Valente, & D. Xerri, Innovative Practices in Early English Language Education (pp. 259-283). Cham: Palgrave Macmillan.

Valtonen, T., Tedre, M., Mäkitalo, K., & Vartiainen, H. (2019). Media Literacy Education in the Age of Machine Learning. *The Journal of Media Literacy Education*, *11*(2), 20–36. doi:10.23860/JMLE-2019-11-2-2

Vasudevan, L. (2010). Literacies in a participatory multimodal world: The arts and aesthetics of Web 2.0. *Language Arts*, *88*(1), 43–50.

Vessey, I. (1994). The effect of information presentation on decision making: A cost-benefit analysis. *Information & Management*, *27*(2), 103–119. doi:10.1016/0378-7206(94)90010-8

Viadero, D., Jones, S., & Baker, L. (2018). *Data: Most states require history but not civics*. Education Week. https://www.edweek.org/teaching-learning/data-most-states-require-history-but-not-civics/2018/10

Vicinus, M., & Eisner, C. (2008). *Originality, imitation, and plagiarism: Teaching writing in the digital age*. University of Michigan Press.

Vishkaei, B. (2022). Metaverse: A New Platform for Circular Smart Cities. In P. De Giovanni (Ed.), *Cases on Circular Economy in Practice* (pp. 51–69). IGI Global. doi:10.4018/978-1-6684-5001-7.ch003

Voithofer, R. (2005). Designing new media education research: The materiality of data, representation, and dissemination. *Educational Researcher*, *34*(3), 3–14. doi:10.3102/0013189X034009003

Walters, M. (2018). *Elementary educators' knowledge, beliefs, and planned and implemented practices for digital citizenship*. ProQuest Dissertations.

Walters, M. G., Gee, D., & Mohammed, S. (2019). A literature review: Digital citizenship and the elementary educator. *International Journal of Technology in Education*, *2*(1), 1–21.

Wang, M., Yu, H., Bell, Z., & Chu, X. (2022). Constructing an Edu-Metaverse Ecosystem: A New and Innovative Framework. *IEEE Transactions on Learning Technologies*, 1–13. doi:10.1109/TLT.2022.3226345

Wang, Y., Su, Z., Zhang, N., Xing, R., Liu, D., Luan, T., & Shen, X. (2022). A survey on metaverse: fundamentals, security, and privacy. *IEEE Communications Surveys and Tutorials*, 1–1. doi:10.1109/COMST.2022.3202047

Wargo, J. M. (2015). Spatial stories with nomadic narrators: Affect, Snapchat, and feeling embodiment in youth mobile composing. *Journal of Language & Literacy Education*, *11*(1), 47–64.

Watts, R. J., & Hipolito-Delgado, C. P. (2015). Thinking ourselves to liberation?: Advancing sociopolitical action in critical consciousness. *The Urban Review*, *47*(5), 847–867. doi:10.100711256-015-0341-x

Watulak, S. L., & Kinzer, C. K. (2013). Beyond technology skills: Toward a framework for critical digital literacies in pre-service technology education. In J. Avila & J. Z. Pandya (Eds.), Critical digital literacies as social praxis: Intersections and challenges (pp. 127-156). Peter Lang.

Webster, J. P. (2018). *Teacher Perceptions and Implementation of Digital Citizenship Curriculum in One-To-One High Schools in Missouri*. ProQuest Dissertations.

Weinstein, E., & James, C. (2022). *School-Based Initiatives Promoting Digital Citizenship and Healthy Digital Media Use. Handbook of Adolescent Digital Media Use and Mental Health*. Cambridge University Press.

Westheimer, J., & Kahne, J. (2004). Educating the "good" citizen: Political choices and pedagogical goals. *PS, Political Science & Politics*, *37*(2), 241–247. doi:10.1017/S1049096504004160

Westheimer, J., & Kahne, J. (2004). What kind of citizen? The politics of educating for democracy. *American Educational Research Journal*, *41*(2), 237–269. doi:10.3102/00028312041002237

West, J., & Gallagher, S. (2006). Challenges of open innovation: the paradox of firm investment in open-source software. *R & D Management*, *36*(3), 319–331. doi:10.1111/j.1467-9310.2006.00436.x

Whalen, J. (2020). Should teachers be trained in emergency remote teaching? Lessons learned from the COVID-19 pandemic. *Journal of Technology and Teacher Education*, *28*(2), 189–199.

What is digital citizenship & how do you teach it ? (2023). Applied Educational Systems. https://www.aeseducation.com/blog/what-is-digital-citizenship#:~:text=Digital%20citizenship%20refers%20to%20the,with%20society%20on%20any%20level

What is digital citizenship ? (2022). Canada's Centre for Digital and Media Literacy. https://mediasmarts.ca/digital-media-literacy/general-information/digital-media-literacy-fundamentals/what-digital-citizenship

Wheeler, T. H. (2002). *Phototruth or photofiction? Ethics and Media Imagery in the Digital Age*. Lawrence Erlbaum Associates.

Whitney, A. E. (2011). In search of the authentic English classroom: Facing the schoolishness of school. *English Education*, *44*(1), 51–62.

Wineburg, S., Breakstone, J., Mcgrew, S., Smith, M. D., & Ortega, T. (2022). Supplemental Material for Lateral Reading on the Open Internet: A District-Wide Field Study in High School Government Classes. *Journal of Educational Psychology*, *114*(5), 893–909. doi:10.1037/edu0000740

Wu, J. Y., & Xie, C. (2018). Using time pressure and note-taking to prevent digital distraction behavior and enhance online search performance: Perspectives from the load theory of attention and cognitive control. *Computers in Human Behavior*, *88*, 244–254. doi:10.1016/j.chb.2018.07.008

Wulandari, E., Winarno, W., & Triyanto, T. (2021). Digital Citizenship Education: Shaping Digital Ethics in Society 5.0. *Universal Journal of Educational Research*, *9*(5), 948–956. doi:10.13189/ujer.2021.090507

Wysocki, A. F., Johnson-Eilola, J., Selfe, C. L., & Sirc, G. (2004). *Writing new media: Theory and applications for expanding the teaching of composition*. Utah State University Press.

Xiong, M., & Wang, H. (2022). Digital twin applications in aviation industry: A review. *International Journal of Advanced Manufacturing Technology*, *121*(9), 5677–5692. doi:10.100700170-022-09717-9

Xu, S., Yang, H. H., & Zu, S. (2019, July). *An investigation of 21st-century digital skills on digital citizenship among college students. International Symposium on Educational Technology*, Czech Republic. 10.1109/ISET.2019.00056

Yalçın-Irmak, A., & Ardıç, A. (2018). Dijital oyunların çocuk ve ergenler üzerindeki etkileri: Literatür inceleme [The effects of digital games on children and adolescents: Literature review]. *Gençlik Araştırmaları Dergisi*, *6*(16), 71–84.

Yang, C., Tu, X., Autiosalo, J., Ala-Laurinaho, R., Mattila, J., Salminen, P., & Tammi, K. (2022). Extended reality application framework for a digital-twin-based smart crane. *Applied Sciences (Basel, Switzerland)*, *12*(12), 6030. doi:10.3390/app12126030

Zaff, J. F., Malanchuk, O., & Eccles, J. S. (2008). Predicting positive citizenship from adolescence to young adulthood: The effects of a civic context. *Applied Developmental Science*, *12*(1), 38–53. doi:10.1080/10888690801910567 PMID:22837638

Zappavinga, M. (2014). CoffeeTweets: bonding around the bean on Twitter. In P. Seargeant & C. Tagg (Eds.), *The language of social media: Identity and community on the internet* (pp. 139–166). Palgrave Macmillan. doi:10.1057/9781137029317_7

Zhang, Z., Wen, F., Sun, Z., Guo, X., He, T., & Lee, C. (2022). Artificial intelligence-enabled sensing technologies in the 5g/internet of things era: from virtual reality/augmented reality to thc digital twin. *Advanced Intelligent Systems*, *4*(7), 2100228. doi:10.1002/aisy.202100228

Zhou, S., & Barbaro, N. (2023). Understanding student expression across higher ed: Heterodox Academy's annual campus expression survey. Heterodox Academy.

Zukin, C., Keeter, S., Andolina, M., Jenkins, K., & Delli Carpini, M. X. (2006). *A new engagement?: Political participation, civic life, and the changing American citizen*. Oxford University Press. doi:10.1093/acprof:oso/9780195183177.001.0001

About the Contributors

Jason D. DeHart is a passionate educator and has served as a middle grades teacher for eight years. He also served as an assistant professor of reading education at Appalachian State University from 2019-2022, and has taught reading education courses at The University of Tennessee, Knoxville and Lee University in Cleveland, Tennessee. DeHart's research interests include multimodal literacy, including film and graphic novels, and literacy instruction with adolescents. His work has recently appeared in SIGNAL Journal, English Journal, and The Social Studies.

* * *

Seda Adıgüzel is a Ph.D. student discovering teachers' Digital Citizenship perceptions. Also, IT teacher at ODTÜ GV schools in Ankara, Turkey. She co-organized several technology workshops for students and teachers. Her interests include social media and internet technologies, computer-based education and learning, and information technologies in education.

Paulo Barroso is an Assistant Professor at the Polytechnic Institute of Viseu (Portugal), College of Education, Department of Communication and Art, teaching Strategic Communication, Sociology of Communication, Semiotics, Theories of Communication, Ethics and Deontology; PhD in Philosophy (University of Santiago de Compostela, Spain), Post-Doctoral Researcher in Communication Sciences (University of Minho, Portugal) between 2007 and 2013 and presently integrated researcher at NOVA Institute of Communication – NOVA University of Lisbon (School of Social Sciences and Humanities) with the ORCID identifier number 0000-0001-7638-5064.

Crystal L. Beach is a part-time assistant professor at the University of Georgia and an academic specialist in Union County Schools. She strives to create meaningful literacy learning opportunities for all as she knows how important literacy is to the overall success of communities. Over the years, she has presented and published on a variety of areas in education, including literacy, multimodality, teacher education, rural education, and technologies in the classroom. She also serves on many editorial review boards, volunteers to mentor new teachers, and enjoys working with community stakeholders to forge powerful connections with schools.

Susana Bernardino is an Adjunct Professor of Management at the Polytechnic of Porto - Institute of Accounting and Administration (ISCAP), S. Mamede de Infesta, Portugal, and researcher at CEOS. PP/ISCAP. She completed her under-graduation degree in Management at the Faculty of Economics of the University of Porto, where also obtained the curricular part of the master's course in Finance. She got her PhD in Management in 2014 from the Portucalense University. Her research interests focus on social entrepreneurship, entrepreneurship, social and solidarity economy, smart cities, urban and regional development. and crowdfunding.

Emine Bozkurt Polat graduated from Necmettin Erbakan University, Department of Preschool Education in 2012 and was appointed as a research assistant at Kahramanmaraş Sütçü İmam University in the same year. After completing his master's degree in Necmettin Erbakan University Preschool Education Department in 2016, he was appointed as a research assistant to Gazi University Preschool Education Department in the same year and started his doctorate education. She worked as the education coordinator at Gazi Practice Kindergarten for 1 year in the 2018-2019 academic year. The author is still working as a research assistant at Gazi University. As an academician, he has publications on technology, augmented reality, robotic coding values education in preschool children. She worked on robotic coding in her doctoral thesis.

Niroj Dahal is currently a lecturer at the Department of STEAM Education at Kathmandu University School of Education, Hattiban, Lalitpur, Nepal. His research interests include ICT in education, qualitative research, mathematics education, open, distance & e-learning, STEAM education, artificial intelligence (AI), research and development, and ICT & e-Research. Mr. Dahal has been teaching graduate and undergraduate students for over a decade. He has also been continuously taking part and presenting his research and practices in more than a dozen plus national and international conferences, workshops, and seminars. He has published articles, research notes, editorials, book reviews, and book chapters in various national and international journals and publication presses in the field of ICT, qualitative research, education in general and mathematics education, and STEAM education in particular.

Mohammad Daradkeh is an Associate Professor of Business Analytics and Data Science at the College of Engineering & IT, University of Dubai. Prior to joining the University of Dubai, Dr. Daradkeh worked at Yarmouk University in Jordan and Lincoln University in New Zealand. His research interests are mainly in the areas of Business Intelligence, Data Analytics, and Innovation Management. He has published numerous research papers with reputed publishers such as Elsevier, Springer, Emerald MDPI and IGI. He has presented research papers at various international conferences. He is also a member of the editorial board of several reputed journals.

Shernette Dunn is the director for first year success and has been working with faculty and staff to provide the best educational opportunities for students to achieve academic success. She has researched and published on this topic extensively. Shernette is also an instructional designer who consults with educators and organizations to ensure that classes are designed to meet a variety of learning needs while using various different instructional technologies. She has been recognized by her work by the Federal Government and continues to support all learners.

Ali Orhan holds a BA degree in English Language and Literature from Ege University, Türkiye. He has received his MA degree and PhD in the Department of Curriculum and Instruction. He is working at Zonguldak Bülent Ecevit University in Türkiye. His research interests focus on higher-order thinking skills such as critical thinking and problem solving, curriculum development, meta-analysis, flipped learning, online learning, mobile learning, digital citizenship, and digital literacy. He is the author of numerous articles published in international peer-reviewed journals like Journal of Educational Research, Journal of Psychoeducational Assessment, Smart Learning Environments, and Learning Environments Research.

Mary Rice is an Associate Professor of Literacy at the University of New Mexico. Her work focuses on how educators engage with technological literacies to make curriculum that supports diverse students. Mary is a fellow at the Michigan Virtual Learning Institute where she participates in research on standards development, designerly thinking in K-12 online courses, and curriculum accessibility.

Vera Rocha graduated in human resources management from ISMAI (2018) and master in entrepreneurship and internationalization from ISCAP (2022). Always open to challenges, with an interest in social entrepreneurship and the well-being of society, helping to meet its needs and in the search for a better future.

Index

F

G

H

I

L

M

N

O

P

R

S

T

U

V

W

www.ingramcontent.com/pod-product-compliance
Lightning Source LLC
LaVergne TN
LVHW080643011026
849160LV00014B/1306